T0339891

# POLYUNSATURATED FATTY ACID METABOLISM

# POLYUNSATURATED FATTY ACID METABOLISM

*Edited by*

GRAHAM C. BURDGE

ELSEVIER

ACADEMIC PRESS
An imprint of Elsevier

AOCS
PRESS

Academic Press and AOCS Press
Academic Press is an imprint of Elsevier
125 London Wall, London EC2Y 5AS, United Kingdom
525 B Street, Suite 1800, San Diego, CA 92101-4495, United States
50 Hampshire Street, 5th Floor, Cambridge, MA 02139, United States
The Boulevard, Langford Lane, Kidlington, Oxford OX5 1GB, United Kingdom

Published in cooperation with American Oil Chemists' Society www.aocs.org
Director, Content Development: Janet Brown

**Library of Congress Cataloging-in-Publication Data**
A catalog record for this book is available from the Library of Congress

**British Library Cataloguing-in-Publication Data**
A catalogue record for this book is available from the British Library

ISBN: 978-0-12-811230-4

For information on all Academic Press publications visit our website at
https://www.elsevier.com/books-and-journals

Working together
to grow libraries in
developing countries

www.elsevier.com • www.bookaid.org

*Publisher:* Gerhard Andre Wolff
*Acquisition Editor:* Nancy Maragioglio
*Editorial Project Manager:* Billie Jean Fernandez
*Production Project Manager:* Omer Mukthar
*Designer:* Vicky Pearson Esser

Typeset by Thomson Digital

# Contents

## 9. Polyunsaturated Fatty Biosynthesis and Metabolism in Reproductive Tissues

D. CLAIRE WATHES, ZHANGRUI CHENG

## 10. The Effect of Dietary Modification on Polyunsaturated Fatty Acid Biosynthesis and Metabolism

BEVERLY S. MUHLHAUSLER

## 11. Omega-3 Polyunsaturated Fatty Acid Metabolism in Vegetarians

GRAHAM C. BURDGE, CHRISTIANI J. HENRY

## 12. Genetic Influences on Polyunsaturated Fatty Acid Biosynthesis and Metabolism

COLETTE O' NEILL, ANNE M. MINIHANE

## 13. Interactions Between Polyunsaturated Fatty Acids and the Epigenome

KAREN A. LILLYCROP, GRAHAM C. BURDGE

# List of Contributors

**Karen P. Best**   Healthy Mothers, Babies and Children Theme, South Australian Health and Medical Research Institute (SAHMRI), Adelaide, SA, Australia

**Graham C. Burdge**   University of Southampton, Southampton, United Kingdom

**Philip C. Calder**   University of Southampton, Southampton, United Kingdom

**Luís Filipe C. Castro**   CIIMAR—Interdisciplinary Centre of Marine and Environmental Research, U. Porto – University of Porto, Porto, Portugal

**Zhangrui Cheng**   Royal Veterinary College, North Mymms, Hatfield, Herts, United Kingdom

**Carmel T. Collins**   Healthy Mothers, Babies and Children Theme, South Australian Health and Medical Research Institute (SAHMRI), Adelaide, SA, Australia

**Michael E.R. Dugan**   Agriculture and Agri-Food Canada, Lacombe Research Centre, Lacombe, AB, Canada

**Jacqueline F. Gould**   Healthy Mothers, Babies and Children Theme, South Australian Health and Medical Research Institute (SAHMRI), Adelaide, SA, Australia

**Christiani J. Henry**   Clinical Nutrition Research Centre, Centre for Translational Medicine, National University of Singapore, Singapore

**Gabriela E. Leghi**   FOODplus Research Centre, The University of Adelaide, Adelaide, SA, Australia

**Karen A. Lillycrop**   Centre for Biological Sciences, University of Southampton, Southampton, United Kingdom

**Cletos Mapiye**   Department of Animal Sciences, Stellenbosch University, Stellenbosch, South Africa

**Anne M. Minihane**   University of East Anglia, Norwich, United Kingdom

**Oscar Monroig**   University of Stirling, Stirling, United Kingdom

**Beverly S. Muhlhausler**   FOODplus Research Centre, The University of Adelaide; Healthy Mothers, Babies and Children Theme, South Australian Health and Medical Research Institute (SAHMRI), Adelaide, SA, Australia

**Colette O' Neill**   Cork Centre for Vitamin D and Nutrition Research, University College Cork, Cork, Ireland

**Woo Jung Park**   Gangneung-Wonju National University, Gangneung, Gangwon, The Republic of Korea

**Andrew J. Sinclair**   Deakin University, Geelong; South China University of Technology, Guangzhou, China; Monash University, Clayton, Victoria, Australia

**Douglas R. Tocher**   University of Stirling, Stirling, United Kingdom

**Payam Vahmani**   Agriculture and Agri-Food Canada, Lacombe Research Centre, Lacombe, AB, Canada

**D.Claire Wathes**   Royal Veterinary College, North Mymms, Hatfield, Herts, United Kingdom

# 1

# Introduction: More Than 50 Years of Research on Polyunsaturated Fatty Acid Metabolism

*Andrew J. Sinclair*

Deakin University, Geelong, Victoria, Australia;
South China University of Technology, Guangzhou,
China; Monash University, Clayton, Victoria, Australia

## OUTLINE

Why would anyone be interested in lipids and fatty acids for more than 50 years? This is a question that I often get asked by family and friends and it is not all that easy to give a simple answer, other than to say that, like many branches of science, there is something new to learn every day. People not involved in science find this hard to believe. In some respects, science can be like the daily news cycle: always something interesting on a daily basis, some of which is totally in the moment and other bits, which remain with you forever. Specifically, this field is fascinating because there is such a great variety of research in biology, medicine, agriculture and nutrition, and many of the discoveries are of fundamental importance to everyday lives.

Two such examples suffice to whet the reader's appetite:

1. Omega 3 polyunsaturated fatty acids (PUFA) can play a significant role in improving the quality of life for those suffering from rheumatoid arthritis, and in addition reduce their antiinflammatory drug use (Proudman et al., 2015).

Polyunsaturated Fatty Acid Metabolism. http://dx.doi.org/10.1016/B978-0-12-811230-4.00001-6

**2.** Chronic headache is a severe debilitating condition. Using foods to significantly reduce linoleic acid and increase long chain omega 3 PUFA have significantly reduced the number of hours of headaches each day, in those suffering from chronic headaches (Ramsden et al., 2013).

This chapter serves as a precursor to the chapters following that deal in detail with specific areas of metabolism of fatty acids. At this point, I make sincere apologies to authors whose work I have not mentioned—there are many legends in this field and their work has been acknowledged previously by me, in publications from the various laboratories where I have worked Melbourne (Australia), London (Canada), London (United Kingdom), Coleraine (Northern Ireland), Dijon (France), Guangzhou and Hangzhou (China), and Geelong (Victoria, Australia).

When I started my PhD in the mid-1960s, Ralph Holman was the king of fatty acids, especially PUFAs. His publications were many and legendary, and with coauthors like Mohrhauer, H., Caster, W.O., Pfeiffer, J.J., Rahm, J.J., Christie, W.W., and others, he published exciting studies on PUFA metabolism, effects of single PUFA on rat tissue fatty acid composition, the competitive effects of two PUFA at a time, the requirements of female rats for essential fatty acids (EFA), positional distribution of fatty acids in lecithin and more (Holman, 1986, 1996). These studies were conducted at the Hormel Institute in Austin, Minnesota, at a time when packed column gas chromatography (GC) was starting to make its mark; remember, this technique was only invented in the late 1950s (James, 1970). Students of today would not believe the difficulties faced by early users of such equipment. Being a semiskilled plumber and mechanic, with a strong body to manipulate 1.7 m tall gas cylinders, were useful attributes.

It was clear from reading the early literature on EFA that there were a number of earlier champions in the field, such as Burr and Burr (1929, 1930) (they discovered the essentiality of linoleic acid [18:2n-6] and alpha-linolenic acid [18:3n-3]), Aaes-Jorgensen (1961; who wrote papers on this topic including a fantastic review of the field), Klenk who conducted seminal studies on metabolic interconversions of fatty acid (Klenk, 1965), Mead (1981) who, later than Klenk, built on these studies. The $\Delta 6$, $\Delta 5$, and $\Delta 4$ pathway was born through all these studies, and refined by the work of Brenner who studied the impact of nutrition and physiological state on the pathway (Peluffo et al., 1984). This metabolic pathway was later revised by Howard Sprecher (Voss et al., 1991), whereby the $\Delta 4$ was replaced by an additional elongase and another $\Delta 6$ step; this metabolic pathway is still the subject of debate (Infante and Huszagh, 1998; Park et al., 2015).

The era of late 1960s to early 1970s was one where it was reasoned that adult humans would not develop EFA deficiency because of the store of these PUFA in adipose tissue; however, during my PhD we were involved in the first identification of EFA deficiency in an adult (Collins et al., 1971); the patient had a significant removal of much of his bowel and was being fed totally by intravenous (IV) drip, with only sporadic use of IV lipid emulsions due to adverse effects of these on patients (increased temperature). The patient developed a wound around the area of his stoma. Prior to using the IV lipids, the patient's plasma fatty acid profiles showed high levels of 20:3n-9 and reduced 20:4n-6 levels; this fatty acid profile was rapidly reversed by using the IV lipid emulsion, and this was associated with improvements in the skin condition around the wound.

Around this time, it was reasoned that linoleic acid was the main (only) EFA for humans, while alpha-linolenic acid (ALA) was the main (only) EFA for fish and other marine species. Things have certainly changed since those days. In fact, infant formulas in Europe (and possibly elsewhere) were being made with very high linoleic acid (LA) to alpha-linolenic acid levels, with little regard to the composition of human milk (Crawford et al., 1973; Gibson and Kneebone, 1981).

A technological revolution occurred with the arrival of capillary columns and one of the earliest users was the great Bob Ackman, working in Halifax, Nova Scotia. What a genius he was in this area, and with an extraordinarily dry sense of humor. He published hundreds of papers on the fatty acid composition of marine species using capillary GC, as well as deriving the retention times of many fatty acids, aided by his strong chemical background (Ackman, 1987). This new technology sped up the GC analysis remarkably as well as sharpening the later-eluting peaks because it was possible to temperature program the GC oven, as the phases used were quite stable at high temperatures (up to 300°C typically). The first capillary columns were stainless steel (support coated open tubular), replaced later by synthetic flexible polymer columns (wall coated open tubular), ranging in length from 25 to 100 m (in contrast to the original glass columns packed with stationery phase, which were about 2 m long and 1 cm diameter, and which were operated isothermally at around 170–200°C).

Today, GC technology and fatty acid analyses from blood have advanced considerably with the use of finger prick blood samples, robotic systems for extracting samples and derivitizing the ester bound fatty acids to fatty acid methyl esters, and using very short columns (Bell et al., 2011; Lin et al., 2012; Stark and Salem, 2005).

When I started working with Michael Crawford at the Nuffield Institute for Comparative Medicine, Zoological Society of London, in 1970, he was still using packed column GC, and the GC charts were often 3 m long and needed to be spread on the lab bench to look at the results. The later eluting peaks (e.g., docosahexaenoic acid (DHA) [22:6n-3]) were only evident as small rises in height but with large widths, so accurate integration was key to calculating compositions; many labs simply ignored these peaks, such as the lab where I completed my PhD, in Melbourne. Fortunately, the Crawford lab converted to using capillary systems soon after, at my insistence, making our lives much easier—having expert lab managers/engineers was also crucial to progress toward new technology.

Professor Michael Crawford is without doubt one of the most amazing characters of the field, always enthusiastic, passionate, and unafraid of thinking on a large scale. Additionally, he is exceedingly good company with a great love of fresh seafood and malt products from Scotland. He is still an active researcher in lipids, and nutrition in general. While the first reports on the human brain fatty acid composition were published by O'Brien and Sampson (1965) and Svennerholm (1968), Crawford came to realize through his studies that the brain of all mammals studied in his lab (more than 35) had the same fingerprint of PUFA (20:4n-6 [arachidonic acid (AA)], 22:4n-6 and DHA), especially in the gray matter (cellular area) (Crawford et al., 1976; Crawford and Sinclair, 1972a). Recognizing the significance of DHA in the human brain, and that the brain and PUFA characteristic fingerprint established itself soon after birth (Ballabriga and Martinez, 1978; Martinez et al., 1988), Crawford and his team became champions for paying attention to both omega 6 and omega 3 fatty acid in human nutrition, particularly in utero and during lactation (Crawford et al., 1973; Crawford and Sinclair 1972a,b; Sinclair, 1975; Sinclair and Crawford, 1972). The Crawford team was also the first to report the effects of a deficiency of alpha-linolenic acid in a primate (Fiennes et al., 1973).

Crawford and I are of the belief that our research in the early 1970s was a major contributing factor, together with the fundamental discoveries on the essentiality of omega 3 PUFA for retinal function by Gene Anderson and colleagues (Wheeler et al., 1975), to the eventual inclusion of AA and DHA into infant formulas, though there is still argument about the need to include AA in some influential regulatory quarters (Richard et al., 2016).

One of the newcomers to the Crawford lab, John Rivers, was given the task of creating an artificial, meat-free diet for cats. The hypothesis was that an animal, which evolved to eat meat, might have lost the capacity to make longer chain fatty acid from the 18-carbon precursors, in much the same way that cats appeared to have lost capacity to convert beta-carotene to vitamin A. The task was finally accomplished and the results showed that cats fed diets rich in linoleic acid or alpha-linolenic acid, but devoid of longer chain PUFA, did not accumulate significant amounts of AA or DHA (Rivers et al., 1975). This work led to a series of studies at the Nuffield Institute of Comparative Medicine, as reported in Sinclair (1994), in a tribute to the life of John Rivers. The initial study by Rivers et al. (1975) suggested that cats lacked the Δ6 desaturase, while there was no apparent impairment in the chain elongation process. This was tentatively confirmed by subsequent work in Melbourne, showing that there was evidence of a delta 8 and a delta 5 desaturase in cats fed diets without any meat products (Sinclair et al., 1979).

It was during my 4 years with Crawford that I developed an interest in 22:5n-3 (DPAn-3). I have no idea why it took me more than 40 years to do anything about this, by way of research. Maybe, it was waiting for an alignment of availability of pure DPAn-3 and money that played a critical role in the delay? Anyway, we have now published a number of papers on this interesting, minor omega 3 PUFA (Kaur et al., 2011; Markworth et al., 2016). The Crawford lab was studying the fatty acids in liver, brain, and muscle tissue from many wild animals (sourced from Africa or the London Zoo), and DPAn-3 was a major omega 3 PUFA in ruminant liver and muscle tissue, and its proportions almost always exceeded those of DHA. This was not the case in the same tissues from carnivores, or smaller animals, such as rabbits, where the proportions of DHA were greater than DPAn-3 (Crawford et al., 1969, 1970). It seemed curious at the time that DPAn-3 showed such species fluctuations in proportions in tissues, and while the reasons for DPAn-3 levels fluctuating between species have not been studied at an enzyme or molecular level, it may have metabolic explanations, which may be species-specific (enzyme activity, and/or synthesis and turnover of DPAn-3 and DHA). It is only in recent times that whole body synthesis and secretion kinetics of long chain omega 3, in free-living rats, has been studied (Metherel et al., 2016). Docosapentaenoic acid (DPA) appears to be a unique long chain omega 3 fatty acid which may function as a reservoir of tissue eicosapentaenoic acid (EPA) and/or DHA.

A great deal has happened since those exciting days of the 1970s, including the development of a great friendship with Norman Salem and his team at National Institute of Alcoholism and Alcohol Abuse (at NIH) in Bethesda, MD. He had assembled the strongest omega 3 research group in the world and although Norman has retired from there, his work continues through the stewardship of several champions, including Hee-Yong Kim, Klaus Gawrisch, and Joe Hibblen (Mihailescu et al., 2011; Ramsden et al., 2016; Sidhu et al., 2016).

That team led calmly, and most ably by Norman, demonstrated the power of interdisciplinary research and big picture thinking on the importance of DHA in the brain and visual function and translational research. Art Spector and Bill Lands, legends in their own right, joined them and they both contributed significantly to the work in that great "mega" lab, particularly adding intellectual stimulation and rigor. Bill Lands, in particular, has been a longtime champion of the inflammatory potential of diets rich in linoleic acid (Bibus and Lands, 2015; Lands, 2012, 2016). No doubt he has played a significant role in influencing the direction of the NIH research on diets rich in linoleic acid in relation to inflammation. It was also fortunate that Stanley Rapoport, also at NIH, soon became interested in AA and DHA in the brain. His group has conducted some of the most elegant studies on turnover of these fatty acid in rodents and, more recently, in humans (Basselin et al., 2010; Kim et al., 2011; Rapoport et al., 2010; Taha et al., 2016a,b). Stanley Rapoport continues to publish fundamental breakthrough data in the area, including a recent review on bipolar disorder (Saunders et al., 2016). Richard Bazinet, one of Rapoport's former students, is now in Toronto and continuing to publish outstanding papers on turnover and metabolism of PUFA (Metherel et al., 2016). To illustrate the importance of bringing different skills to tackle some DHA-related research questions, Norman employed Hee-Yong Kim, a mass spectroscopist. Over the years she has turned herself into a legend in molecular biology of DHA metabolism in the brain (Lee et al., 2016), including the discovery of the importance of DHA in phosphatidylserine and synaptomide (N-docosahexanoylethanolamide) in the development of neurones and cognitive function.

What is exciting about this field is the way it has advanced in different areas (e.g., through multidisciplinary research, such as that being conducted by Tom Brenna's group on the metabolic pathway [Kothapalli et al., 2016], and by Joe Hibblen and Chris Ramsden in areas, such as the effects of significant diet change on chronic headache (Ramsden et al., 2013). What is frustrating, however, is how much published research on PUFA appears to be stuck in the past (black box research: feed A, observe Z, and make a few measurements on mechanisms of action; inferring enzyme activities based on ratios of fatty acids—in 2016, for goodness sake).

It only takes a few minutes of searching PubMed (https://www.ncbi. nlm.nih.gov/pubmed) to realize that nutrition is being overrun by big picture research, much of which involves the microbiome, with huge teams including experts in molecular biology, metabolomics, bioinformatics, medicine, immunology, and the microbiome (Arrieta et al., 2015; Caesar et al., 2015; O'Keefe et al., 2015; Pedersen et al., 2016). It is time for the fatty acid field to broaden its horizons to realize that multidisciplinary research is essential to answer biological questions, which might have a fatty acid involvement. Lipid scientists do work in multidisciplinary teams, and two examples suffice, though there are many other examples: (1) the recent work by Tom Brenna and his team on the discovery of causal FADS2 Indel polymorphism influencing arachidonic acid biosynthesis (Kothapalli et al., 2016). This work was interpreted to mean that South Asians may have an advantage in synthesis of LCn-6 fatty acids because of their plant-based diets, (2) the significant progress in rheumatoid arthritis where Les Cleland, a rheumatologist, worked with Michael James (biomedical expert on eicosanoids) and Bob Gibson (fatty acid specialist) to conduct some of the most highly cited papers in this field, with research clearly at the intersection of medicine, eicosanoids, and fatty acids (Proudman et al., 2015). Not only have these scientists been involved in research on arthritis, but also James and Cleland have written a most incisive commentary on omega 3 coronary heart disease trials (James et al., 2014). In addition, in recent times, Mick James has been publishing on the regulation of elongases (Gregory et al., 2013).

Of course, there are many other examples of exciting interdisciplinary research, which could have been cited. I take the opportunity here to mention interdisciplinary collaborative work I have had the fortune to be involved with, including visual function and omega 3 fatty acids with Algis Vingrys (an expert in visual function and measurement, Weisinger et al., 1996), ingestive physiology and omega 3 fatty acids with Richard Weisinger (an expert on salt and water appetite, Begg et al., 2012; Weisinger et al., 2001), and the interaction between omega 3 fatty acids and zinc with Leigh Ackland (a zinc expert, Suphioglu et al., 2010). Similarly, the work on Australian aborigines and their diets was only possible through working with an expert in that field (Kerin O'Dea) (Mann et al., 1997; Sinclair et al., 1983, 1994; Sinclair and O'Dea, 1993). Finally, it has been a pleasure working with Professor Yonghua Wang in multidisciplinary research on furan fatty acids (Xu et al., 2017).

So far, I have ignored the work of many champions in this field and it is difficult to emphasize their impact in such a short space. I mention but a few, including: Bill Christie (a superb methods guru). I am confident most lipid students and research staff have frequently consulted his many books on techniques in the analysis of lipids (see Oily Press Lipid Library Series, http://store.elsevier.com/Oily-Press-Lipid-Library-Series/EST_SER-8400014/), as well as many publications from him (Christie, 1969; Christie et al., 2007).

Doug Tocher and colleagues for many years have been publishing papers on fatty acid metabolism in fish, and considering broader issues, such as how to sustain levels of long chain omega 3 fatty acids in farmed fish, such as salmon (Sprague et al., 2016).

Michel Lagarde and colleagues first reported that 2-lysophosphatidylcholine (DHA) was a carrier of DHA to the brain (Thiés et al., 1992). His work has progressed to the point that a novel, structured lipid, and 1-acetyl,2-DHA-PC has been reported to be a specific carrier of DHA to the brain, compared with DHA-containing PC and non-esterified DHA (Hachem et al., 2016).

Claudio Galli (longtime researcher on omega 3 and prostaglandins), was the first to report that in omega 3 deficiency, there was a significant accumulation of 22:5n-6 in rodent brains, in an apparent replacement of the lost 22:6n-3 (Galli et al., 1970).

Philippe Legrande (longtime researcher on lipids and fatty acids; Rioux et al., 2015) has become interested in the effects of diets rich in linoleic acid on inflammatory processes, with results that question the current high intakes of linoleic acid in current diets worldwide.

Philip Calder has specialized in lipids and immune function for more than 25 years; however, his work is far broader than that, with considerable emphasis on health benefits of plant and marine omega 3 fatty acids (Calder, 2015).

Graham Burdge has also worked for many years in the field and is known among other things for his work on gender effects on PUFA metabolism, including identifying greater DHA synthesis in females than male subjects (Burdge and Wootton, 2002), and also the impact of maternal diet on offspring epigenetics (Lillycrop and Burdge, 2015).

Tom Clandinin has been involved in lipid research over a 40-year period, with many significant contributions to the field as well as in training scientists. A recent theme of his work has been on the role of gangliosides in immune function (Miklavcic et al., 2012).

Stephen Cunnane has also had a long and distinguished career in the field, with important studies on revisiting EFA requirements (Cunnane and Guesnet, 2011), and more recently exploring the role of ketones as a brain fuel in the elderly (Cunnane et al., 2016).

Giovanni Turchini, an outstanding lipid nutritionist and marine scientist, continues to publish papers that challenge dogma in terms of just what the requirements are for different long chain PUFA in salmon and that EPA and DHA are metabolized differently in this species (Emery et al., 2016).

David Horrobin, a former professor of medicine, was a highly innovative thinker in the PUFA field for more than 20 years. Often scorned for his commercial interests, he was someone who never stopped thinking about possibilities for uses of different PUFA, such as gamma-linolenic acid or EPA. He published many papers, and a review paper published more than 14 years ago is worth reading because it discusses how novel PUFA could be useful treatments for psychiatric disorders (Horrobin, 2002).

Alex Leaf was a pioneer in the understanding of omega 3 in cardiac arrhythmias (Leaf et al., 2005). His seminal studies, which started in the late 1990s, were the catalyst for research on mechanisms of action of long chain omega 3 fatty acids in the prevention of fatal arrhythmias.

When one reads about novel lipid mediators, the first name that springs to mind is that of Charles Serhan. He has discovered many novel mediators from EPA, DPA, and DHA and he and his group regularly publish novel data and reviews on the resolution of inflammation (Chiurchiu et al., 2016).

Jorn Dyerberg, with his collaborator Bang, H.O., opened the eyes of the world to omega 3 and heart disease in Eskimos. This work led to a huge surge in the interest in omega 3 fatty acids from that time (Bang et al., 1971). In recent times, Dyerberg with colleagues has put a lot of effort into public health measures to change the food supply in Denmark, with emphasis on reducing artificial trans fats (Dyerberg et al., 2004).

Maria Makrides and Bob Gibson have been involved in some of the largest multicenter, multidisciplinary intervention trials in pregnant women supplemented with DHA for many years, with the view of examining the impact on infants (Makrides et al., 2014). Their work is highly regarded throughout the area of paediatric nutrition.

The future—a wish perhaps? The brain is the area most understudied in a multidisciplinary sense. The people trying to grapple with this include Stanley Rapoport, Chris Ramsden, and Hee-Yong Kim, who separately are examining important basic research questions on the role of linoleic acid in the diet on brain AA metabolism and brain lipid mediators (Taha et al., 2016a,b). Hee-Yong Kim and her colleagues are providing new insights into mechanisms by which DHA promotes brain development and function (Lee et al., 2016). In general, however, there has been a failure of the lipid field to engage with experts in brain physiology, imaging, and anatomical mapping using gene mapping, and it is my guess that the vast majority of neuroscientists have little understanding of brain PUFA and why this topic is regarded as important by many fatty acid scientists!

A second area that really must be solved is to understand the molecular biochemical and physiological basis, which contribute to the substantial variations in the red blood cell omega 3 index between people given the same dose of omega 3 fatty acids (Von Schacky, 2009). Some factors thought to modify this index have been identified but it seems that we are still a long way short of knowing how to reliably raise the red cell omega 3 levels to the same extent in people from the same ethnic background. It has been suggested that as in other fields, the dose of omega 3 fatty acids should be provided on the basis of body weight rather than the same dose independent of weight (Ghasemifard et al., 2014).

In my view, the future is bright for this field because it has a number of rising stars, already mentioned.

## Acknowledgments

I would like to thank Hyunsin (Hedy) Sung, PhD scholar, for her assistance with the referencing of this chapter, and all my colleagues and graduate students over the course of my working journey, recorded here briefly. It has been such a rewarding life to work with so many fine, bright young students and my fellow scientists. In recent years, I have worked in China at Zhejiang University, Hangzhou and South China University of Technology, Guangzhou. I have been blown away by the talents of both the undergraduate and graduate students, their dedication to their goals and their intellectual capacity. In addition, their language skills are impressive. I have greatly valued my time interacting with them and their supervisors including Professor Duo Li and Professor Yonghua Wang. Thank you for sharing your intellectual endeavors and your friendships over the years.

## References

Aaes-Jorgensen, E., 1961. Essential fatty acids. Physiol. Rev. 41, 1–51.

Ackman, R.G., 1987. Simplification of analyses of fatty acids in fish lipids and related lipid samples. Acta Med. Scand. 222, 99–103.

Arrieta, M.-C., Stiemsma, L.T., Dimitriu, P.A., Thorson, L., Russell, S., Yurist-Doutsch, S., Kuzeljevic, B., Gold, M.J., Britton, H.M., Lefebvre, D.L., 2015. Early infancy microbial and metabolic alterations affect risk of childhood asthma. Sci. Transl. Med. 7, 307ra152.

Ballabriga, A., Martinez, M., 1978. A chemical study on the development of the human forebrain and cerebellum during the brain "growth spurt" period: II. Phosphoglyceride fatty acids. Brain Res. 159, 363–370.

Bang, H.O., Dyerberg, J., Nielsen, A.B., 1971. Plasma lipid and lipoprotein pattern in Greenlandic west-coast Eskimos. Lancet 1 (7710), 1143–1145.

Basselin, M., Kim, H.-W., Chen, M., Ma, K., Rapoport, S.I., Murphy, R.C., Farias, S.E., 2010. Lithium modifies brain arachidonic and docosahexaenoic metabolism in rat lipopolysaccharide model of neuroinflammation. J. Lipid. Res. 51, 1049–1056.

Begg, D.P., Sinclair, A.J., Weisinger, R.S., 2012. Thirst deficits in aged rats are reversed by dietary omega-3 fatty acid supplementation. Neurobiol. Aging 33, 2422–2430.

Bell, J.G., Mackinlay, E.E., Dick, J.R., Younger, I., Lands, B., Gilhooly, T., 2011. Using a fingertip whole blood sample for rapid fatty acid measurement: method validation and correlation with erythrocyte polar lipid compositions in UK subjects. Br. J. Nutr. 106, 1408–1415.

Bibus, D., Lands, B., 2015. Balancing proportions of competing omega-3 and omega-6 highly unsaturated fatty acids (HUFA) in tissue lipids. Prostagland. Leuk. Essent. Fatty Acids 99, 19–23.

Burdge, G.C., Wootton, S.A., 2002. Conversion of alpha-linolenic acid to eicosapentaenoic, docosapentaenoic and docosahexaenoic acids in young women. Br. J. Nutr. 88, 411–420.

Burr, G.O., Burr, M.M., 1929. A new deficiency disease produced by the rigid exclusion of fat from the diet. J. Biol. Chem. 82, 345–367.

Burr, G.O., Burr, M.M., 1930. On the nature and role of the fatty acids essential in nutrition. J. Biol. Chem. 86, 587–621.

Caesar, R., Tremaroli, V., Kovatcheva-Datchary, P., Cani, P.D., Bäckhed, F., 2015. Crosstalk between gut microbiota and dietary lipids aggravates WAT inflammation through TLR signaling. Cell Metab. 22, 658–668.

Calder, P.C., 2015. Marine omega-3 fatty acids and inflammatory processes: effects, mechanisms and clinical relevance. Biochim. Biophys. Acta 1851, 469–484.

Chiurchiu, V., Leuti, A., Dalli, J., Jacobsson, A., Battistini, L., Maccarrone, M., Serhan, C.N., 2016. Proresolving lipid mediators resolvin D1, resolvin D2, and maresin 1 are critical in modulating T cell responses. Sci. Transl. Med. 8, 353ra111.

Christie, W.W., 1969. The glyceride structure of Sapium sebiferum seed oil. Biochim. Biophys. Acta, Lipids Lipid Metab. 187, 1–5.

Christie, W.W., Dobson, G., Adlof, R.O., 2007. A practical guide to the isolation, analysis and identification of conjugated linoleic acid. Lipids 42, 1073–1084.

Collins, F., Sinclair, A.J., Royle, J., Coats, D., Maynard, A., Leonard, R., 1971. Plasma lipids in human linoleic acid deficiency. Ann. Nutr. Metab. 13, 150–167.

Crawford, M., Sinclair, A.J., 1972a. Nutritional influences in the evolution of mammalian brain. Ciba Foundation Symposium 3-Lipids, Malnutrition & the Developing Brain. Wiley Online Library. , pp. 267–292.

Crawford, M.A., Sinclair, A.J., 1972b. The limitations of whole tissue analysis to define linolenic acid deficiency. J. Nutr. 102, 1315–1321.

Crawford, M., Gale, M.M., Woodford, M., 1969. Linoleic acid and linolenic acid elongation products in muscle tissue of *Syncerus caffer* and other ruminant species. Biochem. J. 115, 25–27.

Crawford, M.A., Gale, M., Woodford, M., Casperd, N., 1970. Comparative studies on fatty acid composition of wild and domestic meats. Int. J. Biochem. 1, 295–305.

Crawford, M., Sinclair, A.J., Msuya, P., Munhambo, A., 1973. Structural Lipids and Their Polyenoic Constituents in Human Milk. Raven Press, New York, pp. 41–56.

Crawford, M., Casperd, N., Sinclair, A., 1976. The long chain metabolites of linoleic and linolenic acids in liver and brain in herbivores and carnivores. Comp. Biochem. Physiol. B 54, 395–401.

Cunnane, S.C., Guesnet, P., 2011. Linoleic acid recommendations: a house of cards. Prostagl. Leuk. Essent. Fatty Acids 85, 399–402.

Cunnane, S.C., Courchesne-Loyer, A., St-Pierre, V., Vandenberghe, C., Pierotti, T., Fortier, M., Croteau, E., Castellano, C.A., 2016. Can ketones compensate for deteriorating brain glucose uptake during aging? Implications for the risk and treatment of Alzheimer's disease. Ann. NY Acad. Sci. 1367, 12–20.

Dyerberg, J., Eskesen, D.C., Andersen, P.W., Astrup, A., Buemann, B., Christensen, J.H., Clausen, P., Rasmussen, B.F., Schmidt, E.B., Tholstrup, T., Toft, E., Toubro, S., Stender, S., 2004. Effects of trans- and n-3 unsaturated fatty acids on cardiovascular risk markers in healthy males: an 8 weeks dietary intervention study. Eur. J. Clin. Nutr. 58, 1062–1070.

Emery, J.A., Norambuena, F., Trushenski, J., Turchini, G.M., 2016. Uncoupling EPA and DHA in fish nutrition: dietary demand is limited in Atlantic salmon and effectively met by DHA alone. Lipids 51, 399–412.

Fiennes, R.N., Sinclair, A.J., Crawford, M.A., 1973. Essential fatty acid studies in primates linolenic acid requirements of capuchins. J. Med. Primatol. 2, 155–169.

Galli, C., White, Jr., H.B., Paoletti, R., 1970. Brain lipid modifications induced by essential fatty acid deficiency in growing male and female rats. J. Neurochem. 17, 347–355.

Ghasemifard, S., Turchini, G.M., Sinclair, A.J., 2014. Omega-3 long chain fatty acid bioavailability: a review of evidence and methodological considerations. Prog. Lipid Res. 56, 92–108.

Gibson, R.A., Kneebone, G.M., 1981. Fatty acid composition of human colostrum and mature breast milk. Am. J. Clin. Nutr. 34, 252–257.

Gregory, M.K., Cleland, L.G., James, M.J., 2013. Molecular basis for differential elongation of omega-3 docosapentaenoic acid by the rat Elovl5 and Elovl2. J. Lipid Res. 54, 2851–2857.

Hachem, M., Geloen, A., Van, A.L., Foumaux, B., Fenart, L., Gosselet, F., Da Silva, P., Breton, G., Lagarde, M., Picq, M., Bernoud-Hubac, N., 2016. Efficient docosahexaenoic acid uptake by the brain from a structured phospholipid. Mol. Neurobiol. 53, 3205–3215.

Holman, R.T., 1986. Nutritional and biochemical evidences of acyl interaction with respect to essential polyunsaturated fatty acids. Prog. Lipid Res. 25, 29–39.

Holman, R.T., 1996. How I got my start in lipids, and where it led me. FASEB J. 10, 931–934.

Horrobin, D.F., 2002. A new category of psychotropic drugs: neuroactive lipids as exemplified by ethyl eicosapentaenoate (EE). Progress in Drug Research. Springer, Birkhauser, Basel, pp. 171–199.

Infante, J.P., Huszagh, V.A., 1998. Analysis of the putative role of 24-carbon polyunsaturated fatty acids in the biosynthesis of docosapentaenoic (22: 5n-6) and docosahexaenoic (22: 6n-3) acids. FEBS Lett. 431, 1–6.

James, A.T., 1970. The development of gas-liquid chromatography. Biochem. Soc. Symp. 30, 199–211.

James, M.J., Sullivan, T.R., Metcalf, R.G., Cleland, L.G., 2014. Pitfalls in the use of randomised controlled trials for fish oil studies with cardiac patients. Br. J. Nutr. 112, 812–820.

Kaur, G., Cameron-Smith, D., Garg, M., Sinclair, A.J., 2011. Docosapentaenoic acid (22:5n-3): a review of its biological effects. Prog. Lipid Res. 50, 28–34.

Kim, H.-W., Rao, J.S., Rapoport, S.I., Igarashi, M., 2011. Dietary n-6 PUFA deprivation downregulates arachidonate but upregulates docosahexaenoate metabolizing enzymes in rat brain. Biochim. Biophys. Acta, Mol. Cell. Biol. Lipids 1811, 111–117.

Klenk, E., 1965. The metabolism of polyenoic fatty acids. Adv. Lipid Res. 3, 1–23.

Kothapalli, K.S., Ye, K., Gadgil, M.S., Carlson, S.E., O'Brien, K.O., Zhang, J.Y., Park, H.G., Ojukwu, K., Zou, J., Hyon, S.S., 2016. Positive selection on a regulatory insertion-deletion polymorphism in FADS2 influences apparent endogenous synthesis of arachidonic acid. Mol. Biol. Evol. 33, 1726–1739.

Lands, B., 2012. Consequences of essential fatty acids. Nutrients 4, 1338–1357.

Lands, B., 2016. Benefit-risk assessment of fish oil in preventing cardiovascular disease. Drug Saf. 39, 787–799.

Leaf, A., Xiao, Y.F., Kang, J.X., Billman, G.E., 2005. Membrane effects of the n-3 fish oil fatty acids, which prevent fatal ventricular arrhythmias. J. Membr. Biol. 206, 129–139.

Lee, J.-W., Huang, B.X., Kwon, H., Rashid, M.A., Kharebava, G., Desai, A., Patnaik, S., Marugan, J., Kim, H.-Y., 2016. Orphan GPR110 (ADGRF1) targeted by N-docosahexaenoylethanolamine in development of neurons and cognitive function. Nat. Commun. 7, 13123.

Lillycrop, K.A., Burdge, G.C., 2015. Maternal diet as a modifier of offspring epigenetics. J. Dev. Orig. Health Dis. 6, 88–95.

Lin, Y.H., Salem, Jr., N., Wells, E.M., Zhou, W., Loewke, J.D., Brown, J.A., Lands, W.E., Goldman, L.R., Hibbeln, J.R., 2012. Automated high-throughput fatty acid analysis of umbilical cord serum and application to an epidemiological study. Lipids 47, 527–539.

Makrides, M., Gould, J.F., Gawlik, N.R., Yelland, L.N., Smithers, L.G., Anderson, P.J., Gibson, R.A., 2014. Four-year follow-up of children born to women in a randomized trial of prenatal DHA supplementation. JAMA 311, 1802–1804.

Mann, N., Sinclair, A., Pille, M., Johnson, L., Warrick, G., Reder, E., Lorenz, R., 1997. The effect of short-term diets rich in fish, red meat, or white meat on thromboxane and prostacyclin synthesis in humans. Lipids 32, 635–644.

Markworth, J.F., Kaur, G., Miller, E.G., Larsen, A.E., Sinclair, A.J., Maddipati, K.R., Cameron-Smith, D., 2016. Divergent shifts in lipid mediator profile following supplementation with n-3 docosapentaenoic acid and eicosapentaenoic acid. FASEB J. 30, 3714–3725.

Martinez, M., Ballabriga, A., Gil-Gibernau, J.J., 1988. Lipids of the developing human retina: I. total fatty acids, plasmalogens, and fatty acid composition of ethanolamine and choline phosphoglycerides. J. Neurosci. Res. 20, 484–490.

Mead, J.F., 1981. The essential fatty acids: past, present and future. Prog. Lipid Res. 20, 1–6.

Metherel, A.H., Domenichiello, A.F., Kitson, A.P., Hopperton, K.E., Bazinet, R.P., 2016. Whole-body DHA synthesis-secretion kinetics from plasma eicosapentaenoic acid and alpha-linolenic acid in the free-living rat. Bioch. Biophs. Acta 1861, 997–1004.

Mihailescu, M., Soubias, O., Worcester, D., White, S.H., Gawrisch, K., 2011. Structure and dynamics of cholesterol-containing polyunsaturated lipid membranes studied by neutron diffraction and NMR. J. Membr. Biol. 239, 63–71.

Miklavcic, J.J., Schnabl, K.L., Mazurak, V.C., Thomson, A.B., Clandinin, M.T., 2012. Dietary ganglioside reduces proinflammatory signaling in the intestine. J. Nutr. Metab. 2012, 280286.

O'Keefe, S.J., Li, J.V., Lahti, L., Ou, J., Carbonero, F., Mohammed, K., Posma, J.M., Kinross, J., Wahl, E., Ruder, E., 2015. Fat, fibre and cancer risk in African Americans and rural Africans. Nat. Commun. 6, 6342.

O'Brien, J.S., Sampson, E.L., 1965. Fatty acid and fatty aldehyde composition of the major brain lipids in normal human gray matter, white matter, and myelin. J. Lipid Res. 6, 545–551.

Park, H.G., Park, W.J., Kothapalli, K.S., Brenna, J.T., 2015. The fatty acid desaturase 2 (FADS2) gene product catalyzes Δ4 desaturation to yield n-3 docosahexaenoic acid and n-6 docosapentaenoic acid in human cells. FASEB J. 29, 3911–3919.

Pedersen, H.K., Gudmundsdottir, V., Nielsen, H.B., Hyotylainen, T., Nielsen, T., Jensen, B.A., Forslund, K., Hildebrand, F., Prifti, E., Falony, G., Le Chatelier, E., Levenez, F., Dore, J., Mattila, I., Plichta, D.R., Poho, P., Hellgren, L.I., Arumugam, M., Sunagawa, S., Vieira-Silva, S., Jorgensen, T., Holm, J.B., Trost, K., Kristiansen, K., Brix, S., Raes, J., Wang, J., Hansen, T., Bork, P., Brunak, S., Oresic, M., Ehrlich, S.D., Pedersen, O., 2016. Human gut microbes impact host serum metabolome and insulin sensitivity. Nature 535, 376–381.

Peluffo, R.O., Nervi, A.M., Gonzalez, M.S., Brenner, R.R., 1984. Effect of different amino acid diets on Δ5, Δ6 and Δ9 desaturases. Lipids 19, 154–157.

Proudman, S.M., James, M.J., Spargo, L.D., Metcalf, R.G., Sullivan, T.R., Rischmueller, M., Flabouris, K., Wechalekar, M.D., Lee, A.T., Cleland, L.G., 2015. Fish oil in recent onset rheumatoid arthritis: a randomised, double-blind controlled trial within algorithm-based drug use. Ann. Rheum. Dis. 74, 89–95.

Ramsden, C.E., Faurot, K.R., Zamora, D., Suchindran, C.M., Macintosh, B.A., Gaylord, S., Ringel, A., Hibbeln, J.R., Feldstein, A.E., Mori, T.A., Barden, A., Lynch, C., Coble, R.,

Mas, E., Palsson, O., Barrow, D.A., Mann, J.D., 2013. Targeted alteration of dietary n-3 and n-6 fatty acids for the treatment of chronic headaches: a randomized trial. Pain 154, 2441–2451.

Ramsden, C.E., Zamora, D., Majchrzak-Hong, S., Faurot, K.R., Broste, S.K., Frantz, R.P., Davis, J.M., Ringel, A., Suchindran, C.M., Hibbeln, J.R., 2016. Re-evaluation of the traditional diet-heart hypothesis: analysis of recovered data from Minnesota Coronary Experiment (1968–73). BMJ 353, i1246.

Rapoport, S.I., Igarashi, M., Gao, F., 2010. Quantitative contributions of diet and liver synthesis to docosahexaenoic acid homeostasis. Prostagl. Leukot. Essent. Fatty Acids 82, 273–276.

Richard, C., Lewis, E.D., Field, C.J., 2016. Evidence for the essentiality of arachidonic and docosahexaenoic acid in the postnatal maternal and infant diet for the development of the infant's immune system early in life. Appl. Physiol. Nutr. Metab. 41, 461–475.

Rioux, V., Choque, B., Ezanno, H., Duby, C., Catheline, D., Legrand, P., 2015. Influence of the cis-9, cis-12 and cis-15 double bond position in octadecenoic acid (18:1) isomers on the rat FADS2-catalyzed Delta6-desaturation. Chem. Phys. Lipids 187, 10–19.

Rivers, J.P., Sinclair, A.J., Crawford, M.A., 1975. Inability of the cat to desaturate essential fatty acids. Nature 258, 171–173.

Saunders, E.F., Ramsden, C.E., Sherazy, M.S., Gelenberg, A.J., Davis, J.M., Rapoport, S.I., 2016. Reconsidering dietary polyunsaturated fatty acids in bipolar disorder: a translational picture. J. Clin. Psych. 77, e1342–e1347.

Sidhu, V.K., Huang, B.X., Desai, A., Kevala, K., Kim, H.Y., 2016. Role of DHA in aging-related changes in mouse brain synaptic plasma membrane proteome. Neurobiol. Aging 41, 73–85.

Sinclair, A.J., 1975. Incorporation of radioactive polyunsaturated fatty acids into liver and brain of developing rat. Lipids 10, 175–184.

Sinclair, A.J., 1994. John Rivers (1945–1989): his contribution to research on polyunsaturated fatty acids in cats. J. Nutr. 124, 2513S.

Sinclair, A.J., Crawford, M.A., 1972. The incorporation of linolenic aid and docosahexaenoic acid into liver and brain lipids of developing rats. FEBS Lett. 26, 127–129.

Sinclair, A., O'Dea, K., 1993. The significance of arachidonic acid in hunter-gatherer diets: implications for the contemporary western diet. J. Food Lipids 1, 143–157.

Sinclair, A.J., McLean, J., Monger, E., 1979. Metabolism of linoleic acid in the cat. Lipids 14, 932–936.

Sinclair, A.J., O'Dea, K., Naughton, J.M., 1983. Elevated levels of arachidonic acid in fish from northern Australian coastal waters. Lipids 18, 877–881.

Sinclair, A., Johnson, L., O'Dea, K., Holman, R., 1994. Diets rich in lean beef increase the eicosatrienoic, arachidonic, eicosapentaenoic and docosapentaenoic acid content of plasma phospholipids. Lipids 29, 337–343.

Sprague, M., Dick, J.R., Tocher, D.R., 2016. Impact of sustainable feeds on omega-3 long-chain fatty acid levels in farmed Atlantic salmon, 2006–2015. Sci. Rep. 6, 21892.

Stark, K., Salem, N., 2005. Fast gas chromatography for the identification of fatty acid methyl esters from mammalian samples. Lipid Technol. 17, 181.

Suphioglu, C., Sadli, N., Coonan, D., Kumar, L., De Mel, D., Lesheim, J., Sinclair, A.J., Ackland, M.L., 2010. Zinc and DHA have opposing effects on the expression levels of histones H3 and H4 in human neuronal cells. Br. J. Nutr. 103, 344–351.

Svennerholm, L., 1968. Distribution and fatty acid composition of phosphoglycerides in normal human brain. J. Lipid Res. 9, 570–579.

Taha, A.Y., Blanchard, H.C., Cheon, Y., Ramadan, E., Chen, M., Chang, L., Rapoport, S.I., 2016a. Dietary linoleic acid lowering reduces lipopolysaccharide-induced increase in brain arachidonic acid metabolism. Mol. Neurobiol. 23, 1–13.

Taha, A.Y., Hennebelle, M., Yang, J., Zamora, D., Rapoport, S.I., Hammock, B.D., Ramsden, C.E., 2016b. Regulation of rat plasma and cerebral cortex oxylipin concentrations with increasing levels of dietary linoleic acid. Prostagl. Leuk. Essent. Fatty Acids, pii: S0952-3278(16)30017-5.

Thiés, F., Delachambre, M.C., Bentejac, M., Lagarde, M., Lecerf, J., 1992. Unsaturated fatty acids esterified in 2-acyl-l-lysophosphatidylcholine bound to albumin are more efficiently taken up by the young rat brain than the unesterified form. J. Neurochem. 59, 1110–1116.

Von Schacky, C., 2009. Omega 3 fatty acids vs. cardiac disease: the contribution of the omega 3 index. Cell. Mol. Biol. 56, 93–101.

Voss, A., Reinhart, M., Sankarappa, S., Sprecher, H., 1991. The metabolism of 7, 10, 13, 16, 19-docosapentaenoic acid to 4, 7, 10, 13, 16, 19-docosahexaenoic acid in rat liver is independent of a 4-desaturase. J. Biol. Chem. 266, 19995–20000.

Weisinger, H.S., Vingrys, A.J., Sinclair, A.J., 1996. The effect of docosahexaenoic acid on the electroretinogram of the guinea pig. Lipids 31, 65–70.

Weisinger, H.S., Armitage, J.A., Sinclair, A.J., Vingrys, A.J., Burns, P.L., Weisinger, R.S., 2001. Perinatal omega-3 fatty acid deficiency affects blood pressure later in life. Nat. Med. 7, 258–259.

Wheeler, T.G., Benolken, R.M., Anderson, R.E., 1975. Visual membranes: specificity of fatty acid precursors for the electrical response to illumination. Science 188, 1312–1314.

Xu, L., Sinclair, A.J., Faiza, M., Li, D., Han, X., Yin, H., Wang, Y., 2017. Furan fatty acids–beneficial or harmful to health? Progress in Lipid Research 68, 119–137.

## Further Readings

Comte, C., Bellenger, S., Bellenger, J., Tessier, C., Poisson, J.P., Narce, M., 2004. Effects of streptozotocin and dietary fructose on delta-6 desaturation in spontaneously hypertensive rat liver. Biochimie 86, 799–806.

Park, W.J., Kothapalli, K.S., Lawrence, P., Tyburczy, C., Brenna, J.T., 2009. An alternate pathway to long-chain polyunsaturates: the FADS2 gene product Δ8-desaturates 20: 2n-6 and 20: 3n-3. J. Lipid Res. 50, 1195–1202.

# 2

# Polyunsaturated Fatty Acid Biosynthesis and Metabolism in Adult Mammals

*Graham C. Burdge*

University of Southampton, Southampton, United Kingdom

## INTRODUCTION

Mammals are able to synthesize n-9 series polyunsaturated fatty acids (PUFA) from acetyl-CoA via fatty acid synthase to form 16:0, which can then be converted through chain elongation and stearoyl-CoA desaturase activities to form 18:1n-9. Synthesis of 20:3n-9 then proceeds by one of two pathways involving sequential desaturation and carbon chain elongation to form 20:3n-9 (Fig. 2.1) (Ichi et al., 2014).

Synthesis of 20:3n-9 is important in dietary essential fatty acid (EFA) deficiency. However, mammals lack $\Delta12$ and $\Delta15$ desaturases and,

Polyunsaturated Fatty Acid Metabolism. http://dx.doi.org/10.1016/B978-0-12-811230-4.00002-8

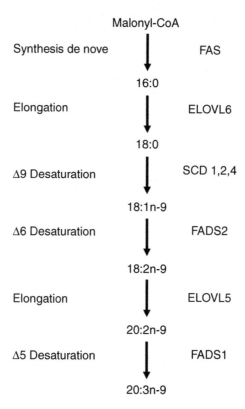

FIGURE 2.1  **Conversion of 18:1n-9 to 20:3n-9 in mammals.** *FADS,* Fatty acid desaturase; *FAS,* fatty acid synthase; *ELOVL,* elongase; *SCD,* stearoyl-CoA deasturase.

consequently, are unable to convert 18:1n-9 to long chain PUFA of either the n-3 or n-6 series from acetyl-CoA. Hence, the products of Δ12 desaturase and Δ15 desaturase activities in plants, 18:2n-6 and 18:3n-3, respectively, are essential in the mammalian diet (Simopoulos, 1999), such that dietary deficiency results in specific symptoms, including skin lesions, infertility, and impaired renal function (Innis, 1991). The majority of the biological effects of 18:3n-3 and 18:2n-6 PUFA have been ascribed to their longer chain metabolites acting via membrane fluidity, production of second messengers, and activation of transcription factors (Sinclair et al., 2002). VLCPUFA are also required for normal development. For example, 22:6n-6 is incorporated into cell membranes of the central nervous system during development and accounts for between 20% and 50% of fatty acids in brain and retinal cells membranes (Uauy et al., 2001). Deficits in 22:6n-3 accumulation can result in impaired neurological function (Uauy et al., 1996; Uauy and Hoffman, 2000). Adequate dietary intake is important for meeting requirements for very long chain polyunsaturated fatty acids (VLCPUFA). The primary source of n-3 VLCPUFA, namely 20:5n-3,

22:5n-3, and 22:6n-6, is oily fish while n-6 VLCPUFA, namely 20:3n-6 and 20:4n-6, are obtained mainly from meat. Nevertheless, mammals can convert n-6 and n-3 EFAs to their respective series of VLCPUFA. However, there are between and within species variations in capacity for conversion of 18:3n-3 and 18:2n-6 to their longer chain, more unsaturated metabolites.

# PARTITIONING OF EFAs TOWARDS β-OXIDATION AND CARBON RECYCLING

A substantial proportion of ingested EFAs are used in energy production. Estimates of the proportion of ingested 18:3n-3 recovered as $^{13}CO_2$ on breath in men have been reported to vary between 24% and 33%, although this variation may reflect differences in the duration of collecting samples after ingestion of labeled 18:3n-3 (Bretillon et al., 2001; Burdge et al., 2002, 2003a; DeLany et al., 2000). Using the same methodology 22% of labeled 18:3n-3 was recovered as $^{13}CO_2$ in women (Burdge and Wootton, 2002a) compared to 33% in men (Burdge et al., 2002). This difference between sexes may reflect relatively lower muscle mass in women compared to men. This has implications for capacity for conversion of 18:3n-3 to VLCPUFA in that the proportion of ingested 18:3n-3 available for conversion would be expected to be greater in women than in men. Such metabolic partitioning does not appear to be influenced by dietary intakes of 18:3n-3 or 20:5n-3 plus 22:6n-3 (Burdge et al., 2003b). The proportion of labeled 18:2n-6 as $^{13}CO_2$ over 9 h was shown to be less (16%) than for 18:3n-3 (24%) in men (DeLany et al., 2000). Others have reported similar findings (18:2n-6, 14%; 18:3n-3, 23%) when excretion of $^{13}CO_2$ on breath was followed for up to 48 h (Bretillon et al., 2001). Moreover, comparison in the same individuals showed that although the proportion of ingested 18:2n-6 that was recovered as $^{13}CO_2$ (14%) was similar to other long chain fatty acids (16:0, 14%; 18:0, 11%; and 18:1n-9, 17%), the proportion of [$^{13}C$]18:3n-3 recovered as $^{13}CO_2$ (24%) was similar to 12:0 (33%). Recovery of ingested 18:3n-3 as $CO_2$ in rats (63%) was also found to be greater than that of 18:2n-3 (48%), similar to 12:0 (63%), and approximate 1.6–2.5-fold greater than 14:0, 16:0, and 18:0 (Leyton et al., 1987). McCloy et al. (2004) found lower recovery of [$^{13}C$]18:2n-6 (12%) as $^{13}CO_2$ than [$^{13}C$]18:3n-3 (19%) in women. Preferential β-oxidation of 18:3n-3 has also been reported in the liver of catfish (Bandyopadhyay et al., 1982). Together findings suggest that preferential partitioning of 18:3n-3 for energy production is conserved between classes as well as between species. If so, this appears to be counterintuitive if 18:3n-3 is an important source of n-3 VLCPUFA.

Carbon from fatty acids undergoing β-oxidation can by incorporated into cholesterol or fatty acids synthesized *de novo* or into ketone bodies (Cunnane et al., 2003). Each cycle of the β-oxidation pathway released two carbons as acetyl-CoA, which is converted to citrate and exported from mitochondria.

Acetyl-CoA can be used as a substrate for fatty acid synthase or converted to acetoacetyl-CoA, which is a substrate for hydroxymethylglutaryl-CoA (HMG-CoA) reductase in the cholesterol biosynthesis pathway or HMG-CoA lyase to form acetoacetate in the ketone body synthesis pathway. Sinclair showed that radioactivity from [$^{14}$C]18:2n-6 and [$^{14}$C]18:3n-3 was incorporated into 18:0 and 18:1n-9 by 22 h after administration in brain and liver from neonatal rats, although the majority of the label was present in the EFAs (Sinclair, 1974). In contrast, others have shown that radioactivity derived from [$^{14}$C]18:3n-3 was enriched primarily in 16:0, 8 h after administration in neonatal rat brain and that the radiolabel was associated initially with brain phosphatidylcholine (PC), but subsequently enriched in cholesterol (Dhopeshwarkar and Subramanian, 1975a,b). Carbon from [$^{13}$C]18:n-3 was enriched in plasma SFA and MUFA from regnant rhesus macaques 5 days after administration of the labeled fatty acid (Sheaff Greiner et al., 1996). Furthermore, enrichment in 16:0 or 18:0 was approximately twice that of 20:5n-3 or 22:6n-6. The same series of experiments showed that carbon from [$^{13}$C]18:2n-6 was enriched in 18:1n-9, but not 16:0, 16:1n-7 or 18:0. Enrichment of 18:1n-9 was approximately ninefold greater than in 20:4n-6. Recycling of carbon from [$^{13}$C]22:6n-3 into saturated (SFA) and monounstuarated (MUFA) fatty acids was approximately eightfold and sevenfold lower than from 18:2n-6 to 18:3n-3, respectively. These findings suggest that there was no selective sparing of 18:3n-3 or 18:2n-6 from $\beta$-oxidation during pregnancy to meet increased fetal and maternal demands for VLCPUFA, although partitioning of preformed 22:6n-3 towards energy production appeared to be lower than for EFAs. Recycling of carbon from [$^{13}$C]18:3n-3 has also been reported in men and women (Burdge and Wootton, 2003). In men, the rank order of enrichment in plasma PC plus triacylglycerol (TAG) was 16:0 > 18:0 = 18:1n-9 = 16:1n-7. However, the rank order in women was 16:0 > 18:0 > 18:1n-9 > 16:1n-7. Furthermore, synthesis of SFA and MUFA fatty acids from 18:3n-3 was 20% greater in men than in women. It has been suggested that partitioning of EFAs towards $\beta$-oxidation may be involved in regulating that fatty acid composition of cell membranes and hence facilitating optimum function by preventing excessive accumulation of these fatty acids in the lipid bilayer if abundant in the diet (Cunnane et al., 2003). However, this does not explain why there are differences between males and females in the extent of such partitioning and assumes that the composition of cells membranes is determined primarily by the diet.

## CONVERSATION OF EFAs TO VLCPUFA

The consensus pathway for conversion of 183n-3 and 18:2n-6 to VLCPUFA is summarized in Fig. 2.2 (Sprecher and Chen, 1999).

The historical background to the elucidation of this pathway has been documented by Sprecher (2000). Although this pathway has been largely accepted, there have been several reports that have suggested alternative or additional reactions may be involved. The conversion of 18 carbon EFAs to VLCPUFA was described originally more than 40 years ago (Klenk and Mohrhauer, 1960). All reactions take place in the endoplasmic reticulum up to the synthesis of 24:6n-3 and 24:5n-6. The initial, rate-limiting reaction is insertion of a double bond into either 9,12–18:2 or 9,12,15–18:3 at the Δ6 carbon to form 6,9,12–18:3 or 6,9,12,15–18:4, respectively. The reaction is catalyzed by Δ6 desaturase (see Chapter 5), which is encoded by the FADS2 gene. 18:3n-6 and 18:4n-3 are converted to 8,11,14–20:3 and 8,11,14,17–20:4, respectively, by addition of 2 carbons by elongase 5 (see Chapter 6) activity (Fig. 2.2). 20:3n-6 and 20:4n-3 are then converted to 5,8,11,14–20:4 and 5,8,11,14,17–20:5, respectively, by addition of a double bond at the Δ5 position by Δ5 desaturase. Two cycles of chain elongation catalyzed by elongase 2 or 5 and then elongase 2 subsequently convert 20:4n-6 to 7,10,13,16–22:4 and then to 9,12,15,18–24:4, and 20:5n-3 to 7,10,13,16,19–22:5, and then 9,12,15,18,21–24:5 (Fig. 2.2). Because there is only limited esterification of these 24 carbon fatty acids, it has been assumed that they are metabolic intermediates (Voss et al., 1991). Synthesis of 4,7,10,13,16–22:5 and 4,7,10,13,16,19–22:6 does not appear to involve Δ4 desaturase activity (Ayala et al., 1973). Instead, a double bond

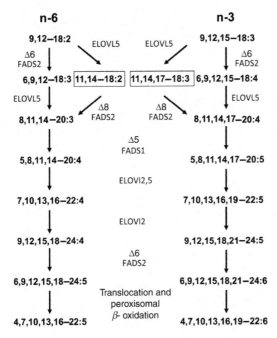

FIGURE 2.2 Pathway for conversion of essential fatty acids to very long chain polyundsaturated fatty acids. *FADS*, Fatty acid desaturase; *ELOVL*, elongase; Δ8,6,5, position of desaturated bond.

is inserted at the Δ6 position by Δ6 desaturase to form 6,9,12,15,18,21–24:5 and 6,9,12,15,18,21–24:6, respectively, followed by translocation from the endoplasmic reticulum to peroxisomes and one cycle of peroxisomal fatty acid β-oxidation (Voss et al., 1991) (Fig. 2.2). The requirement for persoxisomal fatty acid β-oxidation in 22:5n-6 and 22:6n-3 synthesis is supported by the absence of conversion of 24:6n-3 to 22:6n-3 in fibroblasts from patients with peroxisomal disease (Moore et al., 1995; Su et al., 2001) and that conversion of 18:3n-3 to 22:6n-3 only occurred when microsomes and peroxisomes were incubated together (Li et al., 2000). Furthermore, because 22 carbon fatty acids with the first double bond at the Δ4 position are relatively poor substrates for peroxisomal fatty acid β-oxidation (Luthria et al., 1996; Sprecher et al., 1995), this may provide a mechanism to limit the conversion of 24 carbon fatty acids to 22 carbon VLCPUFA rather than complete hydrolysis of the substrate. The translocation of 24 and 22 carbon fatty acids between the endoplasmic reticulum and peroxisomes has been suggested to be a possible locus of metabolic regulation that could facilitate control of 22:5n-6 and 22:6n-3 synthesis independently from the preceding steps in the pathway (Sprecher, 2000).

There is some evidence that suggests the same Δ6 desaturase catalyzes both the conversion of 18:3n-3 to 18:4n-3 and 24:5n-3 to 24:6n-3. Transfection of the FADS2 gene into COS cells facilitated Δ6 desaturation of both 18:3n-3 and 24:5n-4 (D'Andrea et al., 2002). If so, this suggests that flux through the pathway could be regulated by competition between 18 and 24 carbon substrates for Δ6 desaturase. Furthermore, Δ6 desaturase has greater activity against 18:3n-3 than 18:2n-2 (Brenner and Peluffo, 1966), which may have implications for the regulation of n-3 and n-6 VLCP-UFA synthesis at the 24 carbon steps. Coincubation of Jurkat T cells with [$^{14}$C]18:3n-3 and the Δ6 desaturase inhibitor *trans*-9,12–18:2 reduced enrichment of the label in 24:6n-3 + 22:6n-3, indicating an overall suppression of flux through the pathway due to inhibition of the rate limiting reaction (Marzo et al., 1996). However, the suppression of enrichment of the label in 24:6n-3 + 22:6n-3 was less when cells were incubated with [$^{14}$C]20:5n-3. This was interpreted as indicating that the second Δ6 desaturation was carried out by an enzyme that was less sensitive to the inhibitor. However, these findings have yet to be replicated.

## ALTERNATIVE PATHWAYS FOR VLCPUFA BIOSYNTHESIS

The enzyme product of FADS2 has been shown to exhibit both Δ8 and Δ4 desaturase activities as well as Δ6 desaturase activity. Consequently, alternative pathways for VLCPUFA synthesis to that described by Sprecher (2000) have been reported in mammalian cells. Park et al. (2009) have shown that human MCF7 breast carcinoma cells, that lack Δ6 desaturase activity, synthesise 22:6n-3 and 22:5n-6 when transfected with FADS2 and that this synthesis involves direct desaturation of 22:5n-3 and 22:4n-6, respectively, via Δ4 desaturase activity. These findings demonstrate that human cells can exhibit Δ4 desaturase activity, although whether this is present in unmodified cells, as in other organisms (Meyer et al., 2003; Morais et al., 2012; Qiu et al., 2001), has yet to be demonstrated.

FADS2 from baboons exhibited Δ8 desaturase activity when transfected into yeast such that 18:3n-3 was converted to 20:3n-3 via elonagase activity followed by Δ8 desaturation to form 20:4n-3 (Park et al., 2009). However, Chen et al. (2000) failed to detect endogenous Δ8 desaturase activity in rat testes although others have reported Δ8 desaturation in rat testes (Albert and Coniglio, 1977). There is also evidence for Δ8 desaturase activity in human bladder cells and colonocytes. Some studies have suggested that cats are unable to convert 18:2n-6 to 20:4n-6 (Hassam et al., 1977; Rivers et al., 1975) and lack Δ6 desaturase activity (Sinclair et al., 1979). However, there is circumstantial evidence that 20:4n-6 synthesis in cats involves Δ8 desaturase activity (Sinclair et al., 1981), which explained the low but persistent presence of 20:4n-6 in the tissues of cats maintained on a diet, which contained 18:2n-6, but not longer chain n-6 PUFA. Overall, although there is evidence that Δ4 and Δ8 desaturation can occur in mammalian cells, it has yet to be demonstrated whether these alternative PUFA biosynthesis pathways are quantitatively important.

The enzyme product of FADS2 has also been shown to desaturate 16:0 at the Δ6 position to form 16:1n-10 when 16:0 exceeds capacity for Δ9 desaturation of excess 16:0 (Park et al., 2016a).

## RETROCONVERSION OF VLCPUFA

Despite the apparently limited peroxisomal fatty acid $\beta$-oxidation that is involved in conversion of 24 carbon VLCPUFA to 22:5n-6 and 22:6n-3, there is evidence that 22:6n-3 can undergo retroconversion to 22:5n-3 and 20:5n-3. Enrichment of 22:5n-3 and 20:5n-3 was detected after administration of $[^{13}C]$22:6n-3 to pregnant rhesus macaques such that the levels of these fatty acids were approximately 1% of the administered labeled 22:6n-3 (Sheaff Greiner et al., 1996). In isolated rat hepatocytes, approximately 20% of labeled 22:6n-3 was converted to 20:5n-3. This was independent of (−)-carnitine or (+)-deconoylcarnitine, which suggested that such retroconversion involved peroxisomal, rather than mitochondrial, fatty acid $\beta$-oxidation (Gronn et al., 1991). Dietary supplementation with 22:5n-5 in rats increased the concentrations of 20:5n-3 in liver, heart, adipose tissue, and skeletal muscle, but not brain, while supplementation with 22:6n-3 increased the concentration of 20:5n-3, but not 22:5n-3, in liver (Kaur et al., 2010). However, but there was no evidence of retroconversion of 22:6n-3 in brain, adipose tissue, skeletal, or cardiac muscle

(Kaur et al., 2010). Park et al. (2016b) have shown differential retroconversion of 22:6n-3 to 20:5n-3 between different human cell lines. Retroconversion of [$^{13}$C]22:6n-3 was found to be fivefold to sixfold greater in HepG2 hepatocarcinoma and MCF7 breast carcinoma cells that in Y79 or SK-N-SH neuroblastoma cells, although labeled 22:5n-3 and 24:5n-3 were not detected (Park et al., 2016b). This is in contrast to findings in rats (Sheaff Greiner et al., 1996) and suggests some differences in the retroconversion pathway between species. Retroconversion of 22:6n-3 to 20:5n-3 has been reported to be two-fold greater in elderly humans than young subjects (Plourde et al., 2011). Furthermore, dietary supplementation with 22:6n-3 in omnivores and vegetarians increased its proportion in plasma and platelet phospholipids and was accompanied by an increase in the proportion of 20:5n-3, but a decrease in 22:5n-3 (Conquer and Holub, 1997). This is consistent with the retroconversion of 22:6n-3 in cultured cells and is consistent with the suggestion of metabolic channeling of intermediates in the retroconversion pathway (Park et al., 2016b). There was no difference between omnivores and vegetarians in the extent of retroconversion (approximately 10% of the ingested 22:6n-3) (Conquer and Holub, 1997).

## CONVERSION OF EFAs TO VLCPUFA IN MALES

Humans express all of the enzymes required to convert 18:3n-3 and 18:2n-6 to their respective longer chain metabolites, including 22:6n-3 and 22:5n-6 (de Gomez Dumm and Brenner, 1975). However, in contrast to rodents, the capacity of humans to synthesise VLCPUFA from EFAs appears to be limited. A series of studies investigated the extent of whole body conversion of 18:3n-3 labeled with a stable isotope to 20:5n-3, 22:5n-3, and 22:6n-3 in adult humans. Although there are methodological limitations to this approach, including lack of standardization between studies (Burdge, 2004), the findings show consistency between experiments. Mathematic modeling of the appearance and turnover of metabolites of [$d_5$]-18:3n-3 in blood after consuming a single dose of 1g in a mixed group of men and women showed that approximately 0.2% of the label was recovered in 20:5n-3, 0.13% in 22:5n-3 and 0.05% in 22:5n-3 (Pawlosky et al., 2001). Others have shown in men that approximately 8% of label from [U-$^{13}$C]-18:3n-3 was recovered in 20:5n-3 and in 22:5n-3 in blood (based on calculation of the area under the concentration versus time curve), while incorporation of label into 22:6n-3 was below the limit of detection (Burdge et al., 2002). Salem et al. (1999) also found significant synthesis of 20:5n-3 (peak concentration 57 ng/mL), while incorporation of label into 22:6n-3 was <2 ng/mL. Others have reported similar findings (Goyens et al., 2005; Vermunt et al., 2000). Together these finding imply a constraint in conversion of 18:3n-3 to 22:6n-3 downstream of synthesis of

22:5n-3. This is supported by the findings of a dietary supplementation study in which men and postmenopausal women consumed either 18:3n-3, 18:4n-3 (circumventing the first Δ6 desaturation reaction), or 20:5n-5 (circumventing both the first Δ6 desaturation reaction and Δ5 desaturation) (James et al., 2003). The findings showed an increase in the proportions of 20:5n-3 and 22:5n-3, but not 22:6n-3, in erythrocyte and plasma phospholipids. It remains unclear whether this constraint is at the second desaturation at the Δ6 position, the addition of carbons by elongase 2 or the translocation of 24 carbon intermediates (Fig. 2.3). In contrast to the findings of these studies, Emken et al. (1990) reported that, based on estimates of the area under the concentration versus time curve, that conversion of deuterated 18:3n-3 to 20:5n-3 was 6%, to 22:5n-3 was 3.5%, and to 22:6n-3 was 3.8% in men. The substantially higher conversation to 22:6n-3 compared to other studies has not been explained.

The findings of stable isotope tracer studies are supported by the outcomes of trials in which adults consumed increased amounts of 18:3n-3. These have been reviewed recently by Baker et al. (2016). Although 18:3n-3 intake is associated significantly with the concentration of 20:5n-3 and 22:5n-3 in plasma phospholipids, or erythrocyte leukocyte cell membranes, there was no significant increase in the proportion or concentration of 22:6n-3 even at 18:3n-3 intakes of 15 g per day (Fig. 2.3).

In some studies, 22:6n-3 status decreased with increased 18:3n-3 intake (Baker et al., 2016). This may reflect retroconversion of 22:6n-3 in the absence of significant synthesis or dietary intake.

Overall, men have some capacity for conversion of 18:3n-3 to 20:5n-3 and 22:5n-3, but are unable to synthesize sufficient 22:6n-3 to exceed turnover or retroconversion. One implication of these findings is that consuming 18:3n-3 is not sufficient to meet requirements for 22:6n-3. This suggestion supported by the finding that despite equal or higher 18:3n-3 status or intakes, 22:6n-3 levels are consistently lower in vegans than in omnivores (Sanders, 2009).

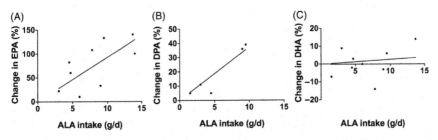

FIGURE 2.3 The effect of increasing 18:3n-3 (ALA) intake on the change in 20:5n-5, 22:5n-3, and 22:6n-3 in human plasma phospholipids. *Source: Reproduced with permission from Baker, E.J., Miles, E.A., Burdge, G.C., Yaqoob, P., Calder, P.C., 2016. Metabolism and functional effects of plant-derived omega-3 fatty acids in humans. Prog. Lipid Res. 64, 30–56.*

There is less information available about the conversion of 18:2n-6 to VLCPUFA than for 18:3n-3. Lin et al. (2005) showed that conversion of 20:3n-6 to 20:4n-6 was 14-fold greater than 18:2n-6 to 20:4n-6 in rats and that the major metabolites of 18:2n-6 were 20:4n-6 and 22:4n-6. Su et al. (1999) showed that fetal baboons derive approximately 50% of 20:4n-6 from 18:2n-6. Following consumption of 1 g [d5]-18:2n-6 by human volunteers (sex mix not reported), peak 20:4n-6 concentration (approximately 0.011 μg/mL) was detected in total plasma lipids at 96 h after the test meal (Salem et al, 1999). Demmelmair et al. (1998) showed that when lactating women consumed 1 mg/kg body weight [U-$^{13}$C]-18:2n-6 on either the 2nd, 6th, or 12th week of breast feeding, recovery of ingested label as $^{13}$CO$_2$ on breath was 18.9%, 24%, and 17.7%, respectively, over 5 days. The proportion of ingested label recovered as 18:2n-6 in breast milk was 12.7%, 13.1%, and 11.7% for each time point studied, respectively. Recovery of ingested labeled 18:2n-6 as 20:3n-6 in breast milk was between 0.17% and 0.2%, and as 0.02% of 20:4n-6 at all time points studied (Demmelmair et al., 1998). The authors estimated that overall, 30% of 18:2n-6 in breast milk was derived directly from the diet, while 11% of 20:3n-6 and 1.2% of 20:4n-6 in breast milk were derived from 18:2n-6 by endogenous synthesis (Demmelmair et al., 1998). In men who consumed a single dose of 400 mg [U-$^{13}$C]-18:2n-6 plus 400 mg [U-$^{13}$C]-18:3n-3, conversion of [$^{13}$C]-18:2n-6 to 20:3n-6 or to 20:4n-6 over 14 days was 0.23% or 0.18% of dose adjusted for precursor pool size, respectively (Hussein et al., 2005). Estimated conversion of [U-$^{13}$C]-18:3n-3 to 20:5n-3 was 0.26%; to 22:5n-3, 0.04%; and to 22:6n-3 <0.01%, adjusted for precursor pool size (Hussein et al., 2005). Thus conversion to 20:5n-3 was similar to 20:4n-6. However, when the same subjects consumed a sunflower oil enriched diet, the proportion of [U-$^{13}$C]-18:2n-6 recovered as 20:4n-6 was 0.26%, while the proportion of [U-$^{13}$C]-18:3n-3 recovered as 20:5n-3 was 0.19%. Conversely, when subjects consumed a diet enriched in flaxseed oil, the proportion of [U-$^{13}$C]-18:2n-6 recovered as 20:4n-6 was 0.12%, while the proportion of [U-$^{13}$C]-18:3n-3 recovered as 20:5n-3 was 0.29% (Hussein et al., 2005). These findings emphasize the competition between n-6 and n-3 PUFA for the PUFA synthesis pathway.

## CONVERSION OF EFAs TO VLCPUFA IN NONPREGNANT AND PREGNANT FEMALES

Comparison of the conversion of [U-$^{13}$C]-18:3n-3 to VLCPUFA in women of reproductive age and similarly aged men using the same experimental design showed that synthesis of 20:5n-3 and 22:6n-3 was substantially greater in women (20:5n-3, 21%; 22:6n-3, 9%) than in men (20:5n-3, 8%; 22:6n-3 undetectable) based on estimated area under the time versus concentration curve in plasma (Burdge et al., 2002; Burdge and Wootton, 2002a). Systematic review of 51 studies that reported fatty acid status in men and women showed that levels of 20:4n-6 and 22:6n-3 were both 20% higher in women than in men (Lohner et al., 2013). Furthermore, higher 22:6n-3 status has also been reported in female rodents (Burdge et al., 2008; Childs et al., 2010) and wild birds (Toledo et al., 2016), compared to males of the species. Such apparent conservation of this sex difference suggests that higher 22:6n-3 status in females may be biologically important.

The developing mammalian fetus requires adequate assimilation of 22:6n-3 in order to support optimal development of the central nervous system. The concentration of 22:6n-3 in plasma phospholipids has been shown to increase significantly from approximately 20 weeks after conception in pregnant women and returns to baseline concentration after delivery (Postle et al., 1995). This involves a switch from PC 18:0/22:6 to PC 16:0/22:6. This increase in the 22:6n-3 content of maternal plasma PC precedes the main period of 22:6n-3 assimilation by the fetus (Kuipers et al., 2012). Similar changes to plasma phospholipid composition have been reported in rats (Burdge et al., 1994). This involves decreased acyl remodeling of sn-1 16:0 molecular species to sn-1 18:0 molecular species, increased PC synthesis de novo, but not phosphatidylethanolamine N-methylation, and increased microsomal choline-phosphotransferase activity. Furthermore, the mRNA expression of FADS2 has been shown to increase in the liver of pregnant rats (Childs et al., 2010, 2012). It is possible that greater 22:6n-3 synthesis and higher status in females together with capacity to upregulate conversion of 18:3n-3 to 22:6n-3 may facilitate an adequate supply of 22:6n-3 to the offspring, although this has not been tested directly. If so, the low DHA status in pregnant vegans (Sanders, 2009) suggests that such physiological adaptations are unable to compensate for exclusion of major foods from the diet.

The mechanism that underlies higher 22:6n-3 status and synthesis in females is not known. Women who took a combined hormone oral contraceptive synthesized 2.5-fold more DHA than those who did not take use hormone therapy (Burdge and Wootton, 2002a). Giltay et al. (2004b) have shown that women who take an oral contraceptive pill had 10% higher plasma DHA concentration than those who did not. The same group also showed that treatment of postmenopausal women with the hepatic estrogen receptor agonist raloxifene did not increase plasma 20:4n-6 concentration, but increased 22:6n-3 concentration by 22.1%, while combined treatment with conjugated equine estrogens plus medroxyprogesterone acetate increased 20:4n-6 concentration by 14% and 22:6n-3 concentration by 14.9% (Giltay et al., 2004a). Thus both estrogen and progesterone appeared to induce VLCPUFA synthesis. Systematic treatment of HepG2 hepatocarcinoma cells or primary human hepatocytes with physiological concentrations of either 17α-ethinylestradiol, progesterone or testosterone showed that while there was no effect of estrogen or testosterone on the conversion of $[^{13}C]$-18:3n-3 to VLCPUFA, progesterone increased synthesis of 20:5n-3, 22:5n-3, and 22:6n-3, the mRNA expression of FADS2, FADS21, and ELOVL2 and 5, which encode elongase 2 and 5, respectively, in a concentration-dependent manner (Sibbons et al., 2014). These findings suggest that progesterone may be the main hormone that is responsible for higher 22:6n-3 synthesis and status in females. This is consistent with the finding that plasma progesterone concentration explained 28% of the variation in FADS2 mRNA expression in nonpregnant rats, although there was no significant association between progesterone and either 20:4n-6 or 22:6n-3 concentrations (Childs et al., 2012). However, oestrogen concentration explained 96% of the variation in FADS2 expression in pregnant rats (Childs et al., 2012). Thus at present, there is no consensus on the regulation of VLCPUFA synthesis by sex hormones in females.

# References

Albert, D.H., Coniglio, J.G., 1977. Metabolism of eicosa-11,14-dienoic acid in rat testes: evidence for delta8-desaturase activity. Biochim. Biophys. Acta 489, 390–396.

Ayala, S., Gaspar, G., Brenner, R.R., Peluffo, R.O., Kunau, W., 1973. Fate of linoleic, arachidonic, and docosa-7,10,13,16-tetraenoic acids in rat testicles. J. Lipid Res. 14, 296–305.

Baker, E.J., Miles, E.A., Burdge, G.C., Yaqoob, P., Calder, P.C., 2016. Metabolism and functional effects of plant-derived omega-3 fatty acids in humans. Prog. Lipid Res. 64, 30–56.

Bandyopadhyay, G.K., Dutta, J., Ghosh, S., 1982. Preferential oxidation of linolenic acid compared to linoleic acid in the liver of catfish (*Heteropneustes fossilis* and *Clarias batrachus*). Lipids 17, 733–740.

Brenner, R.R., Peluffo, R.O., 1966. Effect of saturated and unsaturated fatty acids on the desaturation in vitro of palmitic, stearic, oleic, linoleic, and linolenic acids. J. Biol. Chem. 241, 5213–5219.

Bretillon, L., Chardigny, J.M., Sebedio, J.L., Noel, J.P., Scrimgeour, C.M., Fernie, C.E., Loreau, O., Gachon, P., Beaufrere, B., 2001. Isomerization increases the postprandial oxidation of linoleic acid but not alpha-linolenic acid in men. J. Lipid Res. 42, 995–997.

Burdge, G., 2004. Alpha-linolenic acid metabolism in men and women: nutritional and biological implications. Curr. Opin. Clin. Nutr. Metab. Care 7, 137–144.

Burdge, G.C., Wootton, S.A., 2002a. Conversion of alpha-linolenic acid to eicosapentaenoic, docosapentaenoic and docosahexaenoic acids in young women. Br. J. Nutr. 88, 411–420.

Burdge, G.C., Wootton, S.A., 2003. Conversion of alpha-linolenic acid to palmitic, palmitoleic, stearic and oleic acids in men and women. Prostagl. Leuk. Essent. Fatty Acids 69, 283–290.

Burdge, G.C., Hunt, A.N., Postle, A.D., 1994. Mechanisms of hepatic phosphatidylcholine synthesis in adult rat: effects of pregnancy. Biochem. J. 303 (Pt 3), 941–947.

Burdge, G.C., Jones, A.E., Wootton, S.A., 2002. Eicosapentaenoic and docosapentaenoic acids are the principal products of alpha-linolenic acid metabolism in young men. Br. J. Nutr. 88, 355–363.

Burdge, G.C., Finnegan, Y.E., Minihane, A.M., Williams, C.M., Wootton, S.A., 2003a. Effect of altered dietary n-3 fatty acid intake upon plasma lipid fatty acid composition, conversion of [13C]alpha-linolenic acid to longer-chain fatty acids and partitioning towards beta-oxidation in older men. Br. J. Nutr. 90, 311–321.

Burdge, G.C., Finnegan, Y.E., Minihane, A.M., Williams, C.M., Wootton, S.A., 2003b. Effect of altered dietary n-3 fatty acid intake upon plasma lipid fatty acid composition, conversion of [$^{13}$C]α-linolenic acid to longer-chain fatty acids and partitioning towards beta-oxidation in older men. Br. J. Nutr. 90, 311–321.

Burdge, G.C., Slater-Jefferies, J.L., Grant, R.A., Chung, W.S., West, A.L., Lillycrop, K.A., Hanson, M.A., Calder, P.C., 2008. Sex, but not maternal protein or folic acid intake, determines the fatty acid composition of hepatic phospholipids, but not of triacylglycerol, in adult rats. Prostagl. Leuk. Essent. Fatty Acids 78, 73–79.

Chen, Q., Yin, F.Q., Sprecher, H., 2000. The questionable role of a microsomal delta8 acyl-coA-dependent desaturase in the biosynthesis of polyunsaturated fatty acids. Lipids 35, 871–879.

Childs, C.E., Romeu-Nadal, M., Burdge, G.C., Calder, P.C., 2010. The polyunsaturated fatty acid composition of hepatic and plasma lipids differ by both sex and dietary fat intake in rats. J. Nutr. 140, 245–250.

Childs, C.E., Hoile, S.P., Burdge, G.C., Calder, P.C., 2012. Changes in rat n-3 and n-6 fatty acid composition during pregnancy are associated with progesterone concentrations and hepatic FADS2 expression. Prostagl. Leuk. Essent. Fatty Acids 86, 141–147.

Conquer, J.A., Holub, B.J., 1997. Dietary docosahexaenoic acid as a source of eicosapentaenoic acid in vegetarians and omnivores. Lipids 32, 341–345.

Cunnane, S.C., Ryan, M.A., Nadeau, C.R., Bazinet, R.P., Musa-Veloso, K., McCloy, U., 2003. Why is carbon from some polyunsaturates extensively recycled into lipid synthesis? Lipids 38, 477–484.

D'Andrea, S., Guillou, H., Jan, S., Catheline, D., Thibault, J.N., Bouriel, M., Rioux, V., Legrand, P., 2002. The same rat Delta6-desaturase not only acts on 18- but also on 24-carbon fatty acids in very-long-chain polyunsaturated fatty acid biosynthesis. Biochem. J. 364, 49–55.

de Gomez Dumm, I.N., Brenner, R.R., 1975. Oxidative desaturation of α-linoleic, linoleic, and stearic acids by human liver microsomes. Lipids 10, 315–317.

DeLany, J.P., Windhauser, M.M., Champagne, C.M., Bray, G.A., 2000. Differential oxidation of individual dietary fatty acids in humans. Am. J. Clin. Nutr. 72, 905–911.

Demmelmair, H., Baumheuer, M., Koletzko, B., Dokoupil, K., Kratl, G., 1998. Metabolism of U$^{13}$C-labeled linoleic acid in lactating women. J. Lipid Res. 39, 1389–1396.

Dhopeshwarkar, G.A., Subramanian, C., 1975a. Metabolism of 1-(14)C linolenic acid in developing brain: II. Incorporation of radioactivity from 1-($^{14}$)C linolenate into brain lipids. Lipids 10, 242–247.

Dhopeshwarkar, G.A., Subramanian, C., 1975b. Metabolism of linolenic acid in developing brain: I. incorporation of radioactivity from 1-($^{14}$)C linolenic acid into brain fatty acids. Lipids 10, 238–241.

Emken, E.A., Adlof, R.O., Rakoff, H., Rohwedder, W.K., Gulley, R.M., 1990. Metabolism in vivo of deuterium-labelled linolenic and linoleic acids in humans. Biochem. Soc. Trans. 18, 766–769.

Giltay, E.J., Duschek, E.J., Katan, M.B., Zock, P.L., Neele, S.J., Netelenbos, J.C., 2004a. Raloxifene and hormone replacement therapy increase arachidonic acid and docosahexaenoic acid levels in postmenopausal women. J. Endocrinol. 182, 399–408.

Giltay, E.J., Gooren, L.J., Toorians, A.W., Katan, M.B., Zock, P.L., 2004b. Docosahexaenoic acid concentrations are higher in women than in men because of estrogenic effects. Am. J. Clin. Nutr. 80, 1167–1174.

Goyens, P.L., Spilker, M.E., Zock, P.L., Katan, M.B., Mensink, R.P., 2005. Compartmental modeling to quantify alpha-linolenic acid conversion after longer term intake of multiple tracer boluses. J. Lipid Res. 46, 1474–1483.

Gronn, M., Christensen, E., Hagve, T.A., Christophersen, B.O., 1991. Peroxisomal retrocon-version of docosahexaenoic acid (22:6(n-3)) to eicosapentaenoic acid (20:5(n-3)) studied in isolated rat liver cells. Biochim. Biophys. Acta 1081, 85–91.

Hassam, A.G., Rivers, J.P., Crawford, M.A., 1977. The failure of the cat to desaturate linoleic acid; its nutritional implications. Nutr. Metab. 21, 321–328.

Hussein, N., Ah-Sing, E., Wilkinson, P., Leach, C., Griffin, B.A., Millward, D.J., 2005. Long-chain conversion of [$^{13}$C]linoleic acid and alpha-linolenic acid in response to marked changes in their dietary intake in men. J. Lipid Res. 46, 269–280.

Ichi, I., Kono, N., Arita, Y., Haga, S., Arisawa, K., Yamano, M., Nagase, M., Fujiwara, Y., Arai, H., 2014. Identification of genes and pathways involved in the synthesis of Mead acid (20:3n-9), an indicator of essential fatty acid deficiency. Biochim. Biophys. Acta 1841, 204–213.

Innis, S.M., 1991. Essential fatty acids in growth and development. Prog. Lipid Res. 30, 39–103.

James, M.J., Ursin, V.M., Cleland, L.G., 2003. Metabolism of stearidonic acid in human sub-jects: comparison with the metabolism of other n-3 fatty acids. Am. J. Clin. Nutr. 77, 1140–1145.

Kaur, G., Begg, D.P., Barr, D., Garg, M., Cameron-Smith, D., Sinclair, A.J., 2010. Short-term docosapentaenoic acid (22:5 n-3) supplementation increases tissue docosapentaenoic acid, DHA and EPA concentrations in rats. Br. J. Nutr. 103, 32–37.

Klenk, E., Mohrhauer, H., 1960. Studies on the metabolism of polyenoic fatty acids in the rat. Hoppe Seylers Z. Physiol. Chem. 320, 218–232.

Kuipers, R.S., Luxwolda, M.F., Offringa, P.J., Boersma, E.R., Dijck-Brouwer, D.A., Muskiet, F.A., 2012. Fetal intrauterine whole body linoleic, arachidonic and docosahexaenoic acid contents and accretion rates. Prostagl. Leuk. Essent. Fatty Acids 86, 13–20.

Leyton, J., Drury, P.J., Crawford, M.A., 1987. Differential oxidation of saturated and unsatu-rated fatty acids in vivo in the rat. Br. J. Nutr. 57, 383–393.

Li, Z., Kaplan, M.L., Hachey, D.L., 2000. Hepatic microsomal and peroxisomal docosahexae-noate biosynthesis during piglet development. Lipids 35, 1325–1333.

Lin, Y.H., Pawlosky, R.J., Salem, Jr., N., 2005. Simultaneous quantitative determination of deuterium- and carbon-13-labeled essential fatty acids in rat plasma. J. Lipid Res. 46, 1974–1982.

Lohner, S., Fekete, K., Marosvolgyi, T., Decsi, T., 2013. Gender differences in the long-chain polyunsaturated fatty acid status: systematic review of 51 publications. Ann. Nutr. Metab. 62, 98–112.

Luthria, D.L., Mohammed, B.S., Sprecher, H., 1996. Regulation of the biosynthesis of 4,7,10,13,16,19-docosahexaenoic acid. J. Biol. Chem. 271, 16020–16025.

Marzo, I., Alava, M.A., Pineiro, A., Naval, J., 1996. Biosynthesis of docosahexaenoic acid in human cells: evidence that two different delta 6-desaturase activities may exist. Biochim. Biophys. Acta 1301, 263–272.

McCloy, U., Ryan, M.A., Pencharz, P.B., Ross, R.J., Cunnane, S.C., 2004. A comparison of the metabolism of eighteen-carbon 13C-unsaturated fatty acids in healthy women. J. Lipid Res. 45, 474–485.

Meyer, A., Cirpus, P., Ott, C., Schlecker, R., Zahringer, U., Heinz, E., 2003. Biosynthesis of docosahexaenoic acid in *Euglena gracilis*: biochemical and molecular evidence for the in-volvement of a Delta4-fatty acyl group desaturase. Biochem 42, 9779–9788.

Moore, S.A., Hurt, E., Yoder, E., Sprecher, H., Spector, A.A., 1995. Docosahexaenoic acid syn-thesis in human skin fibroblasts involves peroxisomal retroconversion of tetracosahexae-noic acid. J. Lipid Res. 36, 2433–2443.

Morais, S., Castanheira, F., Martinez-Rubio, L., Conceicao, L.E., Tocher, D.R., 2012. Long chain polyunsaturated fatty acid synthesis in a marine vertebrate: ontogenetic and nu-tritional regulation of a fatty acyl desaturase with Delta4 activity. Biochim. Biophys. Acta 1821, 660–671.

Park, W.J., Kothapalli, K.S., Lawrence, P., Tyburczy, C., Brenna, J.T., 2009. An alternate pathway to long-chain polyunsaturates: the FADS2 gene product Delta8-desaturates 20:2n-6 and 20:3n-3. J. Lipid Res. 50, 1195–1202.

Park, H.G., Kothapalli, K.S., Park, W.J., DeAllie, C., Liu, L., Liang, A., Lawrence, P., Brenna, J.T., 2016a. Palmitic acid (16:0) competes with omega-6 linoleic and omega-3 a-linolenic acids for FADS2 mediated Delta6-desaturation. Biochim. Biophys. Acta 1861, 91–97.

Park, H.G., Lawrence, P., Engel, M.G., Kothapalli, K., Brenna, J.T., 2016b. Metabolic fate of docosahexaenoic acid (DHA; 22:6n-3) in human cells: direct retroconversion of DHA to eicosapentaenoic acid (20:5n-3) dominates over elongation to tetracosahexaenoic acid (24:6n-3). FEBS Lett. 590, 3188–3194.

Pawlosky, R.J., Hibbeln, J.R., Novotny, J.A., Salem, Jr., N., 2001. Physiological compartmental analysis of alpha-linolenic acid metabolism in adult humans. J. Lipid Res. 42, 1257–1265.

Plourde, M., Chouinard-Watkins, R., Vandal, M., Zhang, Y., Lawrence, P., Brenna, J.T., Cunnane, S.C., 2011. Plasma incorporation, apparent retroconversion and beta-oxidation of 13C-docosahexaenoic acid in the elderly. Nutr. Metab. 8, 5.

Postle, A.D., Al, M.D., Burdge, G.C., Hornstra, G., 1995. The composition of individual molecular species of plasma phosphatidylcholine in human pregnancy. Early Hum. Dev. 43, 47–58.

Qiu, X., Hong, H., MacKenzie, S.L., 2001. Identification of a Delta 4 fatty acid desaturase from *Thraustochytrium* sp. involved in the biosynthesis of docosahexanoic acid by heterologous expression in *Saccharomyces cerevisiae* and *Brassica juncea*. J. Biol. Chem. 276, 31561–31566.

Rivers, J.P., Sinclair, A.J., Craqford, M.A., 1975. Inability of the cat to desaturate essential fatty acids. Nature 258, 171–173.

Salem, Jr., N., Pawlosky, R., Wegher, B., Hibbeln, J., 1999. In vivo conversion of linoleic acid to arachidonic acid in human adults. Prostagl. Leuk. Essent. Fatty Acids 60, 407–410.

Sanders, T.A., 2009. DHA status of vegetarians. Prostagl. Leuk. Essent. Fatty Acids 81, 137–141.

Sheaff Greiner, R.C., Zhang, Q., Goodman, K.J., Giussani, D.A., Nathanielsz, P.W., Brenna, J.T., 1996. Linoleate, alpha-linolenate, and docosahexaenoate recycling into saturated and monounsaturated fatty acids is a major pathway in pregnant or lactating adults and fetal or infant rhesus monkeys. J. Lipid Res. 37, 2675–2686.

Sibbons, C.M., Brenna, J.T., Lawrence, P., Hoile, S.P., Clarke-Harris, R., Lillycrop, K.A., Burdge, G.C., 2014. Effect of sex hormones on n-3 polyunsaturated fatty acid biosynthesis in HepG2 cells and in human primary hepatocytes. Prostagl. Leuk. Essent. Fatty Acids 90, 47–54.

Simopoulos, A.P., 1999. Essential fatty acids in health and chronic disease. Am. J. Clin. Nutr. 70, 560S–569S.

Sinclair, A.J., 1974. Fatty acid composition of liver lipids during development of rat. Lipids 9, 809–818.

Sinclair, A.J., McLean, J.G., Monger, E.A., 1979. Metabolism of linoleic acid in the cat. Lipids 14, 932–936.

Sinclair, A.J., Slattery, W., McLean, J.G., Monger, E.A., 1981. Essential fatty acid deficiency and evidence for arachidonate synthesis in the cat. Br. J. Nutr. 46, 93–96.

Sinclair, A.J., Attar-Bashi, N.M., Li, D., 2002. What is the role of alpha-linolenic acid for mammals? Lipids 37, 1113–1123.

Sprecher, H., 2000. Metabolism of highly unsaturated n-3 and n-6 fatty acids. Biochim. Biophys. Acta 1486, 219–231.

Sprecher, H., Chen, Q., 1999. Polyunsaturated fatty acid biosynthesis: a microsomal-peroxisomal process. Prostagl. Leuk. Essent. Fatty Acids 60, 317–321.

Sprecher, H.W., Baykousheva, S.P., Luthria, D.L., Mohammed, B.S., 1995. Differences in the regulation of biosynthesis of 20- versus 22-carbon polyunsaturated fatty acids. Prostagl. Leuk. Essent. Fatty Acids 52, 99–101.

Su, H.M., Corso, T.N., Nathanielsz, P.W., Brenna, J.T., 1999. Linoleic acid kinetics and conversion to arachidonic acid in the pregnant and fetal baboon. J. Lipid Res. 40, 1304–1312.

Su, H.M., Moser, A.B., Moser, H.W., Watkins, P.A., 2001. Peroxisomal straight-chain Acyl-CoA oxidase and d-bifunctional protein are essential for the retroconversion step in docosahexaenoic acid synthesis. J. Biol. Chem. 276, 38115–38120.

Toledo, A., Andersson, M.N., Wang, H.L., Salmon, P., Watson, H., Burdge, G.C., Isaksson, C., 2016. Fatty acid profiles of great tit (*Parus major*) eggs differ between urban and rural habitats, but not between coniferous and deciduous forests. Naturwissenschaften 103, 55.

Uauy, R., Hoffman, D.R., 2000. Essential fat requirements of preterm infants. Am. J. Clin. Nutr. 71, 245S–250S.

Uauy, R., Peirano, P., Hoffman, D., Mena, P., Birch, D., Birch, E., 1996. Role of essential fatty acids in the function of the developing nervous system. Lipids 31 (Suppl.), S167–S176.

Uauy, R., Hoffman, D.R., Peirano, P., Birch, D.G., Birch, E.E., 2001. Essential fatty acids in visual and brain development. Lipids 36, 885–895.

Vermunt, S.H., Mensink, R.P., Simonis, M.M., Hornstra, G., 2000. Effects of dietary alpha-linolenic acid on the conversion and oxidation of [13]C-alpha-linolenic acid. Lipids 35, 137–142.

Voss, A., Reinhart, M., Sankarappa, S., Sprecher, H., 1991. The metabolism of 7,10,13,16,19-docosapentaenoic acid to 4,7,10,13,16,19-docosahexaenoic acid in rat liver is independent of a 4-desaturase. J. Biol. Chem. 266, 19995–20000.

# Polyunsaturated Fatty Acid Biosynthesis and Metabolism in Fish

*Oscar Monroig\*, Douglas R. Tocher\*,*
*Luís Filipe C. Castro\*\**

\*University of Stirling, Stirling, United Kingdom; \*\*CIIMAR–
Interdisciplinary Centre of Marine and Environmental Research,
U. Porto–University of Porto, Porto, Portugal

## INTRODUCTION

Fish and seafood are recognized as important components of a healthy diet, with the n-3 long-chain ($\geq C_{20}$) polyunsaturated fatty acids (LC-PUFA), eicosapentaenoic acid (EPA; 20:5n-3), and docosahexaenoic acid (DHA; 22:6n-3), being the nutrients most associated with the beneficial effects (Lands, 2014). Epidemiological studies, randomized

Polyunsaturated Fatty Acid Metabolism. http://dx.doi.org/10.1016/B978-0-12-811230-4.00003-X

controlled (intervention) trials, and laboratory studies investigating biochemical and molecular mechanisms have all provided key evidence (Gil et al., 2012). Studies have shown that dietary n-3 LC-PUFA can decrease the risk of developing cardiovascular disease (CVD) (Delgado-Lista et al., 2012) and have positive benefits in patients with some disease already (Calder, 2014; Calder and Yaqoob, 2012; Casula et al., 2013; Delgado-Lista et al., 2012). Beneficial effects of dietary n-3 LC-PUFA in inflammatory disease have been obtained in rheumatoid arthritis (Miles and Calder, 2012) and, to some extent, in inflammatory bowel disease, such as Crohn's disease and ulcerative colitis (Cabré et al., 2012). Epidemiological studies indicated that n-3 LC-PUFA may have a protective effect (i.e., decreasing risk) in colorectal, breast, and prostate cancers (Gerber, 2012), and can be beneficial in chemotherapy (Bougnoux et al., 2009). Decreased DHA status can lead to cognitive and visual impairment and DHA supplements have positive outcomes in preterm infants (Campoy et al., 2012; Carlson et al., 1993) and may have beneficial effects in some psychological/behavioral/psychiatric disorders (Ortega et al., 2012), and n-3 LC-PUFA may help prevent some pathological conditions associated with normal aging (Úbeda et al., 2012). Consequently, many recommendations for EPA and DHA intake for humans have been produced by a large number of global and national health agencies and associations, and government bodies (GOED, 2014, 2016). Based on their effects on CVD, many health agencies worldwide recommend up to 500 mg/d of EPA and DHA for reducing CVD risk or 1 g/d for secondary prevention in existing CVD patients, with a dietary strategy for achieving 500 mg/d being to consume two fish meals per week with at least one of oily fish (Aranceta and Pérez-Rodrigo, 2012; EFSA, 2012; Gebauer et al., 2006; ISSFAL, 2004).

Fish, especially "oily" species, such as salmon, are the most important foods in delivering physiologically effective doses of n-3 LC-PUFA to consumers (Henriques et al., 2014; Sprague et al., 2016; Tur et al., 2012). The primary reason for this being that aquatic food webs are n-3 LC-PUFA rich and, as 96% of global water is ocean, EPA and DHA are predominantly of marine origin and fish simply accumulate them from their diet (Bell and Tocher, 2009a; Tocher, 2009). However, global fisheries have been stagnant for the last 20 years and, with increasing global population driving increasing demand, an increasing proportion of fish and seafood are now farmed, reaching more than 50% in 2014 (FAO, 2016). Traditionally, natural diets were replicated in farmed fish by formulating feeds with fishmeal (FM) and fish oil (FO), and so farmed fish were also rich in n-3 LC-PUFA (Tocher, 2015). However, aquaculture has grown by an average of more than 6% annually over the last 15 years or so, outpacing population growth (FAO, 2014). Paradoxically, FM and FO are themselves products of wild capture fisheries that are also at their sustainable limit and, therefore, they are finite resources on an annual basis, which would have limited aquaculture growth and development if they were not increasingly replaced in aquafeeds by alternative ingredients, predominantly terrestrial crop-derived plant meals and vegetable oils (Gatlin et al., 2007; Hardy, 2010; Turchini et al., 2011). As terrestrial plants do not produce LC-PUFA, this has the consequence of reducing the level of n-3 LC-PUFA in farmed fish, potentially compromising their nutritional quality to the human consumer (Sprague et al., 2016; Tocher, 2015). This has prompted considerable research into pathways of LC-PUFA metabolism in fish based on the hypothesis that understanding the molecular basis of LC-PUFA biosynthesis and regulation in fish will allow the pathway to be optimized to enable efficient and effective use of plant-based alternatives in aquaculture while maintaining the n-3 LC-PUFA content of farmed fish for the consumer (Tocher, 2003, 2010).

Fish, like all vertebrates, cannot synthesize PUFA de novo and so they are essential dietary nutrients, but dietary essential PUFA, such as linoleic acid (LOA; 18:2n-6) and α-linolenic acid (ALA; 18:3n-3) can be converted to LC-PUFA in some species (Tocher, 2010). These conversions are carried out by fatty acyl desaturases (Fads) and elongation of very long-chain fatty acid (Elov) proteins as depicted in Fig. 3.1. While a further description is provided later, fish possess Fads and Elovl involved in LC-PUFA biosynthesis and, importantly, it has been established that the extent to

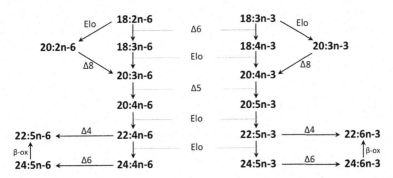

FIGURE 3.1 Biosynthetic pathways of long-chain ($\geq C_{20}$) polyunsaturated fatty acids in fish. Desaturation reactions are catalyzed by fatty acyl desaturases (Fads) and are denoted with "Δ" to indicate the carbon position at which the incipient double bond locates within the carboxylic group end (Δ) of the fatty acyl chain. Elongation reactions, denoted with "Elo," are catalyzed by elongation of very long-chain fatty acid (Elovl) proteins. β-ox, Partial β-oxidation.

which a fish species can convert $C_{18}$ PUFA to LC-PUFA varies, associated with their complement of *fads* and *elovl* genes (Castro et al., 2016; Monroig et al., 2011a; Tocher, 2015). In general, freshwater fish species have a greater ability for conversion of $C_{18}$ PUFA to LC-PUFA than marine species (Tocher, 2010), with the limited capacity of marine fish attributed to deficiencies in one or more key enzymes of the endogenous LC-PUFA biosynthesis pathway (Bell and Tocher, 2009b).

Fish can be classified as cartilaginous fish (e.g., sharks and skates) and teleost (or "bony") fish and this classification has important implications in the repertoire of genes existing in species within these groups (Castro et al., 2012, 2016). For instance, particularly for teleosts, vital processes to consider in the analysis and comprehension of PUFA metabolism are gene duplication and loss, evolutionary events by which teleosts acquired or lost distinctive PUFA biosynthetic functionalities compared to non-teleost vertebrates. This has been particularly well-established for *fads* and *elovl* genes, where a number of studies have described the contribution of tandem (e.g., *fads*) and whole genome (e.g., elongases) duplication events (Castro et al., 2016; Fonseca-Madrigal et al., 2014; Monroig et al., 2016a). The increasingly available genomic and transcriptomic data from fish, as well as the aforementioned interest in the contribution that farmed fish have on a healthy diet for humans, have greatly contributed to the understanding of the molecular mechanisms of PUFA metabolism in fish, particularly with regards to LC-PUFA biosynthesis. Consequently, this chapter focuses primarily on LC-PUFA biosynthesis, arguably the pathway of nutritional physiology most extensively studied in fish. However, this is preceded by discussion of some key processes of PUFA metabolism that can affect tissue fatty acid compositions, such as digestion, absorption, and catabolism, with particular attention to fish-specific mechanisms. Finally, the key roles of PUFA and, especially, LC-PUFA in signaling pathways and the regulation of metabolism in fish are discussed.

## PUFA METABOLISM AND TISSUE FATTY ACID COMPOSITIONS

As in all animals, PUFA and LC-PUFA have important structural roles in fish as constituents of phospholipids that are the major components of cellular biomembranes, and confer various functional properties by affecting both physicochemical properties of the membrane (e.g., fluidity) as well as influencing membrane protein (e.g., receptors, carriers, and enzymes) functions (Tocher, 1995). A review of fish phospholipid and membrane PUFA compositions is without the scope of this chapter, but other reviews provide comprehensive coverage, including dietary influences (Sargent et al., 2002; Tocher, 1995) and environmental effects, especially temperature (Hazel and Williams, 1990; Hochachka and Mommsen, 1995).

The phrase "you are what you eat" may have been used in one form or another for almost 200 years but, in terms of PUFA metabolism, it now has a good basis in science. In short, the fatty acid composition of the diet is arguably the most important factor in determining the fatty acid profiles of fish tissues (Turchini et al., 2011). This, therefore, implies that fish metabolism has only a limited impact on dietary fatty acids. The following is a brief summary of some metabolic pathways and physiological processes that influence tissue fatty acid compositions in fish. These are generally very similar to those in mammals and so only aspects with a particular relevance to fish are discussed.

## Digestion and Absorption

The specificities of digestive enzymes toward different fatty acids and the efficacy of uptake of free fatty acids and other digestion products (e.g., mono- and diacylglycerols, lysophospholipids) are relatively unstudied in fish (Pérez et al., 1999). However, the overall efficiency of digestion and absorption in fish is generally estimated by simply determining the concentrations of nutrients (e.g., fatty acids) in diets and feces in comparison to an indigestible marker (Bell and Koppe, 2011). The apparent digestibility coefficients (ADC) of all PUFA, including LC-PUFA, are usually high, indicating that dietary PUFA are efficiently digested and absorbed in most fish species. However, lipid content, lipid class composition (triacylglyceride, phospholipid) and fatty acid composition of the diet, and water temperature can influence ADC of fatty acids (Bell and Koppe, 2011).

## PUFA Oxidation

As components of triacylglycerol (TAG) and other storage lipids (e.g., wax esters), PUFA can function as an energy reserve, with energy recovered through fatty acid β-oxidation in mitochondria (Tocher, 2003). Based on relative oxidation rates measured in vitro, PUFA and LC-PUFA are poorer substrates for oxidation than saturated or monounsaturated fatty acids, with n-6 PUFA better oxidized than n-3 PUFA (Henderson, 1996). However, comparing dietary and tissue fatty acid compositions showed that the higher the dietary concentration of a fatty acid, the lower its relative deposition, which implies increased oxidation in vivo (Stubhaug et al., 2007). Therefore, fatty acid oxidation is dependent on both dietary fatty acid concentration and enzyme specificity. Generally, there is no preferential retention of PUFA in fish (Stubhaug et al., 2007), but DHA appears to be an exception being usually deposited in tissues of most fish species at higher levels than dietary inclusion (Brodtkord et al., 1997). This appears to be simply due to the Δ4 double bond in DHA being relatively resistant to mitochondrial β-oxidation (Madsen et al., 1998).

A major control point in fatty acid β-oxidation is the mitochondrial carnitine palmitoyltransferase system, which is composed of two enzymatic modules, namely Cpt1 and Cpt2. Cpt1 in particular is vital given its sensitivity and inhibition by malonyl-CoA, an intermediate of fatty acid synthesis (Bonnefont et al., 2004). In fish, clear orthologues of *cpt1a* and *cpt1b* have been recognized (Boukouvala et al., 2010; Lopes-Marques et al., 2015). Two uncharacterized *cpt1* genes are also present in teleost genomes, which most likely represent cryptic orthologues of *cpt1c* (Lopes-Marques et al., 2015). A more complex gene pattern was observed in salmonids due to the additional round of genome duplication. A few studies have detailed the response and involvement of Cpt1b in response to nutritional and hormonal stimuli (Boukouvala et al., 2010; Kolditz et al., 2008; Morash et al., 2010). In gilthead seabream (*Sparus aurata*), *cpt1b* was expressed in the heart, and very strongly in white and red muscle (Boukouvala et al., 2010). Moreover, examination of expression levels of *cpt1b* in response to feeding status showed a rapid postprandial decrease in expression in white and red muscle and heart, followed by a return to preprandial levels at or after 24 h postfeeding (Boukouvala et al., 2010). High-energy (high fat) diets were also shown to increase the expression of *cpt1b* in trout (Kolditz et al., 2008). More recently, *cpt1a* and *cpt1b* also showed a marked increase in expression in liver of seabream *S. aurata*, in response to various diets including those with LC-PUFA derived from genetically modified *Camelina sativa* (Betancor et al., 2016a).

## LC-PUFA BIOSYNTHESIS IN FISH

### Fatty Acyl Desaturases

Current data suggest that teleost fish have lost Fads1 during evolution (Castro et al., 2012), whereas both *fads1* and *fads2* genes were identified in the lesser-spotted dogfish (*Scyliorhinus canicula*). Consistent with mammals, the substrate specificity of *S. canicula* Fads1 and Fads2 enzymes revealed they were Δ5 and Δ6 desaturases, respectively, a gene set more recently confirmed in the genome of the elephant shark (*Callorhinchus milii*) (Castro et al., 2016; Venkatesh et al., 2014). In contrast, while virtually all bony fishes appear to have lost Fads1, the exact number of *fads2* genes varies among species. Thus *fads*-like genes seem to be completely absent in the genomes of pufferfish, including *Takifugu rubripes* and *Tetraodon nigroviridis* (Leaver et al., 2008). Other species possess one (e.g., *Danio rerio*), two (e.g., *Siganus canaliculatus*), three (e.g., *Oreochromis niloticus*), or four (e.g., *Salmo salar*) *fads2* genes (Monroig et al., 2010a). Typically, many teleost

Fads2 are Δ6 desaturases, consistent with the substrate specificities found for mammalian FADS2 (De Antueno et al., 2001). However, there is now a good body of evidence confirming that sub- (acquisition of additional substrate specificities) and neofunctionalization (substitution and/or acquisition of new substrate specificities) events have occurred within the teleost *fads2* gene family (Castro et al., 2016; Fonseca-Madrigal et al., 2014). Thus the zebrafish *D. rerio* Fads2 was the first vertebrate desaturase shown to be a dual Δ6Δ5 fatty acyl desaturase, an enzyme able to introduce double bonds into separate and distinct carbons in the fatty acyl chain (Hastings et al., 2001). At the time, this substrate specificity was unique among vertebrate "front-end" desaturases, but has since been revealed to be a relatively common trait among teleost fish. Thus in addition to the zebrafish bifunctional Δ6Δ5 Fads2, other bifunctional desaturases have been reported in the rabbitfish *S. canaliculatus* (Li et al., 2010), Nile tilapia *O. niloticus* (Tanomman et al., 2013), pike silverside *Chirostoma estor* (Fonseca-Madrigal et al., 2014), African catfish *Clarias gariepinus* (Oboh et al., 2016), and striped snakehead *Channa striata* (Kuah et al., 2016).

Examples of neofunctionalization among teleost fish Fads2 desaturases are also found. For instance, the salmonids Atlantic salmon *S. salar* and rainbow trout *Oncorhynchus mykiss* possess Fads2 that were functionally characterized as monofunctional Δ5 desaturases (Abdul Hamid et al., 2016; Hastings et al., 2005) although the former has been recently found to have also Δ6 activity (Oboh et al., 2017a). Irrespective of whether Δ5 desaturase activity exists as mono- or bifunctional Fads2 enzymes, the acquisition of Δ5 desaturation ability in Fads2 represents a clear advantage in those teleost species because it can compensate the aforementioned loss of Δ5 *fads1* during evolution (Castro et al., 2012). The existence of Fads2 enzymes enabling both Δ6 and Δ5 desaturation is essential to accomplish all the desaturation reactions required to convert $C_{18}$ PUFA into the physiologically important $C_{20-22}$ LC-PUFA including ARA, EPA, and DHA (Fig. 3.1) (Castro et al., 2016). A distinctive trait of the fish LC-PUFA biosynthetic pathway is that, in contrast to mammalian LC-PUFA pathways where Δ6 desaturation is regarded as rate-limiting (Guillou et al., 2010), the lack of *fads1* in teleost genomes suggests that rather it is limited at the Δ5 desaturase level.

One of the most striking findings on the biosynthetic pathways of LC-PUFA in fish has been the discovery of a Fads2 with Δ4 fatty acyl desaturase activity in the marine herbivore rabbitfish *S. canaliculatus*, this representing the first ever report of Δ4 desaturation capability in any vertebrate species (Li et al., 2010). Other Δ4 Fads2 desaturases have been subsequently found in Senegalese sole *Solea senegalensis* (Morais et al., 2012, 2015), pike silverside *C. estor* (Fonseca-Madrigal et al., 2014), and the striped snakehead *C. striata* (Kuah et al., 2015), suggesting that such enzymatic capability appears more widespread than initially appreciated, as previously also found for the $\Delta6\Delta5$ desaturase. Consistently, Oboh et al. (2017a) identified the presence of $\Delta4$ Fads2 in a further 11 species and confirmed the function as $\Delta4$ desaturases of Fads2 from medaka *Oryzias latipes* and Nile tilapia *O. niloticus*. Interestingly, the human FADS2 was recently shown to have the ability to produce DHA by direct $\Delta4$ desaturation of 22:5n-3 (Park et al., 2015). Historically, however, the biosynthesis of DHA in vertebrates had been widely accepted to proceed through the so-called Sprecher pathway (Sprecher, 2000). Briefly, in this pathway, rather than 22:5n-3 being desaturated at the $\Delta4$ position to directly produce DHA (the "$\Delta4$ pathway"), it undergoes elongation to 24:5n-3, which is then $\Delta6$ desaturated to 24:6n-3 by the same enzyme that operates toward $C_{18}$ PUFA and initiates the bioconversion pathways, that is, $\Delta6$ Fads2 (Ferdinandusse et al., 2004; Sprecher, 2000). Finally, 24:6n-3 is catabolized (partial $\beta$-oxidation) to DHA in peroxisomes and thus translocation of metabolic intermediates (24:6n-3) from endoplasmic reticulum and peroxisomes is required. Although the Sprecher pathway, first described in rat (Sprecher et al., 1995; Sprecher, 2000), has been demonstrated to potentially operate in some species of fish including rainbow trout and zebrafish (Buzzi et al., 1996, 1997; Tocher et al., 2003a), it remained unclear whether the ability to desaturate 24:5n-3 to 24:6n-3 was an inherent characteristic of all teleost Fads2. With the exception of the Fads2 from the nibe croaker (*Nibea mitsukurii*), an enzyme with $\Delta6$ desaturase activity toward 18:3n-3 but not 24:5n-3 (Kabeya et al., 2015) all tested Fads2 enzymes from fish species with different evolutionary and ecological backgrounds showed the ability to operate as $\Delta6$ desaturases toward 24:5n-3 enabling them to biosynthesize DHA through the Sprecher pathway (Oboh et al., 2017a). Clearly, the LC-PUFA biosynthetic capabilities of teleosts appear much more varied and adaptive compared to other vertebrates groups, possibly reflecting the diversity of habitats and trophic strategies that fish have occupied during evolution.

With the possible exception of Atlantic salmon, the herbivorous rabbitfish *S. canaliculatus* has arguably become one of the most popular model teleost species for investigating LC-PUFA biosynthesis enzymes (Li et al., 2008, 2010; Monroig et al., 2012) and regulatory mechanisms (Dong et al., 2016, 2018; Wang et al., 2018; Zhang et al., 2014, 2016a,b). Interestingly, the two Fads2 reported in rabbitfish, namely the $\Delta6\Delta5$ and $\Delta4$ Fads2 (Li et al., 2010) and, particularly, their highly similar amino acid sequences (83%), have allowed the identification of the specific residues accounting for the different substrate chain-length specificities and regioselectivities (insertion of a double bond at a specific position in the fatty acyl chain) occurring among teleost Fads2 enzymes. The construction of chimeric proteins from $\Delta6\Delta5$ and $\Delta4$ Fads2 enabled the authors to conclude that four aa residues (FHYQ for $\Delta6\Delta5$ Fads2 and YNYN for $\Delta4$ Fads2) located in the third putative transmembrane domain between the second (HDFGH) and third (QIEHH) histidine boxes were responsible for the activities of these enzymes (Lim et al., 2014). In silico searches using YNYN as query against fish genomic and transcriptomic databases confirmed that these key aa residues for Δ4 function are widespread among teleosts suggesting this activity may also be more widespread in fish (Castro et al., 2016; Oboh et al., 2017a).

The initiation of the LC-PUFA biosynthesis pathway from $C_{18}$ PUFA precursors can proceed, alternative to the Δ6 desaturation described earlier, with an elongation of ALA and LOA to 20:3n-3 and 20:2n-6, respectively (Fig. 3.1). Both 20:3n-3 and 20:2n-6 can be reincorporated into the biosynthetic pathway through desaturation at the Δ8 position, producing 20:4n-3 and 20:3n-6 (Fig. 3.1), and thus are not technically "dead-end" metabolic products as previously thought (Monroig et al., 2011b). In agreement with previous findings in baboon FADS2 (Park et al., 2009), Δ8 activity toward 20:3n-3 and 20:2n-6 was demonstrated to be a characteristic of Fads2 desaturases from a wide range of freshwater, diadromous and marine fish species (Lopes-Marques et al., 2017; Kabeya et al., 2017, 2018; Monroig et al., 2011b; Oboh et al., 2016). Interestingly, functional analyses performed in yeast suggested that the Δ8 activity of Fads2 desaturases varied markedly among species, with marine fish Fads2 generally exhibiting higher Δ8 capability compared to those from freshwater/diadromous fish. It is unclear what advantage that retaining enhanced Δ8 desaturation capability gives to marine species that have natural diets with high availability of preformed EPA and DHA and in which the general absence of a Δ5-desaturase limits the overall activity pathway (Bell and Tocher, 2009b; Castro et al., 2012).

## Elongation of Very Long-Chain Fatty Acid Proteins

Fish, like other vertebrates, possess Elovl enzymes that also play key roles in the LC-PUFA biosynthetic pathways. By far the most extensively investigated type of Elovl in fish has been Elovl5, first reported in zebrafish *D. rerio* (Agaba et al., 2004). Functional characterization assays in yeast revealed that the zebrafish Elovl5 had the ability to elongate both $C_{18}$ (18:4n-3 and 18:3n-6) and $C_{20}$ (20:4n-3 and 20:3n-6) PUFA substrates very efficiently, while $C_{22}$ (22:5n-3 and 22:4n-6) substrates were only marginally elongated in the yeast system (Agaba et al., 2004). This pattern of activity has been repeatedly observed with Elovl5 from a wide variety of species from different habitats and phylogenetic positions (Castro et al., 2016) although in some species, such as *S. canaliculatus*, Elovl5 exhibited relatively high ability to elongate 22:5n-3 compared to other species (Monroig et al., 2012). The elongation abilities demonstrated by fish Elovl5 were largely in agreement with those of mammalian ELOVL5, but some data suggested that teleost Elovl5 were more versatile compared to their mammalian counterparts. For instance, fish Elovl5 have often been reported to have the ability to elongate monounsaturated fatty acids since endogenous 18:1n-7 and 18:1n-9 were converted to 20:1n-7 and 20:1n-9, respectively, in functional assays in yeast (Hastings et al., 2005; Kabeya et al., 2015; Mohd-Yusof et al., 2010; Monroig et al., 2013; Morais et al., 2009). Some fish, including the meagre *Argyrosomus regius* and the nibe croaker *N. mitsukurii*, possess Elovl5 with the ability to elongate the hexadecatrienoic acid (16:3n-3) (Kabeya et al., 2015; Monroig et al., 2013). This is a $C_{16}$ PUFA naturally occurring in some seaweeds like *Caulerpa* sp. and *Codium* sp. (Goecke et al., 2010; Khotimchenko, 1995) and the so-called 16:3 plants, such as *Arabidopsis*, rapeseed, and spinach (Browse and Somerville, 1991; Wallis and Browse, 2002), and thus representing interesting sources of dietary fatty acids for aquafeeds (Monroig et al., 2013). In common with mammalian orthologues (Guillou et al., 2010), teleost fish Elovl5 appear to have a prominent role in the so-called $\Delta 8$ pathway described earlier and thus the initiation of the pathway by an elongation of the dietary essential fatty acids ALA and LOA to 20:3n-3 and 20:2n-6, respectively, was efficiently catalyzed by fish Elovl5 (Gregory et al., 2010; Monroig et al., 2013). A recent study reported the in silico 3D structure model of the Elovl5 from the silver barb (*Puntius gonionotus*) and showed how elongase substrates (e.g., ALA) dock with the catalytic site of this enzyme (Nayak et al., 2018).

In addition to Elovl5, another elongase that has been demonstrated to play a major role in LC-PUFA biosynthesis in fish is Elovl2. The Elovl2 of Atlantic salmon (Morais et al., 2009) and zebrafish (Monroig et al., 2009) were the first of this type of elongase characterized from fish species, with those from rainbow trout and African catfish being reported more recently (Gregory and James, 2014; Oboh et al., 2016). The activities of fish Elovl2 appeared to be largely consistent with those found in mammals (Gregory et al., 2013; Leonard et al., 2002), and thus the preferred substrates being $C_{20}$ (20:4n-3 and 20:3n-6) and $C_{22}$ (22:5n-3 and 22:4n-6) PUFA, with $C_{18}$ PUFA (18:4n-3 and 18:3n-6) being elongated to a notably lesser extent. Whereas $C_{18}$ and $C_{22}$ PUFA can be regarded as specific substrates for Elovl5 and Elovl2 enzymes, respectively, both have ability to elongate $C_{20}$ PUFA substrates, consistent with these two proteins sharing a common (*Elovl2/5*) ancestor in basal chordates (Monroig et al., 2016a). Despite sharing some substrate preference with Elovl5, the ability of Elovl2 to elongate $C_{22}$ to $C_{24}$ LC-PUFA is the characteristic by which this enzyme has been regarded as critical for the biosynthesis of DHA biosynthesis via the Sprecher pathway. Thus, Elovl2 is required for the elongation of 22:5n-3 to 24:5n-3, which is the substrate for $\Delta 6$ desaturation by Fads2 to produce 24:6n-3 that can then be chain-shortened to produce DHA. However, the *elovl2* gene appears to be absent in the vast majority of marine teleost species that are currently produced in aquaculture (Castro et al., 2016; Morais et al., 2009). The absence of Elovl2 along with the apparent lack of $\Delta 5$ desaturase activity associated with the loss of *fads1*, have often been cited as the molecular mechanisms underpinning the low activity of LC-PUFA biosynthesis of marine fish in comparison to freshwater or salmonid species (Castro et al., 2016; Tocher, 2015; Tocher et al., 2003a). Recent evidence has revealed that the situation is far more complex than initially thought and the pattern of gene complement and function is a reflection of, not only ecological factors (e.g., natural diet), but also evolutionary events, shaping these pathways. As mentioned earlier, some marine fish, such as *S. canaliculatus* possess a desaturase with Δ5 activity (Li et al., 2010) and marine species from relatively ancient teleost lineages, such as Clupeiformes (e.g., Atlantic herring *Clupea harengus*) have *elovl2* in their genomes (Monroig's personal communication).

Elovl4 elongases have been also characterized in a range of fish including model species, such as zebrafish (Monroig et al., 2010b) and commercially important species, such as cobia *Rachycentron canadum* (Monroig et al., 2011c), Atlantic salmon *S. salar* (Carmona-Antoñanzas et al., 2011), rabbitfish *S. canaliculatus* (Monroig et al., 2012), nibe croaker *N. mitsukurii* (Kabeya et al., 2015), orange-spotted grouper *Epinephelus coioides* (Li et al., 2017a), African catfish *Clarias gariepinus* (Oboh et al., 2017b), large yellow croaker *Larimichthys crocea* (Li et al., 2017b), black seabream *Acanthopagrus schlegelii* (Jin et al., 2017) and loach *Misgurnus anguillicaudatus* (Yan et al., 2018). Two distinct genes termed Elovl4a and Elovl4b exist in teleost fish (Castro et al., 2016) although Elovl4 isoforms have been most extensively studied (Carmona-Antoñanzas et al., 2011; Kabeya et al., 2015; Monroig et al., 2011c, 2012; Li et al., 2017b). Both isoforms have the ability to efficiently produce very long-chain ($>C_{24}$) saturated fatty acids. Moreover, the Elovl4b was also confirmed to be involved in the biosynthesis of very long-chain ($>C_{24}$) PUFA (VLC-PUFA) because heterologous expression in yeast revealed that they were able to elongate $C_{20}$ (20:5n-3 and 20:4n-6) and $C_{22}$ (22:5n-3 and 22:4n-6) PUFA substrates to produce polyenes of up to $C_{36}$. While such capacity was apparently absent in zebrafish Elovl4a, recent studies on Elovl4a from African catfish *C. gariepinus* (Oboh et al., 2017b) and black seabream *A. schlegelii* (Jin et al., 2017) demonstrated their ability to produce VLC-PUFA from shorter-chain precursors. This was largely consistent with the substrate activities reported for human ELOVL4 (Agbaga et al., 2008). However, whereas the activity of human ELOVL4 indicated it was unlikely to have a major role in the biosynthesis of $C_{20-22}$ LC-PUFA, the capability of fish Elovl4 enzymes to elongate 22:5n-3 to 24:5n-3 implies a role for these enzymes in DHA biosynthesis through the Sprecher pathway, similar to that described above for Elovl2 elongases. These findings further confirmed that, despite the aforementioned lack/loss of Elovl2 in most marine species with commercial interest in fish farming, the biosynthesis of DHA does not appear to be limited at the elongation level since Elovl4 could compensate for the absence of Elovl2. Interestingly, DHA itself does not appear to be a substrate for fish

Elovl4s as it is only poorly elongated (Monroig et al., 2010b, 2011c). This is in contrast with recent evidence that confirmed the presence of the DHA elongation product 32:6n-3 in retinal phosphatidylcholine of gilthead seabream juveniles (Monroig et al., 2016b). It remains unclear whether Elovl4 is responsible for the elongation of DHA to 32:6n-3, and the possibility that other elongases are responsible for such reactions cannot be excluded. Xue et al. (2014) identified 10 genes encoding for Elovl enzymes in Atlantic cod (*Gadus morhua*), and confirmed the presence of *elovl4*-like transcripts termed *elovl4c-1* and *elovl4c-2* that were distinct to typical Elovl4a and Elovl4b. Unfortunately, no functional data are yet available for these novel Elovl4-like elongases although changes in their expression in response to dietary lipid in Atlantic cod and gilthead seabream (Betancor et al., 2016a) suggested they have putative roles in LC-PUFA biosynthesis.

## Tissue Distribution of Fads and Elovl Genes

The aforementioned dichotomy with regard to LC-PUFA biosynthetic capability between freshwater/salmonid fish and marine fish is partly supported by the tissue distribution patterns of genes encoding Fads and Elovl enzymes involved in the pathways. Typically, gene expression analyses by quantitative reverse-transcriptase PCR (qPCR) of a panel of tissues from salmonids, including rainbow trout and Atlantic salmon, showed that *fads2* (Δ6 and Δ5) desaturases and *elovl5* and *elovl2* elongases were highly expressed in intestine, liver, and brain (Abdul Hamid et al., 2016; Geay et al., 2016; Morais et al., 2009; Zheng et al., 2005). In contrast, studies of marine species including Atlantic cod (*G. morhua*), cobia (*R. canadum*), Asian sea bass (*Lates calcarifer*), Senegalese sole (*S. senegalensis*) and chu's croaker (*Nibea coibor*) (Huang et al., 2017) revealed brain as the major metabolic site as indicated by the highest expression of *fads2* (Kabeya et al., 2017; Mohd-Yusof et al., 2010; Morais et al., 2012; Tocher et al., 2006; Zheng et al., 2009). It has been hypothesized that the reason underlying this distinctive expression pattern and the fact that marine fish have retained a functional Δ6 desaturase was in order to guarantee sufficient supply of DHA in neural tissue, particularly in critical early developmental stages. A more restricted tissue distribution appears to exist for fish *elovl4* mRNA. Thus, the Elovl4a isoform in fish is primarily expressed in brain (Monroig et al., 2010b), whereas Elovl4b showed highest expression in eye and pineal gland, both tissues possessing photoreceptors in fish (Carmona-Antoñanzas et al., 2011; Li et al., 2015; Monroig et al., 2010b, 2012; Yan et al., 2018). Therefore, the tissue distribution of fish *elovl4* mRNA largely reflected the distribution of their enzymatic products, that is, the VLC-fatty acids, and thus suggested that their biosynthesis in fish occurs in situ as described for mammals (Agbaga et al., 2010).

# Desaturases and Elongase Expression During Early Development

Early life-cycle stages of vertebrates including fish have high demands for LC-PUFA to satisfy rapidly forming neural tissues in embryos (Tocher, 2010), here defined as developmental stages from zygote to the opening of oesophagus (Gatesoupe et al., 2001). Being lecithotrophic organisms, fish rely on the yolk to fulfil nutritional requirements prior to the onset of the exogenous feeding. In some species, lipids are accumulated in the form of oil droplets, thus constituting an additional source of fatty acids. Irrespective of the presence or absence of oil droplets, the composition of lipid reserves in fish embryos greatly depends on the diet of broodstock fish, as well as genetic factors (Tocher, 2010). Interestingly, there is strong evidence indicating that the LC-PUFA biosynthetic pathway is active, not only during follicle maturation (Ishak et al., 2008), but also in postfertilization stages in the model species zebrafish (Monroig et al., 2009, 2010b; Tan et al., 2010). Study of the temporal expression of genes involved in LC-PUFA biosynthesis during early ontogeny of zebrafish confirmed the presence of transcripts (mRNA) of both desaturases (dual $\Delta6\Delta5$ *fads2*) and elongases (*elovl2* and *elovl5*) throughout the entire embryogenesis, with activation occurring from 24 h postfertilization (hpf) (Tan et al., 2010). It was particularly interesting to note that the three genes investigated were expressed from the zygote stage (0 hpf), suggesting that transcripts of key LC-PUFA biosynthetic genes were transferred maternally to the embryo. Similar results were obtained in the marine species cobia (Monroig et al., 2011c) and Senegalese sole (Morais et al., 2012), and also common carp *Cyprinus carpio* (Ren et al., 2013), the latter study focusing exclusively on the *fads2* gene complement. Hence, in addition to deposition of preformed LC-PUFA in the yolk or lipid globules, the maternal role includes the transfer of key enzyme mRNA transcripts encoding enzymes involved in LC-PUFA biosynthesis. Furthermore, in addition to the temporal expression pattern, the spatial distribution of mRNA of Fads- and Elovl-encoding genes during early development confirmed that LC-PUFA produced by these enzymes were essential compounds in neural tissues, such as brain and retina (Monroig et al., 2009, 2010b; Tan et al., 2010).

# Modulation of the LC-PUFA Biosynthesis in Fish

The LC-PUFA biosynthetic capability of fish can be modulated by dietary fatty acids (nutritional regulation) and environmental factors, primarily temperature and salinity (Vagner and Santigosa, 2011). Nutritional regulation of LC-PUFA in fish has been extensively investigated in farmed species (Tocher, 2015). One study showed that European seabass

(*Dicentrarchus labrax*) fed on either a low or high LC-PUFA diet during larval stages (4–45 days posthatch) had different LC-PUFA biosynthetic capability at later juvenile stages (>7 g) (Vagner et al., 2007a, 2009). The results indicated that fish previously fed the low LC-PUFA diet during larval stages were more capable of utilizing an LC-PUFA-deficient diet later in life as highlighted by increased phospholipid DHA and increased Δ6 *fads2* expression, suggesting a possible epigenetic regulatory mechanism with potential application in aquaculture. The concept of "nutritional programming" has been recently supported by a study that showed early exposure to a low LC-PUFA diet at first feeding enhanced EPA and DHA metabolism including Δ6 *fads2* and *elovl5* expression in Atlantic salmon (Clarkson et al., 2017; Vera et al., 2017). Conversely, absence of clear nutritional conditioning effects have been recently reported in juveniles of rainbow trout *O. mykiss* (Mellery et al., 2017). More often though, nutritional regulation of LC-PUFA biosynthesis has been investigated in juvenile and grow-out life stages in the context of development of sustainable feed formulations with typically reduced levels of LC-PUFA. As mentioned in the Introduction, terrestrial plant ingredients are now widely used in aquafeeds to replace the finite marine ingredients FM and FO. It has been shown that species, such as Atlantic salmon with the full complement of enzymatic activities required for LC-PUFA from $C_{18}$ PUFA, are able to use ALA and LOA derived from VO and satisfy physiological requirements for the key LC-PUFA. Additionally, it is commonly reported in the literature that increased expression and activity of desaturases and elongases occurs in fish fed high VO-rich diets, a mechanism that has been often postulated to at least partially compensates the lower dietary level of preformed LC-PUFA (Bell and Tocher, 2009b; Leaver et al., 2008; Tocher, 2003, 2010). The molecular mechanisms underpinning the increased expression of Fads and Elovl in fish fed VO compared to fish fed FO are being elucidated. Until recently it was not fully clear whether the higher expression in fish fed VO was due to reduced levels of pathway products (e.g., EPA and DHA) or high levels of pathway substrates (e.g., ALA and LOA) (Tocher, 2003). Although there is evidence that substrate level may be influential in rainbow trout (Thanuthong et al., 2011), product inhibition is also involved, and it is now clear that it is DHA rather than EPA that exerts this feedback to suppress the expression of key genes including Δ6 Fads (Betancor et al., 2015a,b, 2016b; Thomassen et al., 2012). It is important to note that, although the LC-PUFA biosynthetic pathway is upregulated in fish fed VO-based diets (Vagner and Santigosa, 2011), the tissue levels of n-3 LC-PUFA achieved are still lower than those of fish fed FO-based diets (Tocher, 2009, 2015). A variety of practical strategies involving functional feeds containing ingredients or supplements with the ability to enhance the biosynthesis/retention of LC-PUFA in fish fed VO-based diets have been investigated. Bioactive fatty acids, such as conjugated linoleic acid (CLA; positional and geometric isomers of LOA), 3-thia fatty acids (e.g., tetradecylthioacetic acid) or petroselinic acid, have been used as potential dietary modulators of LC-PUFA metabolism in fish with some positive results in terms of LC-PUFA levels and expression of Fads-encoding genes (Kennedy et al., 2006, 2007; Randall et al., 2013). Dietary supplementation with micronutrients, such as minerals (iron, zinc, and magnesium) and vitamins (niacin, riboflavin, and biotin) (Giri et al., 2016; Lewis et al., 2013; Senadheera et al., 2012), as well as plant-derived compounds, such as fibrates (Ruyter et al., 1997) and lignans (Schiller Vestergren et al., 2011, 2012; Trattner et al., 2008a,b), have been investigated in salmonids with some degree of success in enhancing the LC-PUFA biosynthesis pathway. Furthermore, dietary cholesterol supplementation induced a significant increase in both gene expression and apparent in vivo activity of Δ6 Fads and elongase in liver in trout fed VO-based diets (Norambuena et al., 2013).

Fish are ectotherms and consequently have minimum control of their body temperature, which ultimately reflects that of the environment. Consequently, exposure of fish to lower temperatures has often been associated with increased unsaturation of cell membrane lipids (e.g., phospholipids) to maintain membrane fluidity through the activation of de novo biosynthesis of monosaturated fatty acids by stearoyl-CoA desaturase (Trueman et al., 2000) or biosynthesis of LC-PUFA (De Torrengo and Brenner, 1976; Hagar and Hazel, 1985; Ninno et al., 1974; Schünke and Wodtke, 1983; Tocher et al., 2004). Therefore, the aforementioned consequences of the limited availability of FO for aquafeeds combined with the potential effects of the expected rise of water temperature due global climate change has prompted interest in evaluation of the dual impacts of these factors in commercially important species (Mellery et al., 2016; Ruyter et al., 2003; Tocher et al., 2004; Vagner et al., 2007b). Generally, these studies have shown that increased temperature leads to decreased activity of the LC-PUFA pathways, which would be contrary to the adaptation to low marine ingredient feeds.

Salinity is another environmental factor that has been reported to influence LC-PUFA production in fish (Vagner and Santigosa, 2011). In particular, diadromous salmonids, such as Atlantic salmon, exhibit increasing activity of the pathway during freshwater with a peak around seawater transfer, with a subsequent reduction during the seawater phase (Tocher et al., 2000, 2003b). However, the parr-smolt transformation in salmon is a preadaptation for a change in salinity, rather than a direct response to salinity. A direct influence of salinity on LC-PUFA biosynthesis has been reported and was associated with osmoregulatory responses required for adaptation to higher salinity, which involved membrane lipid remodeling to ensure normal function of membrane-bound proteins (Fonseca-Madrigal et al., 2012). While the molecular mechanisms of the regulatory process are not fully understood, studies in the marine euryhaline *S. canaliculatus* have confirmed that the expression of *fads2* was regulated by salinity at both transcriptional (Li et al., 2008; Xie et al., 2015; Zhang et al., 2016a) and posttranscription levels (Zhang et al., 2014, 2016b). It is important to note that the response to low or high salinity depends on the fish species and, as described earlier for rabbitfish *S. canaliculatus*, the mullet *Mugil cephalus* also exhibited increased tissue levels of DHA and ARA when reared at reduced salinity (Kheriji et al., 2003). In contrast, lower environmental salinity resulted in reduced tissue levels of EPA and DHA in Japanese seabass (*Lateolabrax japonicus*) (Xu et al., 2010) and European seabass (*D. labrax*) (Hunt et al. 2011), and reduced LC-PUFA biosynthetic activities in the pike silverside *C. estor* (Fonseca-Madrigal et al., 2012).

## Genetic Approaches to Enhance LC-PUFA Biosynthesis in Fish

Genes encoding Fads and Elovl involved in LC-PUFA biosynthesis may be appropriate targets for genetic selection to develop strains of fish with enhanced ability to thrive on more sustainable VO-based feed formulations. Selective breeding programs for commercially important species like Atlantic salmon have largely overlooked traits like flesh n-3 LC-PUFA content despite its high heritability (Leaver et al., 2011) and thus its potential for genetic selection of strains with higher capacity for LC-PUFA biosynthesis (Gjedrem, 2000). Therefore, wild stocks represent a valuable genetic resource for improving this trait as recently shown in landlocked strains of Atlantic salmon that do not migrate to the sea and have an increased capacity for LC-PUFA biosynthesis likely due to the limited dietary supply during their life cycle (Betancor et al., 2016c).

As an alternative to genetic selection, transgenic technology has also been explored to generate fish strains with enhanced capacity to biosynthesise LC-PUFA. Using the model species zebrafish *D. rerio*, studies investigated the effects that overexpression of genes encoding enzymes of LC-PUFA biosynthetic pathway from masu salmon (*Oncorhynchus masou*), namely putative Δ6 (Alimuddin et al., 2005) and Δ5 *fads2* (Alimuddin et al., 2007), and *elovl5* (Alimuddin et al., 2008), had on n-3 LC-PUFA biosynthesis. Transgenic strains of zebrafish carrying the masu salmon Δ6 and Δ5 Fads both resulted in increased production of EPA and DHA compared to nontransgenic fish (Alimuddin et al., 2005, 2007), with similar results obtained when the masu salmon *elovl5* was used as transgene (Alimuddin et al., 2008). More recently, humanized *Caenorhabditis elegans fat1*, an ω3 desaturase, and *fat2*, a Δ12 desaturase, were expressed in zebrafish as an approach to produce a fish strain that was totally independent of dietary PUFA (Pang et al., 2014). These studies have obvious interest to understand the molecular and physiological processes related to LC-PUFA enhancement and the authors suggested that transgenesis was a potential strategy to alleviate and possibly eliminate the need to supply preformed $C_{20-22}$ LC-PUFA in the diet of ongrowing and larval stages of farmed fish. In this regard, similar investigations have been conducted in commercially important species, such as the common carp *C. carpio* (Cheng et al., 2014) and the marine fish nibe croaker (Kabeya et al., 2014, 2016). The applicability of these technologies, however, to fish farming in many parts of the world, not least Europe, is still extremely challenging due in part to sociopolitical issues and food safety regulations, but also to the fact that methodological difficulties might arise when applying transgenic technologies into fish other than model species.

# METABOLIC ROLES OF PUFA AND LC-PUFA

Much of the biological significance of PUFA, and especially LC-PUFA, derives from specific functional effects they have as important regulators of metabolism and physiology through key roles in various signaling pathways. The fatty acids can exert these effects either as themselves or as derivatives, such as eicosanoids. Although much less is known of these roles in fish, the emerging data indicate they are as important in fish as they are in mammals.

## Regulators of Transcription Factors

Various PUFA, LC-PUFA, and their derivatives can exert regulatory effects on cellular metabolism as ligands of transcription factors, including nuclear receptors, such as sterol regulatory element binding proteins (Srebp) that are key regulators of lipogenesis and cholesterol biosynthesis as well as LC-PUFA biosynthesis (Carmona-Antoñanzas et al., 2014; Minghetti et al., 2011), peroxisome proliferator-activated receptors α (Pparα) and γ (Pparγ) that regulate genes of fatty acid oxidation and deposition among other pathways (Leaver et al., 2008), and the liver X receptor (LXR) and G-protein coupled receptors (Gprc) (Gpr120 and Gpr40) (Grygiel-Górniak, 2014; Oh et al., 2010).

The significant roles that Ppar have in regulating fatty acid metabolism have been long-established in mammalian models (Grygiel-Górniak, 2014; Willson et al., 2000). While Pparα is involved fatty acid oxidation, ketone body synthesis and glucose sparing, Pparγ acts as a key regulator inducing the differentiation of preadipocytes into adipocytes, and triglyceride storage (Hihi et al., 2002). Gene orthologues of both *pparα* (two copies in teleosts) and *pparγ* have been isolated and characterized in teleosts (Boukouvala et al., 2004; Leaver et al., 2005; Urbatzka et al., 2015), with a clear indication of a conserved role in LC-PUFA metabolism, including expression tissue patterns and ligand binding profiles (Bouk-ouvala et al., 2004). In the case of Lxr, a critical function was recognized in cholesterol homeostasis (Kalaany and Mangelsdorf, 2006), and its role in LC-PUFA metabolism has been explored in teleosts (Carmona-Antoñan-zas et al., 2014; Cruz García et al., 2009; Zhang et al., 2016a), despite the paucity of data. Gprc have also been linked with LC-PUFA metabolism (Milligan et al., 2015). For example, Gpr120 functions as an n-3 LC-PUFA sensor (Oh et al., 2010). However, these receptors have yet to be charac-terized outside mammals, with their presence in fish genomes still to be firmly established (Castro's personal communication).

## Eicosanoids

Fish produce the same range of highly biologically active LC-PUFA derivatives, that is, eicosanoids, as in mammals including cyclooxygenase (prostaglandins and thromboxanes), and lipoxygenase (hydroperoxy and hydroxy fatty acids, leukotrienes, and lipoxins) products. Furthermore, the eicosanoids appear to have the same physiological roles in inflammatory and immune responses, blood clotting and cardiovascular tone, renal and neural functions, and reproduction in fish as in mammals (Tocher, 2003). In addition, the same competition between ARA and EPA as substrates for eicosanoid synthesis exists in fish as in mammals (Tocher, 1995). Therefore, the fact that ARA is the preferred substrate for eicosanoid synthesis in fish despite the preponderance of EPA in fish tissues is particularly interesting. The molecular mechanism for this is unclear and an early hypothesis that ARA-rich phosphatidylinositol was the source of eicosanoid precursor in fish was never proven (Tocher et al., 1991) and, therefore, it is likely that the fatty acid specificity of phospholipase A enzymes drives this prefer-ence (Tocher, 2003). Whatever the mechanism, eicosanoid production in fish, as in mammals, is influenced by the cellular ratio of ARA:EPA, and an imbalance leading to excessive (strength and duration) inflammatory responses appears to be similarly damaging in fish (Tocher, 2003). In this respect, n-3 LC-PUFA, EPA and DHA, are also the precursors of specialized proresolving lipid mediators (SPM), the D & E series resolvins, (neuro) protectins, and maresins, that prevent excessive inflammation, promote resolution, and expedite the return to tissue homeostasis (Serhan, 2014). While SPM have been reported in fish, specifically resolvins and protec-tins in rainbow trout (Hong et al., 2005) and resolvins in Atlantic salmon (Raatz et al., 2011), little else is known about their roles in fish although it is highly likely that they have similar critical roles in resolving inflamma-tion in fish as in mammals.

The regulatory roles of PUFA and, especially, LC-PUFA emphasize the importance of the balance between the n-3 and n-6 series (Lands, 2014) and highlight the impact that dietary LC-PUFA can have on fish health and disease (Tocher and Glencross, 2015). Therefore, this is an area of current interest in aquaculture due to the development of sustainable feeds based on plant meals and VO that has reduced n-3 LC-PUFA and increased n-6 PUFA levels in farmed fish (Tocher, 2015).

## CONCLUSIONS

Omega-3 LC-PUFA are essential components of the human diet, with fish being the primary source. Thus, research into PUFA metabolism in fish has been thriving in the past decades coinciding with the expansion of finfish aquaculture and the need to supply nutritious products while increasing the use of non-marine ingredients in aquafeeds. In this chapter, we summarized the extent to which our understanding of several key aspects of PUFA metabolism in fish has evolved. This was particularly noteworthy with regards to the LC-PUFA biosynthesis pathway, whereby a plethora of studies combining genomic and functional approaches have illuminated vital aspects of PUFA desaturation and elongation with impact on nutritional strategies in aquaculture. The chapter also covered other areas of research relevant to PUFA metabolism, including digestion, absorption, and catabolism, which largely operate as in mammals. The key roles of PUFA in signaling pathways and the regulation of metabolism in fish are also discussed. Overall, comparative approaches with other vertebrates, namely mammals, provide a clear link between genetic evolutionary background (gene repertoire versus function) and the impact of habitat specific inputs (e.g., diets). Finally, the present-day "omics" revolution will continue to provide valuable insights into our ability to elucidate mechanisms of PUFA metabolism in fish, as will the application of new technologies (e.g., transgenics, CRISP).

### References

Abdul Hamid, N.K., Carmona-Antoñanzas, G., Monroig, Ó., Tocher, D.R., Turchini, G.M., Donald, J.A., 2016. Isolation and functional characterisation of a *fads2* in rainbow trout (*Oncorhynchus mykiss*) with $\Delta 5$ desaturase activity. PLoS One 11, e0150770.

Agaba, M., Tocher, D.R., Dickson, C.A., Dick, J.R., Teale, A.J., 2004. Zebrafish cDNA encoding multifunctional fatty acid elongase involved in production of eicosapentaenoic (20: 5n-3) and docosahexaenoic (22:6n-3) acids. Mar. Biotechnol. 6, 251–261.

Agbaga, M.-P., Brush, R.S., Mandal, M.N.A., Henry, K., Elliott, M.H., Anderson, R.E., 2008. Role of Stargardt-3 macular dystrophy protein (ELOVL4) in the biosynthesis of very long chain fatty acids. Proc. Natl. Acad. Sci. U.S.A. 105, 12843–12848.

Agbaga, M.-P., Mandal, M.N.A., Anderson, R.E., 2010. Retinal very long-chain PUFAs: new insights from studies on ELOVL4 protein. J. Lipid Res. 51, 1624–1642.

Alimuddin, Yoshizaki, G., Kiron, V., Satoh, S., Takeuchi, T., 2005. Enhancement of EPA and DHA biosynthesis by over-expression of masu salmon $\Delta 6$-desaturase-like gene in zebrafish. Transgenic Res. 14, 159–165.

Alimuddin, Yoshizaki, G., Kiron, V., Satoh, S., Takeuchi, T., 2007. Expression of the masu salmon Δ5-desaturase-like gene elevated EPA and DHA biosynthesis in zebrafish. Mar. Biotechnol. 9, 92–100.

Alimuddin, Kiron, V., Satoh, S., Takeuchi, T., Yoshizaki, G., 2008. Cloning and over-expression of a masu salmon (*Oncorhynchus masou*) fatty acid elongase-like gene in zebrafish. Aquaculture 282, 13–18.

Aranceta, J., Pérez-Rodrigo, C., 2012. Recommended dietary reference intakes, nutritional goals and dietary guidelines for fat and fatty acids: a systematic review. Br. J. Nutr. 107, S8–22.

Bell, J.G., Tocher, D.R., 2009a. Farmed fish: the impact of diet on fatty acid compositions. In: Rossell, B. (Ed.), Oils and Fats Handbook Volume 4: Fish Oils. Leatherhead Food International, Leatherhead, pp. 171–184.

Bell, M.V., Tocher, D.R., 2009b. Biosynthesis of fatty acids: general principles and new directions. In: Arts, M.T., Brett, M., Kainz, M. (Eds.), Lipids in Aquatic Ecosystems. Springer-Verlag, Berlin, Germany, pp. 211–236.

Bell, J.G., Koppe, W., 2011. Welfare and health of fish fed vegetable oils as alternative lipid sources to fish oil. In: Turchini, G.M., Ng, W.-K., Tocher, D.R. (Eds.), Fish Oil Replacement and Alternative Lipid Sources in Aquaculture Feeds. Taylor & Francis, CRC Press, Boca Raton, FL, United States, pp. 21–59.

Betancor, M.B., Sprague, M., Usher, S., Sayanova, O., Campbell, P.J., Napier, J.A., Tocher, D.R., 2015a. A nutritionally-enhanced oil from transgenic *Camelina sativa* effectively replaced marine fish oil as a source of eicosapentaenoic acid for farmed Atlantic salmon (*Salmo salar*). Sci. Rep. 5, 8104.

Betancor, M.B., Sprague, M., Sayanova, O., Usher, S., Campbell, P.J., Napier, J.A., Caballero, M.J., Tocher, D.R., 2015b. Evaluation of a high-EPA oil from transgenic *Camelina sativa* in feeds for Atlantic salmon (*Salmo salar* L.): effects on tissue fatty acid composition, histology and gene expression. Aquaculture 444, 1–12.

Betancor, M.B., Sprague, M., Montero, D., Usher, S., Sayanova, O., Campbell, P.J., Napier, J.A., Caballero, M.J., Izquierdo, M., Tocher, D.R., 2016a. Replacement of marine fish oil with de novo omega-3 oils from transgenic *Camelina sativa* in feeds for gilthead sea bream (*Sparus aurata* L.). Lipids 51, 1171–1191.

Betancor, M.B., Sprague, M., Sayanova, O., Usher, S., Metochis, C., Campbell, P.J., Napier, J.A., Tocher, D.R., 2016b. Nutritional evaluation of an EPA-DHA oil from transgenic *Camelina sativa* in feeds for post-smolt Atlantic salmon (*Salmo salar* L.). PLoS One 11, e0159934.

Betancor, M.B., Olsen, R.E., Slostorm, D., Skulstad, D., Tocher, D.R., 2016c. Assessment of a land-locked Atlantic salmon (*Salmo salar* L.) population as a potential genetic resource with a focus on long-chain polyunsaturated fatty acid biosynthesis. Biochim. Biophys. Acta 1861, 227–238.

Bonnefont, J.P., Djouadi, F., Prip-Buus, C., Gobin, S., Munnich, A., Bastin, J., 2004. Carnitine palmitoyltransferases 1 and 2: biochemical, molecular and medical aspects. Mol. Aspects Med. 25, 495–520.

Bougnoux, P., Hajjaji, N., Ferrasson, M.N., Giraudeau, B., Couet, C., Le Floch, O., 2009. Improving outcome of chemotherapy of metastatic breast cancer by docosahexaenoic acid: a phase II trial. Br. J. Cancer 101, 1978–1985.

Boukouvala, E., Antonopoulou, E., Favre-Krey, L., Diez, A., Bautista, J.M., Leaver, M.J., Tocher, D.R., Krey, G., 2004. Molecular characterization of three peroxisome proliferator-activated receptors from the sea bass (*Dicentrarchus labrax*). Lipids 39, 1085–1092.

Boukouvala, E., Leaver, M.J., Favre-Krey, L., Theodoridou, M., Krey, G., 2010. Molecular characterization of a gilthead sea bream (*Sparus aurata*) muscle tissue cDNA for carnitine palmitoyltransferase 1B (CPT1B). Comp. Biochem. Physiol. 157B, 189–197.

Brodtkord, T., Rosenlund, G., Lie, O., 1997. Effects of dietary levels of 20:5n–3 and 22:6n–3 on tissue lipid composition in juvenile Atlantic salmon, *Salmo salar*, with em-phasis on brain and eye. Aquacult. Nutr. 3, 175–187.

Browse, J., Somerville, C., 1991. Glycerolipid synthesis: biochemistry and regulation. Ann. Rev. Plant Physiol. Plant Mol. Biol. 42, 467–506.

Buzzi, M., Henderson, R.J., Sargent, J.R., 1996. The desaturation and elongation of linolenic acid and eicosapentaenoic acid by hepatocytes and liver microsomes from rainbow trout (*Oncorhynchus mykiss*) fed diets containing fish oil or olive oil. Biochim. Biophys. Acta 1299, 235–244.

Buzzi, M., Henderson, R.J., Sargent, J.R., 1997. Biosynthesis of docosahexaenoic acid in trout hepatocytes proceeds via 24-carbon intermediates. Comp. Biochem. Physiol. 116B, 263–267.

Cabré, E., Mañosa, M., Gassulla, M.A., 2012. Omega-3 fatty acids and inflammatory bowel diseases: a systematic review. Br. J. Nutr. 107, S240–S252.

Calder, P.C., 2014. Very long chain omega-3 (n-3) fatty acids and human health. Eur. J. Lipid Sci. Technol. 116, 1280–1300.

Calder, P.C., Yaqoob, P., 2012. Marine omega-3 fatty acids and coronary heart disease. Curr. Opin. Cardiol. 27, 412–419.

Campoy, C., Escolano-Margarit, V., Anjos, T., Szajewska, H., Uauy, R., 2012. Omega 3 fatty acids on child growth, visual acuity and neurodevelopment. Br. J. Nutr. 107, S85–S106.

Carlson, S.E., Werkman, S.H., Rhodes, P.G., Tolley, E.A., 1993. Visual- acuity development in healthy preterm infants: effect of marine-oil supplementation. Am. J. Clin. Nutr. 58, 35–42.

Carmona-Antoñanzas, G., Monroig, Ó., Dick, J.R., Davie, A., Tocher, D.R., 2011. Biosynthesis of very long-chain fatty acids (C ≥ 26) in Atlantic salmon: cloning, functional characterisation, and tissue distribution of an Elovl4. Comp. Biochem. Physiol. 159B, 122–129.

Carmona-Antoñanzas, G., Tocher, D.R., Martinez-Rubio, L., Leaver, M.J., 2014. Conservation of lipid metabolic gene transcriptional regulatory networks in fish and mammals. Gene 534, 1–9.

Castro, L.F.C., Monroig, Ó., Leaver, M.J., Wilson, J., Cunha, I., Tocher, D.R., 2012. Functional desaturase Fads1 (Δ5) and Fads2 (Δ6) orthologues evolved before the origin of jawed vertebrates. PLoS One 7, e31950.

Castro, L.F.C., Tocher, D.R., Monroig, O., 2016. Long-chain polyunsaturated fatty acid biosynthesis in chordates: insights into the evolution of Fads and Elovl gene repertoire. Prog. Lipid Res. 62, 25–40.

Casula, M., Soranna, D., Catapano, A.L., Corrao, G., 2013. Long-term effect of high dose omega-3 fatty acid supplementation for secondary prevention of cardiovascular outcomes: a meta-analysis of randomized, double blind, placebo controlled trials. Atheroscler. Suppl. 14, 243–251.

Cheng, Q., Su, B., Qin, Z., Weng, C.-C., Yin, F., Zhou, Y., Fobes, M., Perera, D.A., Shang, M., Soller, F., Shi, Z., Davis, A., Dunham, R.A., 2014. Interaction of diet and the masou salmon Δ5-desaturase transgene on Δ6-desaturase and stearoyl-CoA desaturase gene expression and N-3 fatty acid level in common carp (*Cyprinus carpio*). Transgenic Res. 23, 729–742.

Clarkson, M., Migaud, H., Metochis, C., Vera, L.M., Leeming, D., Tocher, D.R., Taylor, J.F., 2017. Early nutritional intervention can improve utilisation of vegetable-based diets in diploid and triploid Atlantic salmon (*Salmo salar*). Br. J. Nutr. 118, 17–29.

Cruz García, L., Minghetti, M., Navarro, I., Tocher, D.R., 2009. Molecular cloning, tissue expression and regulation of liver X receptor (LXR) transcription factors of Atlantic salmon (*Salmo salar*) and rainbow trout (*Oncorhynchus mykiss*). Comp. Biochem. Physiol. 153B, 81–88.

De Antueno, R.J., Knickle, L.C., Smith, H., Elliot, M.L., Allen, S.J., Nwaka, S., Winther, M.D., 2001. Activity of human Δ5 and Δ6 desaturases on multiple n-3 and n-6 polyunsaturated fatty acids. FEBS Lett. 509, 77–80.

Delgado-Lista, J., Pérez-Martínez, P., López-Miranda, J., Pérez-Jiménez, F., 2012. Long chain omega-3 fatty acids and cardiovascular disease: a systematic review. Br. J. Nutr. 107, S201–S213.

De Torrengo, M.P., Brenner, R.R., 1976. Influence of environmental temperature on the fatty acid desaturation and elongation activity of fish (*Pimelodus maculatus*) liver microsomes. Biochim. Biophys. Acta 424, 36–44.

Dong, Y., Wang, S., Chen, J., Zhang, Q., Liu, Y., You, C., Monroig, O., Tocher, D.R., Li, Y., 2016. Hepatocyte nuclear factor $4\alpha$ (HNF4$\alpha$) is a transcription factor of vertebrate fatty acyl desaturase gene as identified in marine teleost *Siganus canaliculatus*. PLoS One. doi: 10.1371/journal.pone.0160361.

Dong, Y., Zhao, J., Chen, J., Wang, S., Liu, Y., Zhang, Q., You, C., Monroig, Ó., Tocher, D.R., Li, Y., 2018. Cloning and characterization of $\Delta6/\Delta5$ fatty acyl desaturase (Fad) gene promoter in the marine teleost *Siganuscanaliculatus*. Gene. 647, 174–180.

European Food Safety Authority (EFSA) Panel on Dietetic Products, Nutrition and Allergies (NDA), 2012. Scientific opinion related to the tolerable upper intake level of eicosapentaenoic acid (EPA), docosahexaenoic acid (DHA) and docosapentaenoic acid (DPA). EFSA J. 10, 2815.

FAO, 2014. The State of World Fisheries and Aquaculture 2014. Food and Agriculture Organization of the United Nations, Rome, Italy, p. 223.

FAO, 2016, Italy. The State of World Fisheries and Aquaculture 2014. Food and Agriculture Organization of the United Nations, Rome, p. 204.

Ferdinandusse, S., Denis, S., van Roermund, C.W.T., Wanders, R.J.A., Dacremont, G., 2004. Identification of the peroxisomal-oxidation enzymes involved in the degradation of long-chain dicarboxylic acids. J. Lipid Res. 45, 1104–1111.

Fonseca-Madrigal, J., Pineda-Delgado, D., Martinez-Palacios, C.A., Rodriguez, C., Tocher, D.R., 2012. Effect of salinity on the biosynthesis of n-3 long-chain polyunsaturated fatty acids in silverside *Menidia estor*. Fish Physiol. Biochem. 38, 1047–1057.

Fonseca-Madrigal, J., Navarro, J.C., Hontoria, F., Tocher, D.R., Martínez-Palacios, C., Monroig, Ó., 2014. Diversification of substrate specificities in teleostei Fads2 fatty acyl desaturases: cloning and functional characterisation of $\Delta4$ and bifunctional $\Delta6\Delta5$ desaturases in the freshwater atherinopsid *Chirostoma estor*. J. Lipid Res. 55, 1408–1419.

Gatesoupe, F.J., Zambonino Infante, J.L., Cahu, C., Bergot, P., 2001. Ontogeny, development and digestive physiology of fish larvae. In: Guillaume, J., Kaushik, S., Bergot, P., Métailler, R. (Eds.), Nutrition and Feeding of Fish and Crustacean. Praxis Publishing Ltd., Chichester, pp. 197–212.

Geay, F., Tinti, E., Mellery, J., Michaux, C., Larondelle, Y., Perpète, E., Kestemont, P., 2016. Cloning and functional characterization of $\Delta6$ fatty acid desaturase 2 (FADS2) in Eurasian perch (*Perca fluviatilis*). Comp. Biochem. Physiol. 191B, 112–125.

Gatlin, III, D.M., Barrows, F.T., Brown, P., Dabrowski, K., Gibson, G.T., Hardy, R.W., Elliot, H., Hu, G., Krogdahl, A., Nelson, R., Overturf, K., Rust, M., Sealey, W., Skonberg, D., Souza, E.J., Stone, D., Wilson, R., Wurtele, E., 2007. Expanding the utilization of sustainable plant products in aquafeeds: a review. Aquacult. Res. 38, 551–579.

Gebauer, S.K., Psota, T.L., Harris, W.S., Kris-Etherton, P.M., 2006. n-3 fatty acid dietary recommendations and food sources to achieve essentiality and cardiovascular benefits. Am. J. Clin. Nutr. 83, 1526S–1535S.

Gerber, M., 2012. Omega-3 fatty acids and cancers: a systematic update review of epidemiological studies. Br. J. Nutr. 107, S228–239.

Gil, A., Serra-Majem, L., Calder, P.C., Uauy, R., 2012. Systematic reviews of the role of omega-3 fatty acids in the prevention and treatment of disease. Br. J. Nutr. 107, S1–S2.

Giri, S.S., Graham, J., Hamid, N.K.A., Donal, J.A., Turchini, G.M., 2016. Dietary micronutrients and in vivo n-3 LC-PUFA biosynthesis in Atlantic salmon. Aquaculture 452, 416–425.

Gjedrem, T., 2000. Genetic improvement of cold-water fish species. Aquacult Res. 31, 25–33.

Goecke, F., Hernández, V., Bittner, M., González, M., Becerra, J., Silva, M., 2010. Fatty acid composition of three species of *Codium* (Bryopsidales, Chlorophyta) in Chile. Rev. Biol. Mar. Oceanogr. 45, 325–330.

GOED Global Recommendations for EPA and DHA Intake (Rev 16 April 2014). Available from: http://www.goedomega3.com/healthcare.

GOED. GOED Publishes EPA and DHA Intake Recommendations 2016. Available from: http://www.goedomega3.com/uploads/default/news_photos/press_release_Intake_Recommendations.pdf.

Gregory, M., See, V.H.L., Gibson, R.A., Schuller, K.A., 2010. Cloning and functional characterisation of a fatty acyl elongase from southern bluefin tuna (*Thunnus maccoyii*). Comp. Biochem. Physiol. 155B, 178–185.

Gregory, M.K., Cleland, L.G., James, M.J., 2013. Molecular basis for diferential elongation of omega-3 docosapentaenoic acid by the rat Elovl5 and Elovl2. J. Lipid Res. 54, 2851–2857.

Gregory, M.K., James, M.J., 2014. Rainbow trout (*Oncorhynchus mykiss*) Elovl5 and Elovl2 differ in selectivity for elongation of omega-3 docosapentaenoic acid. Biochim. Biophys. Acta 1841, 1656–1660.

Grygiel-Górniak, B., 2014. Peroxisome proliferator-activated receptors and their ligands: nutritional and clinical implications: a review. Nutr. J. 13, 17.

Guillou, H., Zadravec, D., Martin, P.G.P., Jacobsson, A., 2010. The key roles of elongases and desaturases in mammalian fatty acid metabolism: insights from transgenic mice. Prog. Lipid Res. 49, 186–199.

Hagar, A.F., Hazel, J.R., 1985. Changes in desaturase activity and the fatty acid composition of microsomal membranes from liver tissue of thermally-acclimating rainbow trout. J. Comp. Physiol. 156B, 35–42.

Hardy, R.W., 2010. Utilization of plant proteins in fish diets: effects of global demand and supplies of fishmeal. Aquacult. Res. 41, 770–776.

Hastings, N., Agaba, M., Tocher, D.R., Leaver, M.J., Dick, J.R., Sargent, J.R., Teale, A.J., 2001. A vertebrate fatty acid desaturase with $\Delta 5$ and $\Delta 6$ activities. Proc. Natl. Acad. Sci. U.S.A. 98, 14304–14309.

Hastings, N., Agaba, M.K., Tocher, D.R., Zheng, X., Dickson, C.A., Dick, J.R., Teale, A.J., 2005. Molecular cloning and functional characterization of fatty acyl desaturase and elongase cDNAs involved in the production of eicosapentaenoic and docosahexaenoic acids from α-linolenic acid in Atlantic salmon (*Salmo salar*). Mar. Biotechnol. 6, 463–474.

Hazel, J.R., Williams, E.E., 1990. The role of alterations in membrane lipid composition in enabling physiological adaptation of organisms to their physical environment. Prog. Lipid Res. 29, 167–227.

Henderson, R.J., 1996. Fatty acid metabolism in freshwater fish with particular reference to polyunsaturated fatty acids. Arch. Anim. Nutr. 49, 5–22.

Henriques, J., Dick, J.R., Tocher, D.R., Bell, J.G., 2014. Nutritional quality of salmon products available from major retailers in the UK: content and composition of n-3 long-chain polyunsaturated fatty acids. Br. J. Nutr. 112, 964–975.

Hihi, A.K., Michalik, L., Wahli, W., 2002. PPARs: transcriptional effectors of fatty acids and their derivatives. Cell Mol. Life Sci. 59, 790–798.

Hochachka, P.W., Mommsen, T.P. (Eds.), 1995. Biochemistry and Molecular Biology of Fishes, Environmental and Ecological Biochemistry, vol. 5, Elsevier Press, Amsterdam, Netherlands.

Hong, S., Tjonahen, E., Morgan, E.L., Lua, Y., Serhan, C.N., Rowley, A.F., 2005. Rainbow trout (*Oncorhynchus mykiss*) brain cells biosynthesize novel docosahexaenoic acid-derived resolvins and protectins: mediator lipidomic analysis. Prostaglandins Other Lipid Mediat. 78, 107–116.

Huang, Y., Lin, Z., Rong, H., Hao, M., Zou, W., Li, S., Wen, X., 2017. Cloning, tissue distribution, functional characterization and nutritional regulation of $\Delta 6$ fatty acyl desaturase in chu's croaker *Nibeacoibor*. Aquaculture 479, 208–216.

Hunt, A.O., Oxkan, F.E., Engin, K., Tekelioglu, N., 2011. The effects of freshwater rearing on the whole body and muscle tissue fatty acid profile of the European sea bass (*Dicentrarchus labrax*). Aquacult. Int. 19, 51–61.

Ishak, S.D., Tan, S.H., Khong, H.K., Jaya-Ram, A., Enyu, Y.L., Kuah, M.K., Shu-Chien, A.C., 2008. Upregulated mRNA expression of desaturase and elongase, two enzymes involved in highly unsaturated fatty acids biosynthesis pathways during follicle maturation in zebrafish. Reprod. Biol. Endocrinol. 6, 56–65.

ISSFAL, International Society for the Study of Fatty Acids and Lipids, 2004. Report of the Sub-Committee on Recommendations for Intake of Polyunsaturated Fatty Acids in Healthy Adults. ISSFAL, Brighton.

Jin, M., Monroig, Ó., Navarro, J.C., Tocher, D.R., Zhou, Q., 2017. Molecular and functional characterisation of two elovl4 elongases involved in the biosynthesis of very long-chain (>$C_{24}$) polyunsaturated fatty acids in black seabream *Acanthopagrus schlegelii*. Comp. Biochem. Physiol. 212B, 41–50.

Kabeya, N., Takeuchi, Y., Yamamoto, Y., Yazawa, R., Haga, Y., Satoh, S., Yoshizaki, G., 2014. Modification of the n–3 HUFA biosynthetic pathway by transgenesis in a marine teleost, nibe croaker. J. Biotechnol. 172, 46–54.

Kabeya, N., Yamamoto, Y., Cummins, S.F., Elizur, A., Yazawa, R., Takeuchi, Y., Haga, Y., Satoh, S., Yoshizaki, G., 2015. Polyunsaturated fatty acid metabolism in a marine teleost, Nibe croaker *Nibea mitsukurii*: functional characterization of Fads2 desaturase and Elovl5 and Elovl4 elongases. Comp. Biochem. Physiol. 188B, 37–45.

Kabeya, N., Chiba, M., Haga, Y., Satoh, S., Goro, Y., 2017. Cloning and functional characterization of *fads2* desaturase and elovl5 elongase from Japanese flounder *Paralichthys olivaceus*. Comp. Biochem. Physiol. 214B, 36–46.

Kabeya, N., Yevzelman, S., Oboh, A., Tocher, D.R., Monroig, O., 2018. Essential fatty acid metabolism and requirements of the cleaner fish, ballan wrasse *Labrusbergylta*: Defining pathways of long-chain polyunsaturated fatty acid biosynthesis. Aquaculture 488, 199–206.

Kabeya, N., Takeuchi, Y., Yazawa, R., Haga, Y., Yoshizaki, G., 2016. Transgenic modification of the n-3 HUFA biosynthetic pathway in nibe croaker larvae: improved DPA (docosapentaenoic acid; 22:5n-3) production. Aquacult. Nutr. 22, 472–478.

Kalaany, N.Y., Mangelsdorf, D.J., 2006. LXRS and FXR: the yin and yang of cholesterol and fat metabolism. Annu. Rev. Physiol. 68, 159–191.

Kennedy, S.R., Leaver, M.J., Campbell, P.J., Zheng, X., Dick, J.R., Tocher, D.R., 2006. Influence of dietary oil content and conjugated linoleic acid (CLA) on lipid metabolism enzyme activities and gene expression in tissues of Atlantic salmon (*Salmo salar* L.). Lipids 41, 423–436.

Kennedy, S.R., Bickerdike, R., Berge, R.K., Porter, A.R., Tocher, D.R., 2007. Influence of dietary conjugated linoleic acid (CLA) and tetradecylthioacetic acid (TTA) on growth, lipid composition and key enzymes of fatty acid oxidation in liver and muscle of Atlantic cod (*Gadus morhua* L.). Aquaculture 264, 372–382.

Kheriji, S., El Cafsi, M., Masmoudi, W., Castell, J.D., Romdhane, M.S., 2003. Salinity and temperature effects on the lipid composition of mullet sea fry (*Mugil cephalus*, Linne, 1758). Aquacult. Int. 11, 571–582.

Khotimchenko, S.V., 1995. Fatty acid composition of green algae of the genus *Caulerpa*. Bot. Mar. 38, 509–512.

Kolditz, C., Borthaire, M., Richard, N., Corraze, G., Panserat, S., Vachot, C., Lefèvre, F., Médale, F., 2008. Liver and muscle metabolic changes induced by dietary energy content and genetic selection in rainbow trout (*Oncorhynchus mykiss*). Am. J. Physiol. Regul. Integr. Comp. Physiol. 294, 1154–1164.

Kuah, M.-K., Jaya-Ram, A., Shu-Chien, A.C., 2015. The capacity for long-chain polyunsaturated fatty acid synthesis in a carnivorous vertebrate: functional characterisation and nutritional regulation of a Fads2 fatty acyl desaturase with Δ4 activity and an Elovl5 elongase in striped snakehead (*Channa striata*). Biochim. Biophys. Acta 1851, 248–260.

Kuah, M.-K., Jaya-Ram, A., Shu-Chien, A.C., 2016. A fatty acyl desaturase (*fads2*) with dual Δ6 and Δ5 activities from the freshwater carnivorous striped snakehead *Channa striata*. Comp. Biochem. Physiol. 201A, 146–155.

Lands, B., 2014. Historical perspectives on the impact of n-3 and n-6 nutrients on health. Prog. Lipid Res. 55, 17–29.

Leaver, M.J., Boukouvala, E., Antonopoulou, E., Diez, A., Favre-Krey, L., Ezaz, M.T., Bautista, J.M., Tocher, D.R., Krey, G., 2005. Three peroxisomal proliferator-activated receptor (PPAR) isotypes from each of two species of marine fish. Endocrinology 146, 3150–3162.

Leaver, M.J., Bautista, J.M., Björnsson, T., Jönsson, E., Krey, G., Tocher, D.R., Torstensen, B.E., 2008. Towards fish lipid nutrigenomics: current state and prospects for fin-fish aquaculture. Rev. Fish. Sci. 16, 71–92.

Leaver, M.J., Taggart, J.B., Villeneuve, L.A.N., Bron, J.E., Guy, D.R., Bishop, S.C., Houston, R.D., Matika, O., Tocher, D.R., 2011. Heritability and mechanisms of n-3 long chain polyunsaturated fatty acid deposition in the flesh of Atlantic salmon. Comp. Biochem. Physiol. 6D, 62–69.

Leonard, A.E., Kelder, B., Bobik, E.G., Chuang, L.-T., Lewis, C.J., Kopchick, J.J., Mukerji, P., Huang, Y.-S., 2002. Identification and expression of mammalian long-chain PUFA elongation enzymes. Lipids 37, 733–740.

Lewis, M.J., Hamid, N.K.A., Alhazzaa, R., Hermon, K., Donald, J.A., Sinclair, A.J., Turchini, G.M., 2013. Targeted dietary micronutrient fortification modulates n–3 LC-PUFA pathway activity in rainbow trout (Oncorhynchus mykiss). Aquaculture 412, 215–222.

Li, Y.Y., Hu, C.B., Zheng, Y.J., Xia, X.A., Xu, W.J., Wang, S.Q., Chen, W.Z., Sun, Z.W., Huang, J.H., 2008. The effects of dietary fatty acids on liver fatty acid composition and Δ6 desaturase expression differ with ambient salinities in Siganus canaliculatus. Comp. Biochem. Physiol. 151B, 183–190.

Li, Y., Monroig, Ó., Zhang, L., Wang, S., Zheng, X., Dick, J.R., You, C., Tocher, D.R., 2010. Vertebrate fatty acyl desaturase with Δ4 activity. Proc. Natl. Acad. Sci. U.S.A. 107, 16840–16845.

Li, S., Monroig, Ó., Navarro, J.C., Yuan, Y., Xu, W., Mai, K., Tocher, D.R., Ai, Q., 2017a. Molecular Cloning, functional characterization and nutritional regulation by dietary fatty acid profiles of a putative Elovl4 gene in orange-spotted grouper Epinephelus coioides. Aquacult. Res. 48, 537–552.

Li, S., Monroig, Ó., Wang, T., Yuan, Y., Navarro, J.C., Hontoria, F., Liao, K., Tocher, D.R., Mai, K., Xu, W., Ai Q., 2017b. Functional characterization and differential nutritional regulation of putative Elovl5 and Elovl4 elongases in large yellow croaker (Larimichthyscrocea). Sci. Rep 7, 2303.

Lim, Z., Senger, T., Vrinten, P., 2014. Four amino acid residues influence the substrate chainlength and regioselectivity of Siganus canaliculatus Δ4 and Δ5/6 desaturases. Lipids 49, 357–367.

Lopes-Marques, M., Delgado, I.L., Ruivo, R., Torres, Y., Sainath, S.B., Rocha, E., Cunha, I., Santos, M.M., Castro, L.F., 2015. The origin and diversity of Cpt1 genes in vertebrate species. PLoS One 10, e0138447.

Lopes-Marques, M., Ozório, R., Amaral, R., Tocher, D.R., Monroig, Ó., Castro, L.F.C., 2017. Molecular and functional characterization of a fads2 orthologue in the Amazonian teleost, Arapaima gigas. Comp. Biochem. Physiol. 203B, 84–91.

Madsen, L., Frøyland, L., Dyrøy, E., Helland, K., Berge, R.K., 1998. Docosahexaenoic and eicosapentaenoic acids are differently metabolized in rat liver during mitochondria and peroxisome proliferation. J. Lipid Res. 39, 583–593.

Mellery, J., Geay, F., Tocher, D.R., Kestemont, P., Debier, C., Rollin, X., Larondelle, Y., 2016. Temperature increase negatively affects the fatty acid bioconversion capacity of rainbow trout (Oncorhynchus mykiss) fed a linseed oil-based diet. PLoS One 11, e0164478.

Mellery, J., Brel, J., Dort, J., Geay, F., Kestemont, P., Francis, D.S., Larondelle, Y., Rollin, X., 2017. A n-3 PUFA depletion applied to rainbow trout fry (Oncorhynchus mykiss) does not modulate its subsequent lipid bioconversion capacity. Br. J. Nutr. 117 (2), 187–199.

Miles, E.A., Calder, P.C., 2012. Influence of marine n-3 polyunsaturated fatty acids on immune function and a systematic review of their effects on clinical outcomes in rheumatoid arthritis. Br. J. Nutr. 107, S171–S184.

Milligan, G., Alvarez-Curto, E., Watterson, K.R., Ulven, T., Hudson, B.D., 2015. Characterizing pharmacological ligands to study the long-chain fatty acid receptors GPR40/FFA1 and GPR120/FFA4. Br. J. Pharmacol. 172, 3254–3265.

Minghetti, M., Leaver, M.J., Tocher, D.R., 2011. Transcriptional control mechanisms of genes of lipid and fatty acid metabolism in the Atlantic salmon (*Salmo salar* L.) established cell line, SHK-1. Biochim. Biophys. Acta 1811, 194–202.

Mohd-Yusof, N.Y., Monroig, Ó., Mohd-Adnan, A., Wan, K.-L., Tocher, D.R., 2010. Investigation of highly unsaturated fatty acid metabolism in the Asian sea bass, *Lates calcarifer*. Fish Physiol. Biochem. 3, 827–843.

Monroig, Ó., Rotllant, J., Sánchez, E., Cerdá-Reverter, J.M., Tocher, D.R., 2009. Expression of long-chain polyunsaturated fatty acid (LC-PUFA) biosynthesis genes during zebrafish *Danio rerio* early embryogenesis. Biochim. Biophys. Acta 1791, 1093–1101.

Monroig, Ó., Zheng, X., Morais, S., Leaver, M.J., Taggart, J.B., Tocher, D.R., 2010a. Multiple genes for functional Δ6 fatty acyl desaturases (Fad) in Atlantic salmon (*Salmo salar* L.): gene and cDNA characterization, functional expression, tissue distribution and nutritional regulation. Biochim. Biophys. Acta 1801, 1072–1081.

Monroig, Ó., Rotllant, J., Cerdá-Reverter, J.M., Dick, J.R., Figueras, A., Tocher, D.R., 2010b. Expression and role of Elovl4 elongases in biosynthesis of very long-chain fatty acids during zebrafish *Danio rerio* early embryonic development. Biochim. Biophys. Acta 1801, 1145–1154.

Monroig, Ó., Navarro, J.C., Tocher, D.R., 2011a. Long-chain polyunsaturated fatty acids in fish: recent advances on desaturases and elongases involved in their biosynthesis. In: Cruz-Suarez, L.E., Ricque-Marie, D., Tapia-Salazar, M., Nieto-López, M.G., Villarreal-Cavazos, D.A., Gamboa-Delgado, J., Hernández-Hernández, L.H. (Eds.), In: Proceedings of the XI International Symposium on Aquaculture Nutrition. N.L. Universidad Autónoma de Nuevo León, Monterrey, Nuevo León, México, pp. 257–282.

Monroig, Ó., Li, Y., Tocher, D.R., 2011b. Delta-8 desaturation activity varies among fatty acyl desaturases of teleost fish: high activity in delta-6 desaturases of marine species. Comp. Biochem. Physiol. 159B, 206–213.

Monroig, Ó., Webb, K., Ibarra-Castro, L., Holt, G.J., Tocher, D.R., 2011c. Biosynthesis of long-chain polyunsaturated fatty acids in marine fish: characterization of an Elovl4-like elongase from cobia *Rachycentron canadum* and activation of the pathway during early life stages. Aquaculture 312, 145–153.

Monroig, Ó., Wang, S., Zhang, L., You, C., Tocher, D.R., Li, Y., 2012. Elongation of long-chain fatty acids in rabbitfish *Siganus canaliculatus*: cloning, functional characterisation and tissue distribution of Elovl5- and Elovl4-like elongases. Aquaculture 350-353, 63–70.

Monroig, Ó., Tocher, D.R., Hontoria, F., Navarro, J.C., 2013. Functional characterisation of a Fads2 fatty acyl desaturase with Δ6/Δ8 activity and an Elovl5 with $C_{16}$, $C_{18}$ and $C_{20}$ elongase activity in the anadromous teleost meagre (*Argyrosomus regius*). Aquaculture 412–413, 14–22.

Monroig, Ó., Lopes-Marques, M., Navarro, J.C., Hontoria, F., Ruivo, R., Santos, M.M., Venkatesh, B., Tocher, D.R., Castro, L.F., 2016a. Evolutionary functional elaboration of the Elovl2/5 gene family in chordates. Sci. Rep. 6, 20510.

Monroig, Ó., Hontoria, F., Varó, I., Tocher, D.R., Navarro, J.C., 2016b. Biosynthesis of Very Long-Chain ($>C_{24}$) Polyunsaturated Fatty Acids in Juveniles of Gilthead Seabream (*Sparus aurata*). 17th International Symposium on Fish Nutrition & Feeding, June 5–10, Sun Valley, Idaho, United States. .

Morais, S., Monroig, Ó., Zheng, X., Leaver, M.J., Tocher, D.R., 2009. Highly unsaturated fatty acid synthesis in Atlantic salmon: characterization of Elovl5- and Elovl2-like elongases. Mar. Biotechnol. 11, 627–639.

Morais, S., Castanheira, F., Martínez-Rubio, L., Conceição, L.E.C., Tocher, D.R., 2012. Long-chain polyunsaturated fatty acid synthesis in a marine vertebrate: ontogenetic and nutritional regulation of a fatty acyl desaturase with Δ4 activity. Biochim. Biophys. Acta 1821, 660–671.

Morais, S., Mourente, G., Martínez, A., Gras, N., Tocher, D.R., 2015. Docosahexaenoic acid biosynthesis via fatty acyl elongase and Δ4-desaturase and its modulation by dietary lipid level and fatty acid composition in a marine vertebrate. Biochim. Biophys. Acta 1851, 588–597.

Nayak, M., Pradhan, A., Giri, S.S., Samanta, M., Badireenath Konkimalla, V., Saha, A., 2018. Molecular characterization, tissue distribution and differential nutritional regulation of putative Elovl5 elongase in silver barb (*Puntius gonionotus*). Comp. Biochem. Physiol. 217B, 27–39.

Morash, A.J., Le Moine, C.M.R., McClelland, G.B., 2010. Genome duplication events have led to a diversification in the CPT I gene family in fish. Am. J. Physiol. Regul. Integr. Comp. Physiol. 299, R579–R589.

Ninno, R.E., de Torrengo, M.A.P., Castuma, J.C., Brenner, R.R., 1974. Specificity of 5- and 6-fatty acid desaturases in rat and fish. Biochim. Biophys. Acta 360, 124–133.

Norambuena, F., Lewis, M., Hamid, N.K.A., Hermon, K., Donald, J.A., Turchini, G.M., 2013. Fish oil replacement in current aquaculture feed: is cholesterol a hidden treasure for fish nutrition? PLoS One 8, e81705.

Oboh, A., Betancor, M.B., Tocher, D.R., Monroig, O., 2016. Biosynthesis of long-chain polyunsaturated fatty acids in the African catfish *Clarias gariepinus*: molecular cloning and functional characterisation of fatty acyl desaturase (*fads2*) and elongase (*elovl2*) cDNAs. Aquaculture 462, 70–79.

Oboh, A., Kabeya, N., Carmona-Antoñanzas, G., Castro, L.F.C., Dick, J.R., Tocher, D.R., Monroig, O., 2017a. Two alternative pathways for docosahexaenoic acid (DHA 22:6n-3) biosynthesis are widespread among teleost fish. Sci. Rep. 7, 3889.

Oboh, A., Navarro, J.C., Tocher, D.R., Monroig, O., 2017b. Elongation of very long-chain (>$C_{24}$) fatty acids in *Clarias gariepinus*: Cloning, functional characterization and tissue expression of *elovl4* elongases. Lipids 52, 837–848.

Oh, D.Y., Talukdar, S., Bae, E.J., Imamura, T., Morinaga, H., Fan, W., Li, P., Lu, W.J., Watkins, S.M., Olefsky, J.M., 2010. GPR120 is an omega-3 fatty acid receptor mediating potent anti-inflammatory and insulin-sensitizing effects. Cell 142, 687–698.

Ortega, R.M., Rodriguez-Rodriguez, E., Lopez-Sobaler, A.M., 2012. Effects of omega 3 fatty acids supplementation in behavior and non-neurodegenerative neuropsychiatric disorders. Br. J. Nutr. 107, S261–S270.

Pang, S.-C., Wang, H.-P., Li, K.-Y., Zhu, Z.-Y., Kang, J.X., Sun, Y.-H., 2014. Double transgenesis of humanized *fat1* and *fat2* genes promotes omega-3 polyunsaturated fatty acids synthesis in a zebrafish model. Mar. Biotechnol. 16, 580–593.

Park, W.J., Kothapalli, K.S., Lawrence, P., Tyburczy, C., Brenna, J.T., 2009. An alternate pathway to long-chain polyunsaturates: the FADS2 gene product delta8-desaturates 20:2n-6 and 20:3n-3. J. Lipid Res. 50, 1195–1202.

Park, H.G., Park, W.J., Kothapalli, K.S., Brenna, J.T., 2015. The fatty acid desaturase 2 (FADS2) gene product catalyzes Δ4 desaturation to yield n-3 docosahexaenoic acid and n-6 docosapentaenoic acid in human cells. FASEB J. 29, 3911–3919.

Pérez, J.A., Rodríguez, C., Henderson, R.J., 1999. The uptake and esterification of radiolabelled fatty acids by enterocytes isolated from rainbow trout (*Oncorhynchus mykiss*). Fish Physiol. Biochem. 20, 125–134.

Raatz, S.K., Golovko, M.Y., Brose, S.A., Rosenberger, T.A., Burr, G.S., Wolters, W.R., Picklo, M.J., 2011. Baking reduces prostaglandin, resolvin, and hydroxy-fatty acid content of farm-raised Atlantic salmon (*Salmo salar*). J. Agric. Food Chem. 59, 11278–11286.

Randall, K.M., Drew, M.D., Øverland, M., Østbye, T.K., Bjerke, M., Vogt, G., Ruyter, B., 2013. Effects of dietary supplementation of coriander oil, in canola oil diets, on the metabolism of [1-$^{14}$C] 18:3n-3 and [1-$^{14}$C] 18:2n-6 in rainbow trout hepatocytes. Comp. Biochem. Physiol. 166B, 65–72.

Ren, H.-T., Zhang, G.-Q., Li, J.-L., Tang, Y.-K., Li, H.-X., Yu, J.-H., Xu, P., 2013. Two Δ6-desaturase-like genes in common carp (*Cyprinus carpio* var. Jian): structure characterization, mRNA expression, temperature and nutritional regulation. Gene 525, 11–17.

Ruyter, B., Andersen, Ø., Dehli, A., Östlund Farrants, A.-K., Gjøen, T., Thomassen, M.S., 1997. Peroxisome proliferator activated receptors in Atlantic salmon (*Salmo salar*): effects on PPAR transcription and acyl-CoA oxidase activity in hepatocytes by peroxisome proliferators and fatty acids. Biochim. Biophys. Acta 1348, 331–338.

Ruyter, B., Røsjø, C., Grisdale-Helland, B., Rosenlund, G., Obach, A., Thomassen, M.S., 2003. Influence of temperature and high dietary linoleic acid content on esterification, elongation, and desaturation of PUFA in Atlantic salmon hepatocytes. Lipids 38, 833–840.

Sargent, J.R., Tocher, D.R., Bell, J.G., 2002. The lipids. In: Halver, J.E., Hardy, R.W. (Eds.), Fish Nutrition. third ed. Academic Press, San Diego, pp. 181–257.

Schiller Vestergren, A.L., Trattner, S., Mráz, J., Ruyter, B., Pickova, J., 2011. Fatty acids and gene expression responses to bioactive compounds in Atlantic salmon (*Salmo salar* L.) hepatocytes. Neuroendocrinol. Lett. 32, 41–50.

Schiller Vestergren, A.L., Wagner, L., Pickova, J., Rosenlund, G., Kamal-Eldin, A., Trattner, S., 2012. Sesamin modulates gene expression without corresponding effects on fatty acids in Atlantic salmon (*Salmo salar* L.). Lipids 4, 897–911.

Schünke, M., Wodtke, E., 1983. Cold-induced increase of Δ9- and Δ6-desaturase activities in endoplasmic membranes of carp liver. Biochim. Biophys. Acta 734, 70–75.

Senadheera, S.D., Turchini, G.M., Thanuthong, T., Francis, D.S., 2012. Effects of dietary iron supplementation on growth performance, fatty acid composition and fatty acid metabolism in rainbow trout (*Oncorhynchus mykiss*) fed vegetable oil based diets. Aquaculture 342, 80–88.

Serhan, C.N., 2014. Novel pro-resolving lipid mediators in inflammation are leads for resolution physiology. Nature 510, 92–101.

Sprague, M., Dick, J.R., Tocher, D.R., 2016. Impact of sustainable feeds on omega-3 long-chain fatty acid levels in farmed Atlantic salmon, 2006–2015. Sci. Rep. 6, 21892.

Sprecher, H., Luthria, D.L., Mohammed, B.S., Baykousheva, S.P., 1995. Reevaluation of the pathways for the biosynthesis of polyunsaturated fatty acids. J. Lipid Res. 36, 2471–2477.

Sprecher, H., 2000. Metabolism of highly unsaturated n-3 and n-6 fatty acids. Biochim. Biophys. Acta 1486, 219–231.

Stubhaug, I., Lie, Ø., Torstensen, B.E., 2007. Fatty acid productive value and β-oxidation capacity in Atlantic salmon tissues (*Salmo salar* L.) fed on different lipid sources along the whole growth period. Aquacult. Nutr. 13, 145–155.

Tan, S.-H., Chung, H.-H., Shu-Chien, A.C., 2010. Distinct developmental expression of two elongase family members in zebrafish. Biochem. Biophys. Res. Commun. 393, 397–403.

Tanomman, S., Ketudat-Cairns, M., Jangprai, A., Boonanuntanasarn, S., 2013. Characterization of fatty acid delta-6 desaturase gene in Nile tilapia and heterogenous expression in *Saccharomyces cerevisiae*. Comp. Biochem. Physiol. 166B, 148–156.

Thanuthong, T., Francis, D.S., Senadheera, S.P.S.D., Jones, P.L., Turchini, G.M., 2011. LC-PUFA biosynthesis in rainbow trout is substrate limited: use of the whole body fatty acid balance method and different 18:3n-3/18:2n-6 ratios. Lipids 46, 1111–1127.

Thomassen, M.S., Berge, R.D., Gerd, M., Østbye, T.K., Ruyter, B., 2012. High dietary EPA does not inhibit Δ5 and Δ6 desaturases in Atlantic salmon (*Salmo salar* L.) fed rapeseed oil diets. Aquaculture 360-361, 78–85.

Tocher, D.R., Bell, J.G., Sargent, J.R., 1991. The incorporation of [³H]arachidonic and [¹⁴C] eicosapentaenoic acids into glycerophospholipids and their metabolism via lipoxygenases in isolated brain cells from rainbow trout *Oncorhynchus mykiss*. J. Neurochem. 57, 2078–2085.

Tocher, D.R., 1995. Glycerophospholipid metabolism. In: Hochachka, P.W., Mommsen, T.P. (Eds.), Biochemistry and Molecular Biology of Fishes, Vol. 4 Metabolic and Adaptational Biochemistry. Elsevier Press, Amsterdam, The Netherlands, pp. 119–157.

Tocher, D.R., Bell, J.G., Henderson, R.J., McGhee, F., Mitchell, D., Morris, P.C., 2000. The effect of dietary linseed and rapeseed oils on polyunsaturated fatty acid metabolism in Atlantic salmon (*Salmo salar*) undergoing parr–smolt transformation. Fish. Physiol. Biochem. 23, 59–73.

Tocher, D.R., 2003. Metabolism and functions of lipids and fatty acids in teleost fish. Rev. Fish Sci. 11, 107–184.

Tocher, D.R., Agaba, M., Hastings, N., Teale, A.J., 2003a. Biochemical and molecular studies of the fatty acid desaturation pathway in fish. In: Browman, H.I., Skiftesvik, A.B. (Eds.), The Big Fish Bang: Proceedings of the 26th Annual Larval Fish Conference. Institute of Marine Nutrition, Bergen, Norway, pp. 211–227.

Tocher, D.R., Bell, J.G., McGhee, F., Dick, J.R., Fonseca-Madrigal, J., 2003b. Effects of dietary lipid level and vegetable oil on fatty acid metabolism in Atlantic salmon (*Salmo salar*) over the entire production cycle. Fish Physiol. Biochem. 29, 193–209.

Tocher, D.R., Fonseca-Madrigal, J., Dick, J.R., Ng, W., Bell, J.G., Campbell, P.J., 2004. Effects of water temperature and diet containing palm oil on fatty acid desturation and oxidation in hepatocytes and intestinal enterocytes of rainbow trout (*Onchorhynchus mykiss*). Comp. Biochem. Physiol. 137B, 49–63.

Tocher, D.R., Zheng, X., Schlechtriem, C., Hasting, N., Dick, J.R., Teale, A.J., 2006. Highly unsaturated fatty acid synthesis in marine fish: cloning, functional characterization, and nutritional regulation of fatty acyl Δ6 desaturase of Atlantic cod (*Gadus morhua* L.). Lipids 41, 1003–1016.

Tocher, D.R., 2009. Issues surrounding fish as a source of omega-3 long-chain polyunsaturated fatty acids. Lipid Technol. 21, 13–16.

Tocher, D.R., 2010. Fatty acid requirements in ontogeny of marine and freshwater fish. Aquacult. Res. 41, 717–732.

Tocher, D.R., 2015. Omega-3 long-chain polyunsaturated fatty acids and aquaculture in perspective. Aquaculture 449, 94–107.

Tocher, D.R., Glencross, B.D., 2015. Lipids and fatty acids. In: Lee, C.-S., Lim, C., Webster, C., Gatlin, III, D.M. (Eds.), Dietary Nutrients, Additives, and Fish Health. (Chapter.3). Wiley-Blackwell, Oxford, United Kingdom, pp. 47–94.

Trattner, S., Kamal-Eldin, A., Brännäs, E., Moazzami, A., Zlabek, V., Larsson, P., Ruyter, B., Gjøen, T., Pickova, J., 2008a. Sesamin supplementation increases white muscle docosahexaenoic acid (DHA) levels in rainbow trout (*Oncorhynchus mykiss*) fed high alpha-linolenic acid (ALA) containing vegetable oil: metabolic actions. Lipids 43, 989–997.

Trattner, S., Ruyter, B., Østbye, T.K., Gjøen, T., Zlabek, V., Kamal-Eldin, A., Pickova, J., 2008b. Sesamin increases alpha-linolenic acid conversion to docosahexaenoic acid in Atlantic salmon (*Salmo salar* L.) hepatocytes: role of altered gene expression. Lipids 43, 999–1008.

Trueman, R.J., Tiku, P.E., Caddick, M.X., Cossins, A.R., 2000. Thermal thresholds of lipid restructuring and $\Delta^9$-desaturase expression in the liver of carp (*Cyprinus carpio* L.). J. Exp. Biol. 203, 641–650.

Tur, J.A., Bibiloni, M.M., Sureda, A., Pons, A., 2012. Dietary sources of omega 3 fatty acids: public health risks and benefits. Br. J. Nutr. 107, S23–S52.

Turchini, G.M., Ng, W.-K., Tocher, D.R. (Eds.), 2011. Fish Oil Replacement and Alternative Lipid Sources in Aquaculture Feeds. Taylor & Francis, CRC Press, Boca Raton, FL, United States, p. 533.

Úbeda, N., Achóna, M., Varela-Moreirasa, G., 2012. Omega 3 fatty acids in the elderly. Br. J. Nutr. 107, S137–S151.

Urbatzka, R., Galante-Oliveira, S., Rocha, E., Lobo-da-Cunha, A., Castro, L.F., Cunha, I., 2015. Effects of the PPARα agonist WY-14,643 on plasma lipids, enzymatic activities and mRNA expression of lipid metabolism genes in a marine flatfish, *Scophthalmus maximus*. Aquat. Toxicol. 164, 155–162.

Vagner, M., Zambonino Infante, J.L., Robin, J.H., Person-Le Ruyet, J., 2007a. Is it possible to influence European sea bass (*Dicentrarchus labrax*) juvenile metabolism by a nutritional conditioning during larval stage? Aquaculture 267, 165–174.

Vagner, M., Robin, J.H., Zambonino Infante, J.L., Person-Le Ruyet, J., 2007b. Combined effect of dietary HUFA level and temperature on sea bass (*D. labrax*) larvae development. Aquaculture 266, 179–190.

Vagner, M., Robin, J.H., Zambonino Infante, J.L., Tocher, D.R., Person-Le Ruyet, J., 2009. Ontogenic effects of early feeding of sea bass (Dicentrarchus labrax) larvae with a range of dietary n-3 highly unsaturated fatty acid levels on the functioning of polyunsaturated fatty acid desaturation pathways. Br. J. Nutr. 101, 1452–1462.

Vagner, M., Santigosa, E., 2011. Characterization and modulation of gene expression and enzymatic activity of delta-6 desaturase in teleosts: a review. Aquaculture 315, 131–143.

Venkatesh, B., Lee, A.P., Ravi, V., Maurya, A.K., Lian, M.M., Swann, J.B., Ohta, Y., Flajnik, M.F., Sutoh, Y., Kasahara, M., Hoon, S., Gangu, V., Roy, S.W., Irimia, M., Korzh, V., Kondrychyn, I., Lim, Z.W., Tay, B.-H., Tohari, S., Kong, K.W., Ho, S., Lorente-Galdos, B., Quilez, J., Marques-Bonet, T., Raney, B.J., Ingham, P.W., Tay, A., Hillier, L.W., Minx, P., Boehm, T., Wilson, R., Brenner, S., Warren, W.C., 2014. Elephant shark genome provides unique insights into gnathostome evolution. Nature 505, 174–179.

Vera, L.M., Metochis, C., Taylor, J.F., Clarkson, M., Skjærven, K.H., Migaud, H., Tocher, D.R., 2017. Early nutritional programming affects liver transcriptome in diploid and triploid Atlantic salmon (salmosalar). BMC Genomics 18, 886.

Wallis, J.G., Browse, J., 2002. Mutants of Arabidopsis reveal many roles for membrane lipids. Prog. Lipid Res. 41, 254–278.

Wang, S., Chen, J., Jiang, D., Zhang, Q., Dong, Y., You, C., Tocher, D.R., Monroig, Ó., Li, Y., 2018. Hnf4α is involved in the regulation of vertebrate LC-PUFA biosynthesis: insights into the regulatory role of Hnf4α on expression of liver fatty acyl desaturases in the marine teleost Siganus canaliculatus. Fish Physiol. Biochem. https://doi.org/10.1007/s10695-018-0470-8.

Willson, T.M., Brown, P.J., Sternbach, D.D., Henke, B.R., 2000. The PPARs: from orphan receptors to drug discovery. J. Med. Chem. 43, 527–550.

Xie, D., Wang, S., You, C., Chen, F., Tocher, D.R., Li, Y., 2015. Characteristics of LC-PUFA biosynthesis in marine herbivorous teleost Siganus canaliculatus under different ambient salinities. Aquacult. Nutr. 21, 541–551.

Xu, J., Yan, B., Teng, Y., Lou, G., Lu, Z., 2010. Analysis of nutrient composition and fatty acid profiles of Japanese sea bass Lateolabrax japonicus (Cuvier) reared in seawater and freshwater. J. Food Compost. Anal. 23, 401–405.

Xue, X., Feng, C.Y., Hixson, S.M., Johnstone, K., Anderson, D.M., Parrish, C.C., Rise, M.L., 2014. Characterization of the fatty acyl elongase (elovl) gene family, and hepatic elovl and delta-6 fatty acyl desaturase transcript expression and fatty acid responses to diets containing camelina oil in Atlantic cod (Gadus morhua). Comp. Biochem. Physiol. 175B, 9–22.

Yan, J., Liang, X., Cui, Y., Cao, X., Gao, J., 2018. Elovl4 can effectively elongate C18 polyunsaturated fatty acids in loach Misgurnus anguillicaudatus. Biochem. Biophys. Res. Commun. 495, 2637–2642.

Zhang, Q., Xie, D., Wang, S., You, C., Monroig, O., Tocher, D.R., Li, Y., 2014. miR-17 is involved in the regulation of LC-PUFA biosynthesis in vertebrates: effects on liver expression of a fatty acyl desaturase in the marine teleost Siganus canaliculatus. Biochim. Biophys. Acta 1841, 934–943.

Zhang, Q., You, C., Liu, F., Zhu, W., Wang, W., Xie, D., Monroig, O., Tocher, D.R., Li, Y., 2016a. Cloning and characterization of Lxr and Srebp1, and their potential roles in regulation of LC-PUFA biosynthesis in rabbitfish Siganus canaliculatus. Lipids 51, 1051–1063.

Zhang, Q., Wang, S., You, C., Dong, W., Monroig, O., Tocher, D.R., Li, Y., 2016b. The miR-33 gene is identified in a marine teleost: a potential role in regulation of LC-PUFA biosynthesis in Siganus canaliculatus. Sci. Rep. 6, 32909.

Zheng, X., Tocher, D.R., Dickson, C.A., Dick, J.R., Bell, J.G., Teale, A.J., 2005. Highly unsaturated fatty acid synthesis in vertebrates: new insights with the cloning and characterisation of a Δ6 desaturase of Atlantic salmon. Lipids 40, 13–24.

Zheng, X., Ding, Z., Xu, Y., Monroig, O., Morais, S., Tocher, D.R., 2009. Physiological roles of fatty acyl desaturase and elongase in marine fish: characterisation of cDNAs of fatty acyl Δ6 desaturase and Elovl5 elongase of cobia (Rachycentron canadum). Aquaculture 290, 122–131.

# Polyunsaturated Fatty Acid Biosynthesis and Metabolism in Agriculturally Important Species

*Michael E.R. Dugan\*, Cletos Mapiye\*\*,
Payam Vahmani\**

\*Agriculture and Agri-Food Canada, Lacombe Research Centre,
Lacombe, AB, Canada;
\*\*Department of Animal Sciences, Stellenbosch University, Stellenbosch,
South Africa

## INTRODUCTION

The biosynthesis and metabolism of polyunsaturated fatty acids (PUFA) in agriculturally important species, that is, farm animals, are of interest due to their relationships with animal performance, health, and their influence on the fatty acid composition of animal-derived food products. Farm animals, in similarity to humans, require essential fatty acids, namely linoleic (LA, 18:2n-6) and α-linolenic acid (ALA, 18:3n-3), for normal growth and development. This chapter highlights species differences in terms of PUFA synthesis and metabolism, and the supply and manipulation of PUFA in animal-derived foods.

Polyunsaturated Fatty Acid Metabolism. http://dx.doi.org/10.1016/B978-0-12-811230-4.00004-1

# PRECURSORS FOR PUFA SYNTHESIS

Farm animal diets are largely composed of plant-derived feeds, and their oils are typically dominated by the simplest of the n-6 (LA), n-3 (ALA) and sometimes the n-9 [18:1n-9, oleic acid (OA)] series of fatty acids. Farm animal diets are primarily formulated to meet energy and protein or amino acid requirements, and the fatty acid composition is often not considered unless a problem with food quality results (i.e., excessive softness or oxidative instability). In many developed countries, high-intensity animal production is the norm, and production is coupled with least cost-formulated diets based mainly on grains as energy sources and oilseed meals (i.e., oil extraction by-products) as protein sources. Small grains typically contain less than 4% oil (Liu, 2011), and the composition of their PUFA is dominated by LA (Fig. 4.1).

Corn is considered a large grain, contains upward of 8% oil, and LA is the dominant PUFA (62%). Oil sources with appreciable contents of ALA that may from time to time be included in farm animal diets include whole flaxseed (or linseed), soybean, or canola (or rapeseed), or their extracted oils (Fig. 4.2).

Most other fat/oil sources, including safflower, sunflower, peanut and cotton seed oils have PUFA dominated by LA, while pork lard, palm oil, and beef tallow are noted for their higher contents of saturated fatty acids, primarily palmitic acid (PA, 16:0). There has also been development and use of alternative oilseeds, such as camelina, which are rich in ALA (Waraich et al., 2013). In addition, genetically modified crops are now being produced to accumulate stearidonic acid (SDA, 18:4n-3), which bypasses the rate-limiting step in most animals for long chain (LC) n-3 fatty acid synthesis (Harris, 2012; Whelan and Rust, 2006). Additional genetic engineering is also being investigated for production of plant oils enriched with LCn-3 fatty acids (Kitessa et al., 2014; Murphy, 2014; Yilmaz

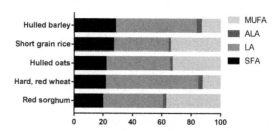

FIGURE 4.1  **Fatty acid proportions of common small grains (percentage of total fatty acids).** *MUFA,* Total monounsaturated fatty acids; *ALA,* α-linolenic acid (18:3n-3); *LA,* linoleic acid (18:2n-6); *SFA,* total saturated fatty acids. *Source: Data adapted from Liu, K., 2011. Comparison of lipid content and fatty acid composition and their distribution within seeds of 5 small grain species. J. Food Sci. 76, C334–C342.*

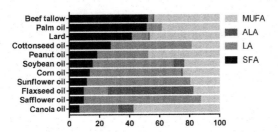

FIGURE 4.2  Fatty acid composition of common oils/fats (percentage of total fatty acids). *MUFA*, Total monounsaturated fatty acids; *ALA*, α-linolenic acid (18:3n-3); *LA*, linoleic acid (18:2n-6); *SFA*, total saturated fatty acids. *Source: Data adapted from Canola Council of Canada (www.canolacouncil.org).*

et al., 2017). In less intensive animal production systems, for example when pasture or forage finishing livestock, diets typically contain 2%–3% crude fat with ALA as the major PUFA (Glasser et al., 2013). Leafy material is rich in chloroplasts, and chloroplasts contain a high proportion of ALA. As a consequence, it is not unusual for lipids from fresh grass or legumes or their hay or silage to contain 50% ALA, but this diminishes as forage quality declines.

## PUFA BIOSYNTHESIS

Biosynthesis of PUFA from LA and ALA in farm animals follows pathways similar to those found in humans involving a series of elongation and desaturation steps (see Chapter 2). Typically there is limited elongation and desaturation beyond arachidonic acid (AA, 20:4n-3) for the n-6 series in farm animals, but in similarity to humans, the n-3 series can proceed to docosahexaenoic acid (DHA, 22:6n-3) via peroxisomal β-oxidation of 24:6n-3 (Sprecher et al., 1999). Humans, however, likely have some of the lowest conversions of ALA to LCn-3 fatty acids, particularly in males (Baker et al., 2016). Some elongation and desaturation of n-9 fatty acids can occur (i.e., from OA, c9-18:1). An increased content of mead acid (20:3n-9), however, is a sign of essential fatty acid deficiency, and rarely found unless specifically induced in farm animals (Holman, 1971). Tissue contents can differ, however, with cartilage being naturally enriched with mead acid (Adkisson et al., 1991). The content and balance of n-6 and n-3 fatty acids in farm animals is dependent on precursors provided in the diet. In developed countries, where LA is the major fatty acid found in diets of intensively produced farm animals, it is not surprising that derived foods contribute to the high n-6 to n-3 fatty acid ratios (10–20:1) currently seen in human diets. In addition, because of the multiple health advantages of including n-3, particularly LCn-3 fatty acids in human diets, there have been calls to rebalance the fatty acid content of animal derived foods (Simopoulos, 1999). Development of animal-derived foods enriched with n-3 fatty acid through to retail have, however, been limited due to issues with product variability coupled with technical, regulatory and market development issues (Dugan et al., 2015; Vahmani et al., 2015a).

## Chicken and Eggs

To date, chicken eggs are the most commercially successful farm animal derived foods enriched with n-3 fatty acids, and their production and effects on health have been extensively reviewed (Cherian, 2012). Reasons for the success of n-3 fatty acid enriched chicken eggs are a single tissue and complications due to contributions of multiple tissues with varying enrichments of n-3 fatty acids can be avoided. Second, although feeding LCn-3 fatty acids can result in their incorporation into chicken eggs, eggs are one of the few tissues, which can effectively accumulate DHA endogenously synthesized from ALA. Consequently, economical sources of ALA, such as flaxseed, can be fed to laying hens, and DHA synthesized in the liver can be deposited in egg yolks. In fact, DHA is typically the most concentrated n-3 fatty acid in chicken eggs (1.2% vs. 0.5% ALA in total fatty acids), and feeding 10% flaxseed to laying hens can increase DHA and ALA by ~2- and 7-fold, respectively (Cherian, 2012). This can be attributed in part to the relatively high content of phospholipids in egg yolk, with highly unsaturated fatty acids (HUFA) being preferred substrates for phospholipid synthesis (Nakamura and Nara, 2004). Chicken carcasses, including the leg, thigh, breast, and wings can incorporate preformed LCn-3 fatty acids, similar to mammals, but their synthesis from ALA is limited (Poureslami and Batal, 2012; Zuidhof et al., 2009). The amount of n-3 fatty acids deposited is also tissue specific. When feeding a diet containing 10% flaxseed to broilers, Zuidhof et al. (2009) found appreciable accumulations of ALA for periods up to 35 days (Fig. 4.3), with more ALA accumulation found in thigh meat (i.e., a tissue with higher lipid content). Breast meat had a lower ALA content, but the amounts of EPA and DHA in breast meat increased slightly over time.

The source of LCn-3 fatty acids for egg yolk is through hepatic synthesis, and unlike muscle, liver n-3 fatty acid contents can change dramatically with age and diet fatty acid composition. Jing et al. (2013) found liver LCn-3 proportions drop immediately posthatching, but gradually increase by day 35 coupled with an increase in the relative abundance of mRNA for FADS1 (fatty acid desaturase 1 or delta-5 desaturase), FADS2 (delta-6 desaturase), ELOVL2 (elongation of very LC fatty acids protein 2), and ELOV5. In addition, when feeding diets containing 6.8% vegetable oil with different LA:ALA ratios, lower ratios led to greater accumulations of n-3 fatty acids (Fig. 4.4), and decreasing the LA:ALA ratio from 11.5 to

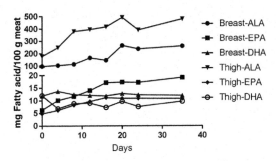

FIGURE 4.3 **The accumulation of n-3 fatty acids in breast and thigh from broilers fed a diet containing 10% flaxseed up to 35 days.** *ALA*, α-Linolenic acid (18:3n3); *EPA*, eicosapentaenoic acid (20:5n-3); *DHA*, docosahexaenoic acid (22:6n-3). *Source: Data adapted from Zuidhof, M.J., Betti, M., Korver, D.R., Hernandez, F.I.L., Schneider, B.L., Carney, V.L., Renema, R.A., 2009. Omega-3-enriched broiler meat: 1. Optimization of a production system. Poult. Sci. 88, 1108–1120.*

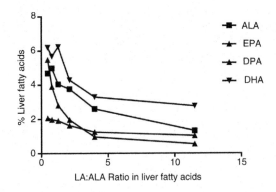

FIGURE 4.4 **Effects of feeding broilers 6.8% vegetable oil with different LA:ALA ratios on the percentage LCn-3 fatty acids in liver.** *LA*, Linoleic acid (18:2n-6); *ALA*, α-linolenic acid (18:3n-3); *EPA*, eicosapentaenoic acid (20:5n-3); *DPA*, docosapentaenoic acid (22:5n-3); *DHA*, docosahexaenoic acid (22:6n-3). *Source: Data adapted from Jing, M., Gakhar, N., Gibson, R., House, J., 2013. Dietary and ontogenic regulation of fatty acid desaturase and elongase expression in broiler chickens. Prostaglandins Leukot. Essent. Fatty Acids 89, 107–113.*

0.47 led to a 1.7–1.8-fold increase in FADS1, FADS2, ELOVL2, and ELOV5 mRNA abundance. Interestingly, liver ELOVL2 and ELOVL5 in chickens differ compared to mammals in that they can efficiently elongate DPA to 24:5n-3 (Gregory et al., 2013), and may provide some advantages in terms of overall DHA synthesis/deposition. Understanding how to increase the amount of LCn-3 fatty acids in the majority of edible tissue (i.e., muscle) will, however, require a greater understanding of transport/deposition from hepatic to extrahepatic tissues (Gregory et al., 2013), and how to influence extrahepatic capacity for LCn-3 fatty acid synthesis/deposition.

When designing production systems for PUFA enriched chicken, particularly LCn-3 fatty acids, precursor metabolism over time also has to be considered. Whole body accumulation (percentage of net intake) of LA and ALA increases with age and this is coupled with reduced rates of β-oxidation, with both β-oxidation and elongation/desaturation being greater for ALA (Fig. 4.5; Poureslami et al., 2010). In chickens, however, gender only marginally affect PUFA metabolism (Poureslami et al., 2010; Zuidhof et al., 2009), which is similar to cattle (Knight et al., 2003; Malau-Aduli et al., 2000), but in pigs (Ntawubizi et al., 2009) and humans (Baker et al., 2016), females have been found to have a greater capacity for DHA synthesis.

## Pigs

In pig production, inclusion of PUFA into diets has been studied as a means to increase their contents in edible tissues, and as a means to try and improve animal health and productivity. Under commercial production conditions, pig life spans are limited and chronic diseases, which affect humans later in life (e.g., cardiovascular disease), and influenced by dietary PUFA, are typically not encountered. Supplementing diets with n-3 fatty acids to improve reproductive and piglet performance has, however, been investigated (Tanghe and De Smet, 2013). In general, pregnant females supplemented with n-3 fatty acids do not have increased numbers of embryos or total number of piglets born, but may prolong gestation. Feeding n-3 PUFA also had no effect on piglet birth weight, although piglets were more viable with improved pre- and postweaning growth rates. Feeding low amounts of n-3 fatty acids during lactation diet may also increase litter size in subsequent gestations.

During fetal development, hepatic delta-5 desaturase and delta-6 desaturase enzyme activities increase, but rates of essential fatty acid conversion to LC-PUFA remains very low (~5%). Fetal development and growth, therefore, relies on placental transfer of LC-PUFA from sow plasma (Tanghe and De Smet, 2013). After birth, hepatic synthesis of DHA

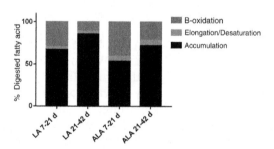

FIGURE 4.5  **Age dependant whole body metabolism of linoleic acid (LA, 18:2n-6) and α-linolenic acid (ALA, 18:3n-3) in broiler chickens.** *d, Day. Source: Data adapted from Poureslami, R., Raes, K., Turchini, G., Huyghebaert, G., De Smet, S., 2010. Effect of diet, sex and age on fatty acid metabolism in broiler chickens: n-3 and n-6 PUFA. Br. J. Nutr. 104, 189–197.*

from ALA increases by 2 weeks of age (~3.5 kg live weight; Malau-Aduli et al., 2000). For the most part, supplementing pig diets with ALA later in life, however, has not been found to appreciably increase DHA in the carcass. When Martínez-Ramírez et al. (2014) fed a control (beef tallow supplemented) diet compared to diets supplying 5 kg of flaxseed from 25 to 50 kg or 85 to 110 kg live weight, feeding flaxseed increased ALA contents in pork belly and loin muscle at slaughter (110 kg), with more ALA accumulating when flaxseed was fed later in the finishing period (Fig. 4.6). Feeding flaxseed also increased LCn-3 fatty acids relative to control, except for DHA. Deposition of LCn-3 fatty acids in finished pigs (110 kg) was independent of when the flaxseed was fed. Retention of LCn-3 fatty acids from early feeding of flaxseed was remarkable, which may in part be related to greater early rates of LCn-3 fatty acid synthesis, preferential deposition in phospholipids and their low rates of turnover. It is of interest to note that when feeding the control diet (i.e., supplemented with

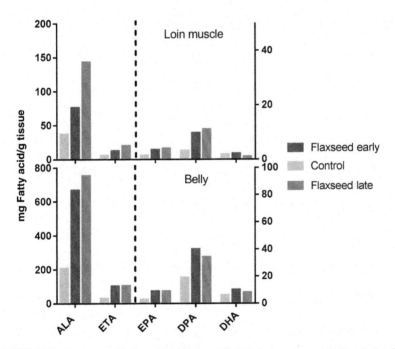

FIGURE 4.6 The accumulation of n-3 fatty acids in loin muscle and belly of pigs fed a control (beef tallow supplemented) diet compared to diets supplying 5 kg of flaxseed from 25 to 50 kg (flaxseed early) or 85–110 kg (flaxseed late). ALA, α-Linolenic acid (18:3n-3); ETA, eicosatrienoic acid (20:3n-33); EPA, eicosapentaenoic acid (20:5n-3); DPA, docosapentaenoic acid (22:5n-3); DHA, docosahexaenoic acid (22:6n-3). Source: Data adapted from Martínez-Ramírez, H.R., Kramer, J.K.G., de Lange, C.F.M., 2014. Retention of n-3 polyunsaturated fatty acids in trimmed loin and belly is independent of timing of feeding ground flaxseed to growing-finishing female pigs. J. Anim. Sci. 92, 238–249.

beef tallow), the most abundant LCn-3 fatty acid in loin and belly was docosapentaenoic acid (DPA, 22:5n-3), whereas when flaxseed was added to the diet, eicosatrienoic acid (ETA, 20:3n-3) became the most abundant LCn-3 fatty acid (Juárez et al., 2011; Martínez-Ramírez et al., 2014). Elongation of ALA to ETA has been thought to be a dead-end pathway with the majority being retroconverted back to ALA. Some evidence, however, suggests limited delta-8 desaturation to 20:4n-3 may occur, which bypasses SDA (18:4n-3) in the synthesis of LCn-3 fatty acids (Park et al., 2009; Schenck et al., 1996). In humans, when ALA supplementation studies are conducted, intakes are typically <10 g/day (Baker et al., 2016), and tissue contents of ETA are low and often not reported. When trying to enrich pork with n-3 fatty acids, however, it is not unusual to feed diets containing 10%–15% flaxseed, which can supply 72–108 g of ALA per day (i.e., 3 kg feed/day, 10%–15% flaxseed in diet, 40% oil in flaxseed oil, and 60% ALA in oil). At such high rates of ALA supplementation, the major elongation and desaturation pathway used for LC-PUFA synthesis may become saturated, and elongation prior to delta-6 desaturation may occur resulting in greater deposition of ETA. At low levels of flaxseed supplementation (providing 3 g ALA/kg diet, or ~1% flaxseed in the diet), small increases in muscle DHA have been found (Enser et al., 1996), but most studies feeding greater amounts of flaxseed have either shown no change or a slight decrease in muscle DHA (Juárez et al., 2011). In liver, short term (25 days) supplementation of increasing levels of flaxseed (0%–15% of the diet) in finisher pig diets increased LCn-3 fatty acids except DHA (Romans et al., 1995). Increases in liver DHA were, however, found when flaxseed was fed for a longer period (~75–80 days) at increasing levels (0%–25%), but DHA contents did not change in other tissues (skeletal muscle, heart, and backfat) (Cherian and Sim, 1995).

The overall metabolism of PUFA appears to be influenced by animal management, age and level of dietary PUFA. Pork and chicken are considered homolipoid as their fatty acid profiles generally reflect their diet composition (Shorland, 1950). This is not to say, however, that bacteria in the digestive tract have no influence on essential fatty acid metabolism. Bazinet et al. (2003) found 21-day-old piglets weaned into a conventional nursery, compared to 14-day-old piglets weaned into a "clean" nursery, had substantially increased rates of LA and ALA oxidation. As a consequence, at 49 days of age, conventionally weaned pigs had 15%–25% lower carcass LA and 20%–30% lower carcass ALA. In grower pigs, whole body disappearance of ALA was found to increase from 2.0% in 37 kg pigs to 23.7% in 46 kg pigs when feeding a 10% flaxseed diet supplying ~4% oil. The low rate of ALA disappearance in lighter pigs (8.8%) was confirmed using direct oxidation of uniformly labelled $^{13}$C-ALA (Martinez-Ramirez et al., 2014). When Duran-Montgé et al. (2010) fed finisher pigs (i.e., ~60–105 kg) diets containing 10% fat/oil including either tallow, sunflower, or flaxseed oil, they found that when dietary LA or ALA levels were low, their whole body depositions were high (80%–100%), and when dietary LA or ALA levels were high, whole body depositions of LA and ALA dropped to 60%–70%. Such a drop in fatty acid deposition (and apparent increased oxidation) could relate to the rather extreme intakes of LA and ALA, with the sunflower oil diet providing 209 g LA and the flaxseed oil diet providing 137 g of ALA per day. Once the need for phospholipid synthesis is met, surplus PUFA are likely diverted toward triacylglycerol synthesis and enter into pools more available for β-oxidation. In support of this theory, when Juárez et al. (2010) fed pigs diets containing 0%, 5%, 10%, and 15% flaxseed for 0, 4, 8, and 12 weeks before slaughter, the proportion of ALA in backfat increased with flaxseed level and duration of feeding (Fig. 4.7). The slope for ALA enrichment in backfat was lowest when feeding 5% ALA, which may be due to its preferential deposition in phospholipid. The slope for ALA enrichment more than doubled (2.6-fold) when feeding 10% flax, indicating a shift towards backfat (i.e., rich in triacylglycerol) deposition, and when 15% flaxseed was fed, the slope increased by only 3.7-fold (i.e., not a doubling of the 2.6-fold increase seen from 5% to 10% flaxseed) indicating more ALA may be going toward β-oxidation. Further studies are required to determine if changes in older pigs are related to actual age effects, differences in diet PUFA contents and to what extent environmental interactions may influence PUFA metabolism.

Feeding sources of LA (e.g., maize) or ALA (e.g., linseed) have not been found to influence lipogenic enzyme activities in pig muscle (Kouba et al., 2003; Kouba and Mourot, 1999). In backfat, however, feeding linseed was found to reduce delta-9 desaturase activity, and feeding maize increased acetyl-CoA carboxylase and malic enzyme activity. The increase in lipogenic enzyme activity in backfat when feeding maize, however,

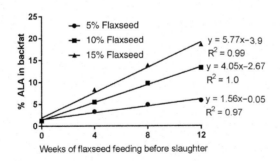

FIGURE 4.7 Effects of duration and feeding level of flaxseed on the accumulation of α-linolenic acid (ALA, 18:3n-3) backfat of grower-finisher pigs. *Source: Data adapted from Juárez, M., Dugan, M.E.R., Aldai, N., Aalhus, J.L., Patience, J.F., Zijlstra, R.T., Beaulieu, A.D., 2010. Feeding co-extruded flaxseed to pigs: effects of duration and feeding level on growth performance and backfat fatty acid composition of grower-finisher pigs. Meat Sci. 84, 578–584.*

may have been due to inhibitory effects of feeding beef tallow [i.e., rich in saturated fatty acids (SFA)] in the control diet, which has been shown to reduce carcass subcutaneous fat content (Dugan et al., 2001; Duran-Montgé et al., 2010). In comparison, feeding PUFA versus SFA sources to poultry has been found to decrease body fat content (Crespo and Esteve-Garcia, 2002). This may relate to differences in the site of fatty acid synthesis, with lipogenesis in pigs mainly occurring in adipose tissue, while poultry are similar to humans in that most lipogenesis occurs in the liver.

Understanding the synthesis, metabolism, and deposition of PUFA is important in developing strategies to enrich their contents in animal derived foods. Important considerations are to determine the degree of enrichment necessary to make the meat or meat product a good source of PUFA to enable a source or health claim, how different tissues respond to dietary treatments, which tissues are included in a serving, and how enrichments may affect product quality (Dugan et al., 2015). Because of their positive health effects, most studies conducted to enrich PUFA are designed to try and enrich n-3 fatty acids. In Canada, for example, 300 mg of total n-3 fatty acids per serving is required for an enrichment claim (CFIA, 2014). Finishing pigs fed a diet containing 10% extruded flaxseed for 11 weeks yielded 207 mg of n-3 fatty acids per 100 g of the major loin muscle (i.e., *Longissimus thoracis*; Juárez et al., 2011). Consequently, an n-3 fatty acid enrichment claim could not be made. When adding all the tissues found in a commercial loin chop together (all muscles including epimysium, seam fat and 5 mm of backfat), however, there was 3362 mg n-3 fatty acids per 100 g serving (Turner et al., 2014). This is well in excess of the requirement for an enrichment claim, and associated with soft fat and reduced sensory quality, particularly in reheated chops and ground pork (Juárez et al., 2011). In addition, although consuming flaxseed is not considered to be an effective means to increase the content of LCn-3 fatty acids in humans (Williams and Burdge, 2007), pigs appear to have somewhat greater conversion efficiency (~1/3 of retained n-3 fatty acids; Kloareg et al., 2007). In addition, when feeding extruded flaxseed to pigs for 11 weeks, and all tissues in commercial pork chops are included, consuming a single serving (100 g) of pork could contribute substantially to consumer intakes of LCn-3 fatty acids (Fig. 4.8; Turner et al., 2014).

Changing pig diets to increase the n-3 fatty acid content of meat and meat products will require a change or modification of feedstuffs, and potentially greater costs of production. A transgenic approach has, therefore, also been investigated. Pigs and cattle have had the *fat-1* gene inserted, which is an n-3 desaturase capable of converting LA to ALA (Liu et al., 2016). Consequently, if these transgenic animals have comparable performance to conventional animals, and the practice is deemed socially acceptable, this could prove an effective way to enrich animal derived foods with n-3 fatty acids.

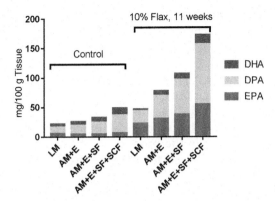

FIGURE 4.8 **Deposition of long-chain n-3 fatty acids in loin tissues from figs fed a control diet of 10% flaxseed (Flax) for 11 weeks.** *EPA*, Eicosapentaenoic acid (20:5n-3); *DPA*, docosapentaenoic acid (22:5n-3); *DHA*, docosahexaenoic acid (22:6n-3); *LM*, *longissimus* muscle. *AM* + *E*, all loin muscles + epimysium; *AM* + *E* + *SF*, AM + E + seam fat; *AM* + *E* + *SF* + *SCF*, AM + E + SF + subcutaneous fat. *Source: Data adapted from Dugan, M.E., Vahmani, P., Turner, T.D., Mapiye, C., Juárez, M., Prieto, N., Beaulieu, A.D., Zijlstra, R.T., Patience, J.F., Aalhus, J.L., 2015. Pork as a source of omega-3 (n-3) fatty acids. J. Clin. Med. 4, 1999–2011.*

## Ruminants

In ruminant animals, such as cattle, sheep, and goats, microbes in the large first stomach (i.e., the rumen) can have an overriding effect on the composition of fatty acids flowing to the small intestine for digestion. As such, ruminants are considered heterolipoid, as the fatty acid composition of their tissues can differ markedly from their diet (Shorland, 1950). After feed consumption, fatty acids are rapidly hydrolyzed from their parent lipids, and free PUFA released are toxic to rumen microbes. To cope, microbes biohydrogenate PUFA and their rumen metabolism and digestion have been extensively reviewed (Bauchart, 1993; Doreau and Ferlay, 1994; Dugan et al., 2011; Harfoot and Hazelwood, 1997; Jenkins et al., 2008). In ruminants, approximately 80% and 92% of LA and ALA are biohydrogenated in the rumen (Doreau and Ferlay, 1994). The degree of biohydrogenation is, however, dependant on the rate of passage and rumen residence time, with smaller rumens in sheep and goats resulting in reduced complete biohydrogenation. Remarkably, it is very rare for ruminant animals to develop essential fatty acid deficiency. Cattle typically also have 50% more LA and 400% more ALA in blood circulation than pigs, however, pigs have 50% more circulating triacylglycerol, and cattle have 290% more phospholipid (Caldari-Torres et al., 2016). Cattle phospholipids in circulation also have more LA and ALA than pigs, while pigs have more LA and ALA in triglyceride. Fat processing postabsorption in cattle also differs from pigs in that small intestinal cells (enterocytes) typically produce more very low density lipoprotein (VLDL) particles rather than large chylomicrons (Bauchart, 1993), and VLDL have a higher phospholipid to triacylglycerol ratio than chylomicrons. With greater PUFA incorporation into phospholipids than triacylglycerol, and fewer fatty acids in phospholipids channelled towards β-oxidation, this serves as a mechanism for conserving PUFA. Recently it has been found, however, that the preferential incorporation of PUFA into phospholipid may be at the level of the liver and muscle, rather than in the intestine (Caldari-Torres et al., 2016).

Biohydrogenation involves a series of isomerization and saturation steps culminating mainly in the production of stearic acid (18:0). Major pathways for biohydrogenation for both LA and ALA have been summarized by Harfoot and Hazelwood (1997) and involve the initial isomerization of the *cis(c)* 12 double bond to *trans* 11 for both LA and ALA, and this is typically found in ruminants raised on pasture or when fed diets containing a greater proportion of forage (Fig. 4.9). When greater amounts of highly fermentable grains (i.e., concentrate) are added to the diet, however, there can be a drop in rumen pH and a change in bacterial population resulting in a shift in isomerization from *c* 9 to *trans(t)* 10 instead of *c* 12 to *t* 11 (Griinari and Bauman, 1999). *Trans* 10 shifted pathways were tentatively proposed by Griinari and Bauman (1999) when feeding highly fermentable carbohydrate, and a key intermediate (*t*10, *c*15-18:2) in ALA biohydrogenation was recently positively identified (Alves and Bessa, 2014). Interestingly, when feeding steers rolled flaxseed in a barley grain based diet, Juarez et al. (2011) found *t*13-/*t*14-18:1 were the most concentrated *t*18:1 isomers in beef, suggesting the *t*10 and *t*11

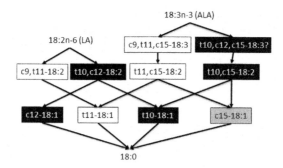

FIGURE 4.9 Ruminal biohydrogenation intermediates of linoleic acid (LA, 18:2n-6) and α-linolenic acid (ALA, 18:3n-3) when feeding high grain/low forage (*black boxes*) or low grain/high forage (*white boxes*) diets. *Source: Data adapted from Alves, S.P., Bessa, R.J., 2014. The trans-10, cis-15 18:2: a missing intermediate of trans-10 shifted rumen biohydrogenation pathway? Lipids 49, 527–541; Griinari, J.M., Bauman, D.E., 1999. Biosynthesis of conjugated linoleic acid and its incorporation into meat and milk in ruminants. In: Yurawecz, M.P., Mossoba, M., Kramer, J.K.G., Nelson, G., Pariza, M.W. (Eds.), Advances in Conjugated Linoleic Acid Research, vol. 1. AOCS Press, Champaign, IL, USA, pp. 180–200; Harfoot, C.G., Hazelwood, G.P., 1997. Lipid metabolism in the rumen. In: Hobson, P.N., Stewart, C.S. (Eds.), The Rumen Microbial Ecosystem. Chapman and Hall, London, UK, pp. 382–426.*

biohydrogenation pathways might not always predominate. In support of this, both $c9$, $t11$, $c15$-18:3 and $c9$, $t13$, $c15$-18:3 have been positively identified in milk (i.e., bovine) fat (Destaillats et al., 2005), but to date $t10$, $c12$, $c15$-18:3 has not been found. Many other minor monoenoic (e.g., $c$ and $t18:1$) and dienoic (e.g., $c/c$-, $c/t$-, $t/c$-, and $t/t$-18:2) biohydrogenation intermediates can also be produced, resulting in complex profiles requiring extraordinary efforts for comprehensive analyses (Aldai et al., 2012).

## Ruminant Fatty Acid Analyses

Most ruminant fatty acid analyses are conducted using gas chromatography (GC), and fatty acids are typically methylated prior to analyses. Acid catalyzed methylation can be used to methylate most lipid classes, but fatty acids with conjugated double bonds (i.e., double bonds separated by two carbons vs. the more typical three carbons including a $-CH_2$–methylene group) are sensitive to acid catalyzed methylation (Kramer et al., 1997). Conjugated linoleic (CLA) and conjugated linolenic acid (CLnA) isomers are intermediates formed during PUFA biohydrogenation. Base catalyzed methylation can be used for conjugated fatty acids, but cannot methylate all lipid classes, for example, free fatty acids (Kramer et al., 1997). Thus, for comprehensive analyses, a combination of acid and base catalyzed methylation often needs to be used and results combined. Analyses of ruminant fatty acids is further complicated, as many isomers generated during biohydrogenation have the same carbon chain length and only differ by double bond position and configuration (i.e., cis or trans). Consequently, resolution of fatty acid methyl esters can be challenging, particularly for 18:1 isomers, as lowering the oven temperature to increase resolution helps separate cis from cis and trans from trans isomers, but cis and trans isomers start overlapping and coeluting. Consequently, classes of fatty acid methyl esters either have to be fractionated using silver-ion chromatography (thin layer chromatography or solid phase extraction) prior to GC analysis, or more than one GC analysis has to be conducted using complementary temperature programs, and results combined (Kramer et al., 2008). In addition, some major conjugated fatty acid isomers (i.e., $t7$, $c9$-18:2 and $c9$, $t11$-18:2) do not separate using highly polar GC columns (i.e., SP2560 or CP-Sil88) typically used for analyses and require the use of silver-ion HPLC (Cruz-Hernandez et al., 2004), or GC using an ionic (SLB-IL111) column (Turner et al., 2011).

### Ruminal Versus Chemical Biohydrogenation

Consumption of trans fatty acids produced through the chemical hydrogenation of PUFA rich vegetable oils has been linked to development of cardiovascular amongst other diseases (Mozaffarian et al., 2009). As a result, there have been concerted efforts to limit human consumption of *trans* fatty acids, and this leaves ruminant meat and milk as the major source of fatty acids containing *trans* double bonds in the human diet (Aldai et al., 2013). Ruminant *trans* fatty acids differ from *trans* fatty acids produced through chemical hydrogenation in that ruminant *trans* fatty acids are produced mostly through enzymatic processes and specific isomers are enriched, while chemical hydrogenation can lead to a broad spectrum of isomers. In addition, whereas industrially produced *trans* fatty acids are associated with ill-health, ruminant production systems can be tailored to enrich *trans* fatty acids with potential health benefits.

### PUFA Biohydrogenation Product Bioactivity and Enrichment

In the late 1970s in Dr. Michael Pariza's lab at the University of Wisconsin, studies were undertaken to determine which compounds in cooked ground beef were carcinogenic. An extract from pan-fried ground beef was found that inhibited mutagenesis (Pariza et al., 1979), and the active compound in the extract was later characterized as CLA (Ha et al., 1987). Since this time, $c9$, $t11$-18:2 (rumenic acid; the main CLA isomer in ruminant derived foods) and its precursor $t11$-18:1 (vaccenic acid) have been studied *in vitro*, and using a number of animal models to examine their effects on several diseases including, but not limited to, various forms of cancer, cardiovascular disease, osteoporosis, asthma, type 2 diabetes, arthritis, and other diseases affected by inflammatory processes (Bhattacharya et al., 2006; Fuke and Nornberg, 2017). Estimated human intakes of rumenic acid thought to be protective against disease range from 0.60 to 3.2 g/day (Siurana and Calsamiglia, 2016). The modes of CLA action have and continue to be extensively studied. To mention a few, CLA can act as a ligand for nuclear receptors (Moya-Camarena et al., 1999), change prostaglandin levels and induce apoptosis in cancer cells (Park et al., 2004), bind to nitric oxide and form nitrofatty acids with potent antiinflammatory effects (Bonacci et al., 2012), and can inhibit foam cell formation by increasing PGC-1α gene expression (Bruen et al., 2017).

Although ruminant derived foods are the main natural source of vaccenic and rumenic acids, their enrichment poses some challenges. Adding PUFA sources to diets as precursors for rumenic and vaccenic acid synthesis does not guarantee their enrichment in ruminant tissues. On one hand, if high concentrate (grain-based) diets are fed, $t10$-18:1 can accumulate instead of vaccenic acid (Vahmani et al., 2015a), and second, if the final step in biohydrogenation is not inhibited, most $t18$:1 can be irreversibly hydrogenated to stearic acid. To date, the most successful method to enrich ruminant products with vaccenic and rumenic acids have been through feeding PUFA sources in forage-based diets. Data in Table 4.1 includes beef muscle (*L. thoracis*) fatty acid compositions of retail samples (Canadian retail survey, barley grain finished beef; Aldai et al., 2009), and when feeding finisher steers extruded flaxseed in a ground alfalfa/grass hay-based diet

**TABLE 4.1** Fatty acid compositions of muscle (*Longissimus thoracis*) from Canadian retail beef (barley grain finished) and from steers fed extruded flaxseed and hay (25:75) as a total mixed ration (TMR) or sequentially (non-TMR)

|  | Canadian retail beef[a] | TMR[b] | Non-TMR[b] |
|---|---|---|---|
| Total fatty acids (%) | 4.3 | 5.2 | 5.5 |
| *% FATTY ACID COMPOSITION[c]* | | | |
| 14:0 | 2.22 | 1.29 | 1.43 |
| 15:0 | 0.44 | 0.21 | 0.20 |
| 16:0 | 24 | 23.08 | 23.71 |
| 17:0 | 1.33 | 0.44 | 0.43 |
| 18:0 | 11.8 | 16.40 | 14.80 |
| Total SFA | 41.30 | 43.50 | 42.54 |
| *t*6-*t*8 18:1 | 0.14 | 0.48 | 0.48 |
| *t*9 18:1 | 0.22 | 0.45 | 0.45 |
| *t*10 18:1 | 1.01 | 0.54 | 0.57 |
| *t*11 18:1 | 0.45 | 4.98 | 7.43 |
| *t*12 18:1 | 0.11 | 0.58 | 0.43 |
| *t*13, *t*14 18:1 | 0.23 | 1.44 | 1.08 |
| *t*15 18:1 | 0.18 | 0.52 | 0.34 |
| *t*16 18:1 | 0.08 | 0.47 | 0.31 |
| Total *t* MUFA | 2.49 | 9.73 | 11.40 |
| *c*9 14:1 | 0.61 | 0.44 | 0.41 |
| *c*7 16:1 | 0.15 | 0.10 | 0.10 |
| *c*9 16:1 | 3.68 | 2.13 | 2.24 |
| *c*9 17:1 | 1.37 | 0.39 | 0.37 |
| *c*9 18:1 | 39.60 | 31.00 | 29.80 |
| *c*11 18:1 | 1.89 | 1.01 | 1.07 |
| *c*12 18:1 | 0.11 | 0.50 | 0.29 |
| *c*13 18:1 | 0.49 | 0.28 | 0.25 |
| *c*14 18:1 | 0.03 | 0.08 | 0.06 |
| *c*15 18:1 | 0.14 | 0.43 | 0.32 |
| *c*16 18:1 | NR[e] | 0.07 | 0.04 |
| *c*9 20:1 | 0.08 | 0.08 | 0.08 |

*(Continued)*

**TABLE 4.1**    Fatty acid compositions of muscle (*Longissimus thoracis*) from Canadian retail beef (barley grain finished) and from steers fed extruded flaxseed and hay (25:75) as a total mixed ration (TMR) or sequentially (non-TMR) (*Cont.*)

| | Canadian retail beef[a] | TMR[b] | Non-TMR[b] |
|---|---|---|---|
| *c*11 20:1 | 0.22 | 0.16 | 0.16 |
| Total *c* MUFA | 48.9 | 36.88 | 35.38 |
| *t*11, *t*15 18:2 | NR | 0.28 | 0.48 |
| *c*9, *t*14 18:2[d] | 0.17 | 0.19 | 0.14 |
| *c*9, *t*13 18:2 | | 0.23 | 0.18 |
| *c*9, *t*15 18:2 | 0.09 | 0.20 | 0.13 |
| *t*9, *c*12 18:2 | 0.03 | 0.10 | 0.13 |
| *t*11, *c*15 18:2 | 0.09 | 1.58 | 2.20 |
| Total atypical dienes | 0.66 | 3.15 | 3.60 |
| *c*9, *t*11 18:2 | 0.26 | 0.48 | 0.64 |
| *t*11, *c*13 18:2 | 0.01 | 0.14 | 0.28 |
| Total CLA | 0.46 | 0.78 | 1.05 |
| *c*9, *t*11, *c*15 18:3 | NR | 0.10 | 0.17 |
| Total CLnA | NR | 0.16 | 0.25 |
| 18:2n-6 | 2.54 | 2.09 | 2.10 |
| 20:4n-6 | 0.77 | 0.26 | 0.23 |
| Total n-6 PUFA | 3.75 | 2.49 | 2.45 |
| 18:3n-3 | 0.26 | 1.13 | 1.26 |
| 20:5n-3 | 0.10 | 0.15 | 0.12 |
| 22:5n-3 | 0.27 | 0.23 | 0.22 |
| 22:6n-3 | 0.03 | 0.02 | 0.01 |
| Total n-3 PUFA | 0.69 | 1.60 | 1.70 |

[a] *Data from Aldai et al. (2009).*
[b] *Data from Vahmani et al. (2017b).*
[c] *Total SFA, sum of 10:0, 12:0, 14:0, 15:0, 16:0, 17:0, 18:0, 19:0, 20:0, 22:0; c, cis; t, trans; total t-18:1, sum of t6-/t7-/t8-, t9-, t10-, t11-, t12-, t13-/t14-, t15-, t16-; total c-MUFA, sum of c9-14:1, c7–16:1, c9–16:1, c11–16:1, c9–17:1, c9–18:1, c11–18:1, c12–18:1, c13–18:1, c14–18:1, c15–18:1, c16–18:1, c9–20:1, c11–20:1; total atypical dienes, sum of t11, t15-, c9,t13-/c9, t14-, c9,t15-, c9, t12–18:1, c9,t12-, t11, c15-, c9, c15-, c12, c15–18:2; total CLA, sum of t7, c9-, c9, t11-, c11, t13-, t11, t13, t, t-18:2; total CLnA, sum of c9, t11, t15- and c9, t11, c15–18:3; total n-6 PUFA, sum of 18:2n-6, 20:2n-6, 20:3n-6, 20:4n-6; total n-3 PUFA, sum of 18:3n-3, 20:3n-3, 22:5n-3.*
[d] *For retail beef, c9, t14- and c9, t13–18 were combined.*
[e] *NR, Not reported.*

(25% and 75% of dietary dry matter, respectively) (Vahmani et al., 2017b). Interestingly, sequential feeding of flaxseed and hay (i.e., feeding flaxseed on its own followed by hay) further increased the amounts of vaccenic and rumenic acids in beef compared to feeding the diet as a total mixed ration. When feeding PUFA sources to ruminants it is important to not exceed 5%–7% fat or oil in the diet, as this has a suppressive effect on fibre digestion (Doreau and Chilliard, 1997). Intermittent feeding of a concentrated PUFA source (e.g., flaxseed), however, appears to inhibit or select against ruminal bacteria which carry out the final step in biohydrogenation (e.g., conversion of vaccenic acid to 18:0), resulting in the accumulation of vaccenic acid (the precursor of rumenic acid).

Although ruminant derived foods are natural sources of rumenic acid, there are concerns that the amounts might not be enough to provide health benefits (i.e., 0.6–3.2 g). In the retail survey, Canadian strip loin beef steak (Aldai et al., 2009) would have supplied 0.012 g of rumenic and 0.022 g of vaccenic acid per 114 g (4 oz) serving. Given the vaccenic to rumenic acid conversion via delta-9 desaturation in humans is ~19% (Turpeinen et al., 2002), the steak would have only provided 0.017 g of rumenic acid equivalents per 114 g serving. Strip loin steak often comes with subcutaneous fat attached, and subcutaneous fat from the same survey would have supplied 0.050 g rumenic acid equivalents per 114 g of fat. Thus, even if 100% subcutaneous fat is consumed, the amount of rumenic acid equivalents would be >10-fold less than the minimum thought to be of potential health benefit (0.60 g). Using the optimized feeding strategy (e.g., sequential feeding of flaxseed and hay; Table 4.1), however, rumenic acid equivalents in 114 g of strip loin steak or back fat would be 0.124 and 3.85 g, respectively. Thus some combination of meat and trim fat could meet minimal rumenic acid needs for health benefit (Vahmani et al., 2017b), and indeed, hamburger in this trial made from 80% sirloin (*Gluteus medius*) and 20% perirenal fat would have provided 0.662 g of rumenic acid equivalents per 114 g serving. Initial research has indicated PUFA biohydrogenation product enriched beef fat can improve insulin sensitivity without altering dyslipidemia in insulin resistant JCR:LA-cp rats (Diane et al., 2016), but more research, including clinical trials, is required to determine how effective ruminant derived foods enriched with rumenic and vaccenic acids are in combating of chronic diseases.

When trying to enhance the fatty acid profile of ruminant derived foods it is important to consider that when increasing the content of one or two fatty acids (i.e., vaccenic and rumenic acids), the composition of the whole food may change. For animal derived foods with enhance fatty acid profiles, therefore, the impact of fatty acid changes in whole diets needs to be considered, and crucially, there needs to be detailed human intervention studies before judgements concerning reduced chronic disease risk can be properly made (Givens, 2015). Indeed, increases in vaccenic and rumenic acids when feeding extruded flaxseed in a hay-based diet (Table 4.1) led to increases in the content of several biohydrogenation products other than vaccenic and rumenic acids, and the health effects of many of these are only beginning to be understood. For example, Vahmani et al. (2015b) found $t$18:1 isomers with double bonds at carbons 6, 11, 12, 13, 14, 15, and 16 undergo delta-9 desaturation in 3T3-L1 cells, but isomers with double bonds at carbons 9 and 10 do not. In addition when culturing HepG2 cells with $t$9 and $t$10-18:1, triacylglycerol and cholesterol synthesis was upregulated, but not for $t$18:1 isomers with double bonds from carbons 11–15 (Vahmani et al., 2017a). In addition, although $t$10, $c$12-18:2 (a CLA isomer primarily found in synthetic supplements) can promote fatty acid mobilization, oxidation, and reductions in body fat, an ALA biohydrogenation product containing a $t$10 double bond (i.e., $t$10, $c$15-18:2) has not been found to have same anti-adipogenic activities in fat cells (Vahmani et al., 2016a). Several other minor dienoic biohydrogenation products in beef have also recently been identified as delta-9 desaturation products of $t$18:1 isomers, including $c$9, $t$12-, $c$9, $t$13-, $c$9, $t$14-, $c$9, $t$15- and $c$9, $t$16-18:2 (Vahmani et al., 2016b), but health effects of most remain unstudied.

### PUFA Biosynthesis and Metabolism in Ruminants

Ruminal biohydrogenation is the main factor limiting the content of PUFA in tissues of ruminant animals. When oils are infused postruminally (i.e., into the fourth stomach or abomasum), tissue contents of PUFA can be increased dramatically (Jenkins and Bridges, 2007; Kliem and Shingfield, 2016). Efforts to increase PUFA contents in ruminant derived foods have been mostly centred on developing ways to protect PUFA from biohydrogenation. Free fatty acids are required for biohydrogenation, and lipase activity needed for their release can be inhibited by reducing rumen pH (Jenkins et al., 2008). Feeding highly fermentable diets (i.e., grain-based) can reduce rumen pH and PUFA biohydrogenation, but can also result in *trans* 10 shifted biohydrogenation pathways (Griinari and Bauman, 1999; Mapiye et al., 2012). As a consequence, PUFA contents of tissues may go up, but their biohydrogenation products may not improve the health value of whole foods. Methods to physically protect PUFA from biohydrogenation have also been investigated by feeding whole or processed oilseeds, or by feeding calcium or other fatty acid salts or amides, but only limited success has been achieved (Jenkins and Bridges, 2007). The most promising technology to date to increase ruminal PUFA bypass, and availability for digestion has been feeding oils encapsulated in formaldehyde treated casein (Scott et al., 1971). The use of formaldehyde is, however, regulated in some jurisdictions (Doreau et al., 2011), and as a consequence, other alternatives, such as using plant extracts rich in polyphenol oxidase activity to emulsify PUFA are being explored (Gadeyne et al., 2015).

As mentioned previously, the PUFA composition of ruminant derived foods is dependent on the diet, with finishing animals on grain-based diets leading to greater deposition of LA and LCn-6 fatty acids, while finishing animals on pasture or feeding forage based diets results in greater deposition of ALA and LCn-3 fatty acids. For example, when Warren et al. (2008) fed steers a grass silage based diet, this resulted in deposition of 43 mg ALA/100 g of loin steak and 59 mg LCn-3 (includes 35 mg DPA). This included a 5–6-fold increase in DHA compared to concentrate feeding (i.e., an increase from ~1 to 5 mg/100 g of muscle). The ability to increase DHA, however, appears to be restricted to when animals are raised on pasture or fed forage based diets (Wood et al., 2008), as feeding other ALA sources, such as flaxseed can either lead to no change in DHA in cattle (Gruffat et al., 2013; Juarez et al., 2011), or in one instance decreased DHA in lambs (Bessa et al., 2007). The amount of LCn-3 in ruminant-derived foods would not be considered excessive considering it is generally recommended that healthy people consume between 250 and 500 mg of combined EPA and DHA per day (Kitessa et al., 2014). In countries, such as Australia, however, where there is limited fish consumption, beef and lamb can contribute close to one third of LCn-3 fatty acid consumed, mainly in the form of DPA (Howe et al., 2006). Consequently, even small changes in LCn-3 fatty acids in ruminant derived food can substantially influence overall intakes, and developing ways to increase these have been a research priority. De Smet et al. (2004) reviewed how meat fatty acid composition is affected by degree of fatness and genetics, and found diet is the main effector of fatty acid composition. Breed, however, has been found to influence the LC-PUFA composition of meat independent of the confounding effect of overall fatness. For example, Choi et al. (2000) found a higher content of EPA in Welsh Black compared to Holstein Friesian steers, and the deposition of n-3 fatty acids was increased with and without n-3 PUFA supplementation. In addition, increased contents of EPA and DPA have been found in Belgian Blue bulls (Raes et al., 2001) and Asturiana de los Valles bulls (Aldai et al., 2008) carrying homozygous for the gene responsible for double muscling.

When trying to change the PUFA content and composition of ruminant derived foods by diet, it is important to understand their underlying metabolism. When liver slices from Normande cows were incubated with LA or ALA in free fatty acid form (i.e., not sequestered in phospholipid), rates of ALA uptake were higher than LA independent of animal diet (control vs. supplementation with extruded flaxseed and rapeseed). The rate of form β-oxidation was also greater in liver slices for ALA (50%) versus LA (27%), and ALA was not converted to LCn-3 fatty acids, but about 14% of LA was converted to arachidonic acid (20:4n-6). Both LA and ALA were also efficiently taken up by adipose tissue slices and mainly esterified to neutral lipids, with 9.5% of LA converted to 20:4n-6 and 1.5% of ALA to

20:5n-3. Trying to explain increases in DHA when animals graze on pasture or when fed high proportions of forage with fatty acid compositions dominated by ALA remains a challenge. Most *de novo* fatty acid synthesis in ruminants occurs in adipose tissue, not the liver. However, mRNA expression of genes involved in LC-PUFA synthesis is greater in liver than muscle or adipose tissue (Cherfaoui et al., 2013). This corresponds with higher contents of LCn-3 fatty acids in liver, and points toward a central role for liver in LC-PUFA metabolism in cattle (da Costa et al., 2014). The mRNA expression of ELOVL5 was noted to be particularly low in muscle (Blonde d'Aquitaine, Limousin, and Angus bulls), but the presence of EPA and DPA in muscle raised the question of where these originate (i.e., possibly the liver as in other species) (Cherfaoui et al., 2013; Gruffat et al., 2013). In Charolais steers, mRNA expression of ELOVL5 did not, however, appear to be as restricted in muscle (Cherfaoui et al., 2013), but muscle contents of LCn-3 fatty acids were not remarkably different than in other breeds of bulls. In addition, in Charolais bulls, feeding grass versus maize silage (i.e., sources of ALA and LA, respectively) increased muscle content of ALA and all LCn-3 (EPA, DPA, and DHA), but mRNA expression of most genes involved in PUFA biosynthesis and peroxisomal oxidation (i.e., needed for conversion of 24:6n-3 to 22:6n-3) were unaffected. In addition, mRNA expression of genes involved in cellular uptake of LCn-3 fatty acids were also unaffected by diet, leading authors speculate that passive diffusion rather than direct transport of LCn-3 fatty acids into muscle cells may be occurring. Interestingly, several mutations in ELOVL genes have been found in Japanese Black cattle, and in the promotor region of ELOVL5 (Matsumoto et al., 2013). Clearly more research is necessary to unravel the key factors involved in LC-PUFA synthesis in ruminants, and how to control these to manipulate their contents.

# References

Adkisson, 4th, H.D., Risener, Jr., F.S., Zarrinkar, P.P., Walla, M.D., Christie, W.W., Wuthier, R.E., 1991. Unique fatty acid composition of normal cartilage: discovery of high levels of n-9 eicosatrienoic acid and low levels of n-6 polyunsaturated fatty acids. FASEB 5, 344–353.

Aldai, N., de Renobales, M., Barron, L.J.R., Kramer, J.K., 2013. What are the trans fatty acids issues in foods after discontinuation of industrially produced trans fats? Ruminant products, vegetable oils, and synthetic supplements. Eur. J. Lipid Sci. Technol. 115, 1378–1401.

Aldai, N., Dugan, M., Nájera, A., Osoro, K., 2008. N-6 and n-3 fatty acids in different beef adipose tissues depending on the presence or absence of the gene responsible for double-muscling. Czech. J. Anim. Sci. 53, 515–522.

Aldai, N., Dugan, M.E.R., Rolland, D.C., Kramer, J.K.G., 2009. Survey of the fatty acid composition of Canadian beef: backfat and longissimus lumborum muscle. Can. J. Anim. Sci. 89, 315–329.

Aldai, N., Kramer, J.K.G., Cruz-Hernandez, C., Santercole, V., Delmonte, P., Mossaba, M.M., Dugan, M.E.R., 2012. Appropriate extraction and methylation techniques for lipid analysis. Cherian, G., Poureslami, R. (Eds.), Fats and Fatty Acids in Poultry Nutrition and Health, vol. 4, Context Products Ltd, Leicestershire, United Kingdom, pp. 249–278.

Alves, S.P., Bessa, R.J., 2014. The trans-10, cis-15 18:2: a missing intermediate of trans-10 shifted rumen biohydrogenation pathway? Lipids 49, 527–541.

Baker, E.J., Miles, E.A., Burdge, G.C., Yaqoob, P., Calder, P.C., 2016. Metabolism and functional effects of plant-derived omega-3 fatty acids in humans. Prog. Lipid Res. 64, 30–56.

Bauchart, D., 1993. Lipid absorption and transport in ruminants. J. Dairy Sci. 76, 3864–3881.

Bazinet, R.P., McMillan, E.G., Seebaransingh, R., Hayes, A.M., Cunnane, S.C., 2003. Whole-body β-oxidation of 18:2ω6 and 18:3ω3 in the pig varies markedly with weaning strategy and dietary 18:3ω3. J. Lipid Res. 44, 314–319.

Bessa, R.J.B., Alves, S.P., Jeronimo, E., Alfaia, C.M., Prates, J.A.M., Santos-Silva, J., 2007. Effect of lipid supplements on ruminal biohydrogenation intermediates and muscle fatty acids in lambs. Eur. J. Lipid Sci. Technol. 109, 868–878.

Bhattacharya, A., Banu, J., Rahman, M., Causey, J., Fernandes, G., 2006. Biological effects of conjugated linoleic acids in health and disease. J. Nutr. Biochem. 17, 789–810.

Bonacci, G., Baker, P.R.S., Salvatore, S.R., Shores, D., Khoo, N.K.H., Koenitzer, J.R., Vitturi, D.A., Woodcock, S.R., Golin-Bisello, F., Cole, M.P., Watkins, S., Croix, St.C., Batthyany, C.I., Freeman, B.A., Schopfer, F.J., 2012. Conjugated linoleic acid is a preferential substrate for fatty acid nitration. J. Biol. Chem. 287, 44071–44082.

Bruen, R., Fitzsimons, S., Belton, O., 2017. Atheroprotective effects of conjugated linoleic acid. Br. J. Clin. Pharmacol. 83, 46–53.

Caldari-Torres, C., McGilliard, M.L., Corl, B.A., 2016. Esterification of essential and non-essential fatty acids into distinct lipid classes in ruminant and non-ruminant tissues. Comp. Biochem. Physiol. B 200, 1–5.

Canadian Food Inspection Agency (CFIA), 2014. Omega-3 and omega-6 polyunsaturated fatty acid claims. Available from: http://www.inspection.gc.ca/food/labelling/food-labelling-for-industry/nutrient-content/specific-claim-requirements/eng/1389907770176/1389907817577?chap=7

Cherfaoui, M., Durand, D., Bonnet, M., Bernard, L., Bauchart, D., Ortigues-Marty, I., Gruffat, D., 2013. A grass-based diet favours muscle n-3 long-chain PUFA deposition without modifying gene expression of proteins involved in their synthesis or uptake in Charolais steers. Animal 7, 1833–1840.

Cherian, G., 2012. Modifying egg lipids for enhancing human health. Cherian, G., Poureslami, R. (Eds.), Fats and Fatty Acids in Poultry Nutrition and Health, vol. 4, Context Products Ltd, Leicestershire, United Kingdom, pp. 57–68.

Cherian, G., Sim, J.S., 1995. Dietary alpha-linolenic acid alters the fatty acid composition of lipid classes in swine tissues. J. Agric. Food Chem. 43, 2911–2916.

Choi, N.J., Enser, M., Wood, J.D., Scollan, N.D., 2000. Effect of breed on the deposition in beef muscle and adipose tissue of dietary n-3 polyunsaturated fatty acids. Anim. Sci. 71, 509–519.

Crespo, N., Esteve-Garcia, E., 2002. Dietary polyunsaturated fatty acids decrease fat deposition in separable fat depots but not in the remainder carcass. Poult. Sci. 81, 512–518.

Cruz-Hernandez, C., Deng, Z., Zhou, J., Hill, A.R., Yurawecz, M.P., Delmonte, P., Mossoba, M.M., Dugan, M.E., Kramer, J.K., 2004. Methods for analysis of conjugated linoleic acids and trans-18:1 isomers in dairy fats by using a combination of gas chromatography, silver-ion thin-layer chromatography/gas chromatography, and silver-ion liquid chromatography. J. AOAC Int. 87, 545–562.

da Costa, A.S.H., Bessa, R.J.B., Pires, V.M.R., Rolo, E.A., Pinto, R.M.A., Fontes, C.M.G.A., Prates, J.A.M., 2014. Is hepatic lipid metabolism of beef cattle influenced by breed and dietary silage level? BMC Vet. Res. 10, 65.

De Smet, S., Raes, K., Demeyer, D., 2004. Meat fatty acid composition as affected by fatness and genetic factors: a review. Anim. Res. 53, 81–98.

Destaillats, F., Trottier, J., Galvez, J.G., Angers, P., 2005. Analysis of α-linolenic acid biohydrogenation intermediates in milk fat with emphasis on conjugated linolenic acids. J. Dairy Sci. 88, 3231–3239.

Diane, A., Borthwick, F., Mapiye, C., Vahmani, P., David, R.C., Vine, D.F., Dugan, M.E.R., Proctor, S.D., 2016. Beef fat enriched with polyunsaturated fatty acid biohydrogenation products improves insulin sensitivity without altering dyslipidemia in insulin resistant JCR:LA-cp rats. Lipids 51, 821–831.

Doreau, M., Bauchart, D., Chilliard, Y., 2011. Enhancing fatty acid composition of milk and meat through animal feeding. Anim. Prod. Sci. 51, 19–29.

Doreau, M., Chilliard, Y., 1997. Digestion and metabolism of dietary fat in farm animals. Br. J. Nutr. 78 (Suppl. 1), S15–S35.

Doreau, M., Ferlay, A., 1994. Digestion and utilisation of fatty acids by ruminants. Anim. Feed Sci. Technol. 45, 379–396.

Dugan, M., Aalhus, J., Lien, K., Schaefer, A., Kramer, J., 2001. Effects of feeding different levels of conjugated linoleic acid and total oil to pigs on live animal performance and carcass composition. Can. J. Anim. Sci. 81, 505–510.

Dugan, M.E., Vahmani, P., Turner, T.D., Mapiye, C., Juárez, M., Prieto, N., Beaulieu, A.D., Zijlstra, R.T., Patience, J.F., Aalhus, J.L., 2015. Pork as a source of omega-3 (n-3) fatty acids. J. Clin. Med. 4, 1999–2011.

Dugan, M.E.R., Aldai, N., Aalhus, J.L., Rolland, D.C., Kramer, J.K.G., 2011. Review: transforming beef to provide healthier fatty acid profiles. Can. J. Anim. Sci. 91, 545–556.

Duran-Montgé, P., Realini, C.E., Barroeta, A.C., Lizardo, R.G., Esteve-Garcia, E., 2010. De novo fatty acid synthesis and balance of fatty acids of pigs fed different fat sources. Livest. Sci. 132, 157–164.

Enser, M., Hallett, K., Hewitt, B., Fursey, G., Wood, J., 1996. Fatty acid content and composition of English beef, lamb and pork at retail. Meat Sci. 42, 443–456.

Fuke, G., Nornberg, J.L., 2017. Systematic evaluation on the effectiveness of conjugated linoleic acid in human health. Crit. Rev. Food Sci. Nutr. 57, 1–7.

Gadeyne, F., Van Ranst, G., Vlaeminck, B., Vossen, E., Van der Meeren, P., Fievez, V., 2015. Protection of polyunsaturated oils against ruminal biohydrogenation and oxidation during storage using a polyphenol oxidase containing extract from red clover. Food Chem. 171, 241–250.

Givens, D.I., 2015. Special issue: lipids and their manipulation in animal-derived foods. Eur. J. Lipid Sci. Technol. 117, 1303–1305.

Glasser, F., Doreau, M., Maxin, G., Baumont, R., 2013. Fat and fatty acid content and composition of forages: a meta-analysis. Anim. Feed Sci. Technol. 185, 19–34.

Gregory, M.K., Geier, M.S., Gibson, R.A., James, M.J., 2013. Functional characterization of the chicken fatty acid elongases. J. Nutr. 143, 12–16.

Griinari, J.M., Bauman, D.E., 1999. Biosynthesis of conjugated linoleic acid and its incorporation into meat and milk in ruminants. Yurawecz, M.P., Mossoba, M., Kramer, J.K.G., Nelson, G., Pariza, M.W. (Eds.), Advances in Conjugated Linoleic Acid Research, vol. 1, AOCS Press, Champaign, IL, USA, pp. 180–200.

Gruffat, D., Cherfaoui, M., Bonnet, M., Thomas, A., Bauchart, D., Durand, D., 2013. Breed and dietary linseed affect gene expression of enzymes and transcription factors involved in n-3 long chain polyunsaturated fatty acids synthesis in Longissimus thoracis muscle of bulls. J. Anim. Sci. 91, 3059–3069.

Ha, Y., Grimm, N., Pariza, M., 1987. Anticarcinogens from fried ground beef: heat-altered derivatives of linoleic acid. Carcinogenesis 8, 1881–1887.

Harfoot, C.G., Hazelwood, G.P., 1997. Lipid metabolism in the rumen. In: Hobson, P.N., Stewart, C.S. (Eds.), The Rumen Microbial Ecosystem. Chapman and Hall, London, UK, pp. 382–426.

Harris, W.S., 2012. Stearidonic acid—enhanced soybean oil: a plant-based source of (n-3) fatty acids for foods. J. Nutr. 142, 600S–604S.

Holman, R.T., 1971. Essential fatty acid deficiency. Prog. Chem. Fats Other Lipids 9, 275–348.

Howe, P., Meyer, B., Record, S., Baghurst, K., 2006. Dietary intake of long-chain omega-3 polyunsaturated fatty acids: contribution of meat sources. Nutrition 22, 47–53.

Jenkins, T.C., Bridges, W.C., 2007. Protection of fatty acids against ruminal biohydrogenation in cattle. Eur. J. Lipid Sci. Technol. 109, 778–789.

Jenkins, T.C., Wallace, R.J., Moate, P.J., Mosley, E.E., 2008. Recent advances in biohydrogenation of unsaturated fatty acids within the rumen microbial ecosystem. J. Anim. Sci. 86, 397–412.

Jing, M., Gakhar, N., Gibson, R., House, J., 2013. Dietary and ontogenic regulation of fatty acid desaturase and elongase expression in broiler chickens. Prostaglandins Leukot. Essent. Fatty Acids 89, 107–113.

Juarez, M., Dugan, M.E., Aalhus, J.L., Aldai, N., Basarab, J.A., Baron, V.S., McAllister, T.A., 2011. Effects of vitamin E and flaxseed on rumen-derived fatty acid intermediates in beef intramuscular fat. Meat Sci. 88, 434–440.

Juárez, M., Dugan, M.E.R., Aldai, N., Aalhus, J.L., Patience, J.F., Zijlstra, R.T., Beaulieu, A.D., 2010. Feeding co-extruded flaxseed to pigs: effects of duration and feeding level on growth performance and backfat fatty acid composition of grower-finisher pigs. Meat Sci. 84, 578–584.

Juárez, M., Dugan, M.E.R., Aldai, N., Aalhus, J.L., Patience, J.F., Zijlstra, R.T., Beaulieu, A.D., 2011. Increasing omega-3 levels through dietary co-extruded flaxseed supplementation negatively affects pork palatability. Food Chem. 126, 1716–1723.

Kitessa, S.M., Abeywardena, M., Wijesundera, C., Nichols, P.D., 2014. DHA-containing oilseed: a timely solution for the sustainability issues surrounding fish oil sources of the health-benefitting long-chain omega-3 oils. Nutrients 6, 2035–2058.

Kliem, K.E., Shingfield, K.J., 2016. Manipulation of milk fatty acid composition in lactating cows: opportunities and challenges. Eur. J. Lipid Sci. Technol. 118, 1661–1683.

Kloareg, M., Noblet, J., van Milgen, J., 2007. Deposition of dietary fatty acids, de novo synthesis and anatomical partitioning of fatty acids in finishing pigs. Br. J. Nutr. 97, 35–44.

Knight, T.W., Knowles, S., Death, A.F., West, J., Agnew, M., Morris, C.A., Purchas, R.W., 2003. Factors affecting the variation in fatty acid concentrations in lean beef from grass-fed cattle in New Zealand and the implications for human health. N. Z. J. Agric. Res. 46, 83–95.

Kouba, M., Enser, M., Whittington, F.M., Nute, G.R., Wood, J.D., 2003. Effect of a high-linolenic acid diet on lipogenic enzyme activities, fatty acid composition, and meat quality in the growing pig. J. Anim. Sci. 81, 1967–1979.

Kouba, M., Mourot, J., 1999. Effect of a high linoleic acid diet on lipogenic enzyme activities and on the composition of the lipid fraction of fat and lean tissues in the pig. Meat Sci. 52, 39–45.

Kramer, J.K., Fellner, V., Dugan, M.E., Sauer, F.D., Mossoba, M.M., Yurawecz, M.P., 1997. Evaluating acid and base catalysts in the methylation of milk and rumen fatty acids with special emphasis on conjugated dienes and total trans fatty acids. Lipids 32, 1219–1228.

Kramer, J.K., Hernandez, M., Cruz-Hernandez, C., Kraft, J., Dugan, M.E., 2008. Combining results of two GC separations partly achieves determination of all cis and trans 16:1, 18:1, 18:2 and 18:3 except CLA isomers of milk fat as demonstrated using Ag-ion SPE fractionation. Lipids 43, 259–273.

Liu, K., 2011. Comparison of lipid content and fatty acid composition and their distribution within seeds of 5 small grain species. J. Food Sci. 76, C334–C342.

Liu, X., Pang, D., Yuan, T., Li, Z., Li, Z., Zhang, M., Ren, W., Ouyang, H., Tang, X., 2016. N-3 polyunsaturated fatty acids attenuates triglyceride and inflammatory factors level in h fat-1 transgenic pigs. Lipids Health Dis. 15, 89.

Malau-Aduli, A.E.O., Siebert, B.D., Bottema, C.D.K., Pitchford, W.S., 2000. Heterosis, sex and breed differences in the fatty acid composition of muscle phospholipids in beef cattle. J. Anim. Physiol. Anim. Nutr. 83, 113–120.

Mapiye, C., Aldai, N., Turner, T.D., Aalhus, J.L., Rolland, D.C., Kramer, J.K., Dugan, M.E., 2012. The labile lipid fraction of meat: from perceived disease and waste to health and opportunity. Meat Sci. 92, 210–220.

Martinez-Ramirez, H.R., Cant, J.P., Shoveller, A.K., Atkinson, J.L., de Lange, C.F.M., 2014. Whole-body retention of alpha-linolenic acid and its apparent conversion to other n-3 PUFA in growing pigs are reduced with the duration of feeding alpha-linolenic acid. Br. J. Nutr. 111, 1382–1393.

Martínez-Ramírez, H.R., Kramer, J.K.G., de Lange, C.F.M., 2014. Retention of n-3 polyunsaturated fatty acids in trimmed loin and belly is independent of timing of feeding ground flaxseed to growing-finishing female pigs. J. Anim. Sci. 92, 238–249.

Matsumoto, H., Shimizu, Y., Tanaka, A., Nogi, T., Tabuchi, I., Oyama, K., Taniguchi, M., Mannen, H., Sasazaki, S., 2013. The SNP in the promoter region of the bovine *ELOVL5* gene influences economic traits including subcutaneous fat thickness. Mol. Biol. Rep. 40, 3231–3237.

Moya-Camarena, S.Y., Heuvel, J.P.V., Blanchard, S.G., Leesnitzer, L.A., Belury, M.A., 1999. Conjugated linoleic acid is a potent naturally occurring ligand and activator of PPARα. J. Lipid Res. 40, 1426–1433.

Mozaffarian, D., Aro, A., Willett, W.C., 2009. Health effects of trans-fatty acids: experimental and observational evidence. Eur. J. Clin. Nutr. 63 (Suppl. 2), S5–S21.

Murphy, D.J., 2014. Using modern plant breeding to improve the nutritional and technological qualities of oil crops. OCL 21, D607.

Nakamura, M.T., Nara, T.Y., 2004. Structure, function, and dietary regulation of $\Delta 6$, $\Delta 5$, and $\Delta 9$ desaturases. Ann. Rev. Nutr. 24, 345–376.

Ntawubizi, M., Raes, K., Buys, N., De Smet, S., 2009. Effect of sire and sex on the intramuscular fatty acid profile and indices for enzyme activities in pigs. Livest. Sci. 122, 264–270.

Pariza, M.W., Ashoor, S.H., Chu, F.S., Lund, D.B., 1979. Effects of temperature and time on mutagen formation in pan-fried hamburger. Cancer Lett. 7, 63–69.

Park, H.S., Cho, H.Y., Ha, Y.L., Park, J.H., 2004. Dietary conjugated linoleic acid increases the mRNA ratio of Bax/Bcl-2 in the colonic mucosa of rats. J. Nutr. Biochem. 15, 229–235.

Park, W.J., Kothapalli, K.S., Lawrence, P., Tyburczy, C., Brenna, J.T., 2009. An alternate pathway to long-chain polyunsaturates: the FADS2 gene product $\Delta 8$-desaturates 20:2n-6 and 20:3n-3. J. Lipid Res. 50, 1195–1202.

Poureslami, R., Batal, A.B., 2012. Modifying meat lipids for human health. Cherian, G., Poureslami, R. (Eds.), Fats and Fatty Acids in Poultry Nutrition and Health, vol. 4, Context Products Ltd, Leicestershire, United Kingdom, pp. 71–80.

Poureslami, R., Raes, K., Turchini, G., Huyghebaert, G., De Smet, S., 2010. Effect of diet, sex and age on fatty acid metabolism in broiler chickens: n-3 and n-6 PUFA. Br. J. Nutr. 104, 189–197.

Raes, K., Smet, S.d., Demeyer, D., 2001. Effect of double-muscling in Belgian Blue young bulls on the intramuscular fatty acid composition with emphasis on conjugated linoleic acid and polyunsaturated fatty acids. Anim. Sci. 73, 253–260.

Romans, J.R., Johnson, R.C., Wulf, D.M., Libal, G.W., Costello, W.J., 1995. Effects of ground flaxseed in swine diets on pig performance and on physical and sensory characteristics and omega-3 fatty acid content of pork: I. Dietary level of flaxseed. J. Anim. Sci. 73, 1982–1986.

Schenck, P.A., Rakoff, H., Emken, E.A., 1996. δ8 Desaturationin vivo of deuterated eicosatrienoic acid by mouse liver. Lipids 31, 593–600.

Scott, T.W., Cook, L.J., Mills, S.C., 1971. Protection of dietary polyunsaturated fatty acids against microbial hydrogenation in ruminants. JAOCS 48, 358–364.

Shorland, F.B., 1950. Effect of the dietary fat on the composition of depot fats of animals. Nature 165, 766.

Simopoulos, A.P., 1999. New products from the agri-food industry: the return of n-3 fatty acids into the food supply. Lipids 34 (Suppl. 1), S297–301.

Siurana, A., Calsamiglia, S., 2016. A metaanalysis of feeding strategies to increase the content of conjugated linoleic acid (CLA) in dairy cattle milk and the impact on daily human consumption. Anim. Feed Sci. Technol. 217, 13–26.

Sprecher, H., Chen, Q., Yin, F.Q., 1999. Regulation of the biosynthesis of 22:5n-6 and 22:6n-3: a complex intracellular process. Lipids 34, S153–S156.

Tanghe, S., De Smet, S., 2013. Does sow reproduction and piglet performance benefit from the addition of n-3 polyunsaturated fatty acids to the maternal diet? Vet. J. 197, 560–569.

Turner, T., Rolland, D.C., Aldai, N., Dugan, M.E.R., 2011. Short communication: rapid separation of cis9, trans11- and trans7, cis9-18:2 (CLA) isomers from ruminant tissue using a 30 m SLB-IL111 ionic column. Can. J. Anim. Sci. 91, 711–713.

Turner, T.D., Mapiye, C., Aalhus, J.L., Beaulieu, A.D., Patience, J.F., Zijlstra, R.T., Dugan, M.E.R., 2014. Flaxseed fed pork: n-3 fatty acid enrichment and contribution to dietary recommendations. Meat Sci. 96, 541–547.

Turpeinen, A.M., Mutanen, M., Aro, A., Salminen, I., Basu, S., Palmquist, D.L., Griinari, J.M., 2002. Bioconversion of vaccenic acid to conjugated linoleic acid in humans. Am. J. Clin. Nutr. 76, 504–510.

Vahmani, P., Mapiye, C., Prieto, N., Rolland, D.C., McAllister, T.A., Aalhus, J.L., Dugan, M.E.R., 2015a. The scope for manipulating the polyunsaturated fatty acid content of beef: a review. J. Anim. Sci. Biotechnol., 6.

Vahmani, P., Meadus, W.J., Duff, P., Rolland, D.C., Dugan, M.E.R., 2017a. Comparing the lipogenic and cholesterolgenic effects of individual trans-18:1 isomers in liver cells. Eur. J. Lipid Sci. Technol. 119, 1600162.

Vahmani, P., Meadus, W.J., Mapiye, C., Duff, P., Rolland, D.C., Dugan, M.E., 2015b. Double bond position plays an important role in delta-9 desaturation and lipogenic properties of trans 18:1 isomers in mouse adipocytes. Lipids 50, 1253–1258.

Vahmani, P., Meadus, W.J., Rolland, D.C., Duff, P., Dugan, M.E., 2016a. Trans10, cis15 18:2 isolated from beef fat does not have the same anti-adipogenic properties as trans10, cis12-18:2 in 3T3-L1 adipocytes. Lipids 51, 1231–1239.

Vahmani, P., Rolland, D.C., Gzyl, K.E., Dugan, M.E., 2016b. Non-conjugated cis/trans 18:2 in beef fat are mainly delta-9 desaturation products of trans-18:1 isomers. Lipids 51, 1427–1433.

Vahmani, P., Rolland, D.C., McAllister, T.A., Block, H.C., Proctor, S.D., Guan, L.L., Prieto, N., Aalhus, J.L., Dugan, M.E.R., 2017b. Effects of feeding steers extruded flaxseed mixed with hay or before hay on animal performance, carcass quality, and meat and hamburger fatty acid composition. Meat Sci. 131, 9–17.

Waraich, E.A., Ahmed, Z., Ahmad, R., Saifullah, M.Y.A., Naeem, M.S., Rengel, Z., 2013. Camelina sativa, a climate proof crop, has high nutritive value and multiple-uses: a review. Aust. J. Crop Sci. 7, 1551.

Warren, H.E., Scollan, N.D., Enser, M., Hughes, S.I., Richardson, R.I., Wood, J.D., 2008. Effects of breed and a concentrate or grass silage diet on beef quality in cattle of 3 ages. I: animal performance, carcass quality and muscle fatty acid composition. Meat Sci. 78, 256–269.

Whelan, J., Rust, C., 2006. Innovative dietary sources of n-3 fatty acids. Annu. Rev. Nutr. 26, 75–103.

Williams, C.M., Burdge, G., 2007. Long-chain n-3 PUFA: plant v. marine sources. Proc. Nutr. Soc. 65, 42–50.

Wood, J.D., Enser, M., Fisher, A.V., Nute, G.R., Sheard, P.R., Richardson, R.I., Hughes, S.I., Whittington, F.M., 2008. Fat deposition, fatty acid composition and meat quality: a review. Meat Sci. 78, 343–358.

Yilmaz, J.L., Lim, Z.L., Beganovic, M., Breazeale, S., Andre, C., Stymne, S., Vrinten, P., Senger, T., 2017. Determination of substrate preferences for desaturases and elongases for production of docosahexaenoic acid from oleic acid in engineered canola. Lipids 52 (3), 1–16.

Zuidhof, M.J., Betti, M., Korver, D.R., Hernandez, F.I.L., Schneider, B.L., Carney, V.L., Renema, R.A., 2009. Omega-3-enriched broiler meat: 1. Optimization of a production system. Poult. Sci. 88, 1108–1120.

# The Biochemistry and Regulation of Fatty Acid Desaturases in Animals

*Woo Jung Park*

Gangneung-Wonju National University, Gangneung, Gangwon,
The Republic of Korea

## INTRODUCTION

Dietary fat is highly important for human health, the prevention of chronic diseases, and development (Roche, 1999). In particular, unsaturated fatty acids (UFA) containing more than one double bond [polyunsaturated fatty acids, (PUFA)] have been considered major component in cellular structures and to be nutritionally critical. Only a small proportion

Polyunsaturated Fatty Acid Metabolism. http://dx.doi.org/10.1016/B978-0-12-811230-4.00005-3

of these fatty acids is changed into diverse PUFA consisting of more than 20 carbons in humans (Venegas-Caleron et al., 2010) (Chapter 2). Moreover, 20-carbon PUFAs can be substrates for the synthesis of eicosanoids and other PUFA-derived mediators that modulate inflammatory and immune responses (Calder, 2013; Wolfe, 1982) (Chapter 8).

Desaturases catalyze the insertion of double bonds in fatty acids chain and are the main enzymes for the biosynthesis of monounsaturated fatty acids (MUFA) and PUFA (Nakamura and Nara, 2004). These enzymes are also known to play an important role in the regulation of physiological and pathogenic processes, such as immune and inflammatory responses, including asthma and arthritis (Calder, 2013; Wolfe, 1982), cardiovascular disease (Superko et al., 2014), diabetes (Fasano et al., 2015), and some cancers (Zheng et al., 2012).

This chapter describes the main functions of desaturases in animals in the context of PUFA biosynthesis (Chapter 2). Recent findings about desaturases from studies involving techniques, such as genomics and transgenic systems, are also discussed.

# DESATURASES AND UNSATURATED FATTY ACIDS

Fatty acid desaturases are named according to the location of the double bond, which is inserted into a fatty acid chain relative to the carboxylic acid group. These enzymes were first reported in cyanobacteria (Reddy et al., 1993) and have been found in many species of plants, algae, and fungi as well as animals.

## Stearoyl-CoA Desaturases

Δ9 desaturase, also known as stearoyl-CoA desaturase (SCD), catalyzes the insertion of a double bond at the 9th carbon from the carboxylic end in a fatty acid chain to convert 16 or 18 carbon saturated fatty acids to MUFAs of the same length (Fig. 5.1). In animals, such as insects (Eigenheer et al., 2002), nematode worms (Watts and Browse, 2000, 2002), and vertebrates (Miyazaki et al., 2005, 2006; Ntambi et al., 2002) this reaction is dependent on NADH, chytochrome b5, and NADH-cytochrome b5 reductase activity for the synthesis of MUFA (Nakamura and Nara, 2004; Sampath and Ntambi, 2014). Δ9 desaturase in mammalians is also named stearoyl-CoA desaturase (SCD), which catalyzes the desaturation of saturated fatty acids, such as palmitic (16:0) or stearic acid (18:0) to generate palmitoleic acid (PA) (16:1n-7) or oleic acid (OA) (18:1n-9), respectively. SCD was first purified from rat liver (Strittmatter et al., 1974) and subsequently its gene was identified in murine cells (Ntambi et al., 1988). SCDs, including first SCD (Scd1),

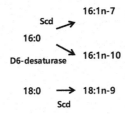

FIGURE 5.1  **Synthesis of monounsaturated fatty acids.** Palmitic acid (16:0) can be converted to either 16:1n-7 by stearoyl-CoA desaturase (Scd) or to sapienic acid (16:1n-10) by Δ6 desaturase. Sapienic acid is a major component of sebum which is unique to humans, while palmitoleic acid fulfills an equivalent role in mice. Stearic acid (18:0) can be converted to oleic acid (18:1n-9) by Scd activity.

incorporate one double bond in fatty acids of chain length greater than 12 carbons (Nakamura and Nara, 2004) and are rate-limiting enzymes in the synthesis of MUFA from saturated fatty acids (Ntambi, 1999). The ratio of MUFA to saturated fatty acid (SFA) can modify the fluidity of cell membranes and hence the activities of integral membrane proteins (Ntambi, 1995).

Humans express two SCD isoforms, SCD1 and SCD5 (Sinner et al., 2012; Zhang et al., 1999), while mice express four; *Scd1*, *Scd2*, *Scd3*, and *Scd4*. Scd1 is the murine homologue of human SCD1, while SCD5 is unique to primates (Miyazaki et al., 2006). Human *SCD1* encodes 359 amino acids and is located on human chromosome 10q24.31. SCD5, which is highly expressed in brain and pancreas, is located on chromosome 4q21.22. The lengths of encoded protein differ considerably between primates; orangutan SCD5 involves 256 amino acids, chimpanzee SCD5 contains 199 amino acids, and human SCD5 has 132 amino acids (Wang et al., 2005). SCD proteins are composed of three histidine boxes, which are assumed to act as catalytic domain chelating an iron (Guillou et al., 2010) and they are known to be integrated with NADH, cytochrome b5, as an electron donor, flavonprotein reductase, and molecular oxygen in the endoplasmic reticulum (Heinemann and Ozols, 2003).

Ntambi et al. (1988) has shown that murine Scd1, containing 352 amino acids, is expressed upon differentiation of 3T3-L1 preadipocytes (Ntambi et al., 1988). Moreover, the pattern of SCD expression is modified by dietary fat intake. Mice fed a diet containing unsaturated fatty acids express SCD only in adipose tissue, while mice fed a fat free diet express SCD in both liver and adipose tissue. Scd2 was reported expressed in 3T3-L1 preadipocytes and this Scd2 amino acid sequence showed about 87% identity with Scd1 (Kaestner et al., 1989). Scd3 is expressed only in mouse skin, especially in sebaceous glands. Scd3 shares approximately, 89% and 85% of its amino acid sequenced with Scd1 and Scd2, respectively (Zheng et al., 2001). Scd4 encoding 342 amino acids has been shown to share approximately 70% sequence identity with Scd1, Scd2, and Scd3 amino acids and is expressed exclusively in murine heart (Miyazaki et al., 2003).

Several SCDs have been reported to be associated with human diseases, such as obesity, atherosclerosis, and skin diseases (Flowers and Ntambi, 2008; Sampath and Ntambi, 2014). *Scd1* null mice maintained on a low fat diet for 10 days developed severe loss of body weight, hypoglycemia, hypercholesterolemia, and a cholestasis-like phenotype involving endoplasmic reticulum stress and altered regulation of several transcription factors involved in lipid homeostasis in liver (Flowers and Ntambi, 2008; Ntambi et al., 2002). Scd2 null mice show increased skin permeability, which was associated with repartitioning of linoleic acid from epidermal acylceramide species into phospholipids. Lipid synthesis was changed in skin and liver development of *Scd2* knockout mice. *Scd2* was associated with the synthesis of MUFA in neonates, which is different from *Scd1* roles in adult mice (Miyazaki et al., 2005). Scd1 and 2 show specific temporal changes in expression during development. Scd1 is induced after weaning, while Scd2 is highly expressed in liver between embryonic day 18.5 and 21 days of age. Neonatal Scd2 null mice exhibited lower liver and plasma triglyceride concentrations than wild-type mice, while there was no significant difference between knockout and wild-type mice in triglyceride concentration in adults (Miyazaki et al., 2005). The primary activity of *Scd3* is to convert palmitic acid to PA rather than synthesis of OA from stearic acid, which suggests that *Scd3* may be regarded as a palmitoyl-CoA desaturase in contrast to other members of the Scd family (Miyazaki et al., 2006).

Individual Scd isoforms exhibit differential expression between tissues such that the highest expression of *Scd1* is in adipose tissue (Guillou et al., 2010) and liver, although the level of expression is dependent on dietary fat intake (Ntambi et al., 1988). *Scd2* expression is high in brain and neuronal tissues in mice (Kaestner et al., 1989) and it has been reported that Scd5 might be involved in growth and differentiation of neuronal cells in humans (Sinner et al., 2012). *Scd3* was expressed in sebocytes, and in harderian and preputial glands, whereas *Scd4* is specifically expressed in the heart (Zheng et al., 2001).

## Desaturases Acting on Polyunsaturated Fatty Acids

### Δ12 and Δ15 Desaturases

Plants are able to synthesise PUFA from carbohydrate precursors via stearic acid and its subsequent conversion by the action of SCD, which inserts a double bond at the Δ9 position to form OA (see later). Δ12 and Δ15 desaturases serve to add a double bond between methyl end and the preexisting double bond between carbons 9 and 10 to form linoleic and alpha-linolenic acid, respectively (Nakamura and Nara, 2004). Prokaryotic Δ12 and/or Δ15 desaturases were first cloned from *Synechocystis* sp., a cyanobacterium (Sakamoto et al., 1994) and the eukyaryotic Δ12 and Δ15 desaturases from *Saccharomyces kluyveri* (Oura and Kajiwara, 2004). Phylogenic studies suggested that Δ12 desaturase is an ancestor gene of Δ15 desaturase (Wang et al., 2013).

These enzymes are not found in mammals and hence their products are essential in the mammalian diet. However, cockroaches *Periplaneta americana* and the House cricket *Acheta domesticus* have been shown to express Δ12 and/or Δ15 desaturases (Borgeson et al., 1990; Cripps et al., 1990). The nematode worm *Caenorhabditis elegans*, also expresses two genes that encode enzymes with Δ12 desaturase (*fat-2*) activity, which acts on 6- and 18-carbon fatty acids to produce n-6 PUFA, and Δ15 desaturase activity (*fat-1*), which acts on 18- and 20-carbon fatty acids to produce n-3 PUFA (Zhou et al., 2011). Figure 5.2 shows the integration of MUFA and PUFA synthesis. The preferred chain length of *Fat* genes is similar to other eukaryotes' Δ12 and/or Δ15 desaturases (Meesapyodsuk et al., 2000). The structures of *fat-1* and *fat-2* have been shown to differ from plant Δ12 and Δ15 desaturases in that they contain three histidine motifs and without an N-terminal cytochrome b$_5$ domain (Wang et al., 2013).

## Mammalian Polyunsaturated Fatty Acid Desaturases

Mouse chromosome 19 and human chromosome 11 contain three fatty acid desaturase genes: *FADS1*, *FADS2*, and *FADS3* (Marquardt et al., 2000). *FADS1* and *FADS2* encode Δ5 and Δ6 desaturases, respectively, while the role of the *FADS3* product in long-chain polyunsaturated fatty acid (LCPUFA) synthesis is unclear. These *FADS* genes in humans are located on a different chromosome to *SCD1*, whereas in mice the *fads* genes are located on the same chromosome as *Scd1*. In humans, *FADS* are located at 11q12.2–11q13.1 and consist of 12 exons and 11 introns, however, all the three enzymes have common features of N-terminal cytochrome b5 domain and three histidine boxes (Marquardt et al., 2000). In contrast SCD does not have a cytochrome b5 domain. Expression of fatty acid desaturase genes alternative splicing. For example, a splice variant of FADS2 (FADS2 AT1), and seven alternative transcripts of FADS3 (FADS3 AT1-AT7) have been identified in various baboon tissues and in human neuroblastoma, SK-N-SH cells (Park et al., 2009b, 2010). Similar to *SCD* genes, the putative FADS2 AT1 contains only three histidine boxes without cytochrome b5 domain, suggesting that the role of FADS2 AT1 could be related to the desaturation of non-methylene interrupted PUFA (Park et al., 2010). However, alternative transcripts of FADS3 display more structural diversity

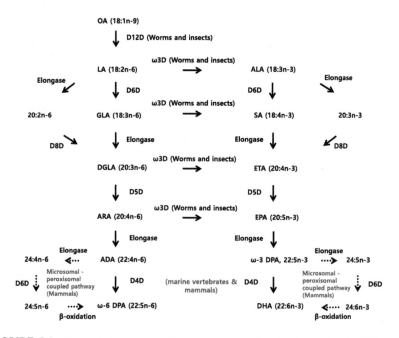

**FIGURE 5.2** **Integrated summary of monounsaturated and polyunsaturated fatty acid synthesis.** In plants and animals stearic acid (SA, 18:0) synthesized from carbohydrate metabolites is converted to oleic acid (OA, 18:1n-9) by stearoyl-CoA desaturase (Scd) activity. In plants, *Caenorhabditis elegans* and some insects, OA is converted sequentially to linoleic acid (LA, 18:2n-6) and then to alpha-linolenic acid (ALA, 18:3n-3) by Δ12 and Δ15 desaturase activities, respectively. 18:2n-6 and 18:3n-3 are the primary PUFA synthesized by plants, although some species covert 18:3n-3 to stearidonic acid (18:4n-3). Conversion of 18:2n-6 and 18:3n-3 to longer chain, more unsaturated PUFA in vertebrates involves sequential desaturation and carbon chain elongation reactions, which are illustrated briefly here. (These reactions are described in more detail in Chapter 2. The consensus view is that synthesis of the longest chain, most unsaturated PUFA involves translocation of 24 carbon internediates between the endoplamic reticulum and peroxisomes, and limited beta oxidation. There is some evidence that direct comversion of 22:4n-6 to 22:5n-6 and 22:5n-3 to 22:6n-6 can also occur, thus bypassing the synthesis of 24 carbon PUFA and beta oxidation. Furthermore, conversion of ALA to eicosatrienoic acid (ETA, 20:4n-3) and LA to DGLA (20:3n-3) has been reported in some cancer cells. DPA(22:5n-6) and DHA(22:6n-3) had been known to be produced via microsomal-peroxisomal coupled pathway, however, recent evidence showed that these pathway are synthesized by D4D.

such that FADS3 AT1 and AT3 include a conserved *N*-terminal cytochrome and three histidine domains, whereas FADS3 AT5 uniquely retains intron 5 (Park et al., 2009b). FADS3 transcript variants are expressed in 12 baboon neonate tissues including liver, kidney, retina, occipital lobes, hippocampus, lung, thymus, spleen, pancreas, skeletal muscle, ovary, and heart and in a tissue-dependent manner. These variants are expressed in human SK-N-SH neuroblastoma cells and their expression showed reciprocal increases and decreases according to the differentiation of those cells

(Brenna et al., 2010; Park et al., 2009b). Three isoforms of FADS3 with 37kD, 51kD, and 75kD were also found in rodent tissues, including lung, heart, kidney, liver, and skeletal muscle. These FADS3 isoforms were differently expressed depending on the tissues and species suggesting that they might have specific functions (Pedrono et al., 2010). However, the function of FADS3 in omega-3 and -6 fatty acid synthesis has not been reported (Marquardt et al., 2000), although FADS3 may catalyze Δ13-desaturation of trans-vaccenate (Rioux et al., 2013). In addition to FADS2 and FADS3, an alternative transcript of FADS1 was identified by using 5'- and 3'-RACE analysis. In silico analysis of the 5'/3' RACE sequences showed the FADS1 gene contains at least four 5' untranslated regions (UTRs) and several 3'UTR. Open reading frame (ORF) finder found FADS1 isoform (FADS1 AT1) containing 360 amino acids and lacking the N-terminal cytochrome b5 domain, in addition to the major isoform of FADS1 consisting of 444 amino acids (Park et al., 2012). Moreover, it was reported that FADS1 AT1 regulates the desaturation of PUFA biosynthesis by potentiating FADS2 Δ6 and Δ8 desaturation activities distinct from the Δ5 desaturase activity of the major FADS1 isoform (Park et al., 2012). This finding was the first report that one gene's function can be modulated by another gene's splice variant in a gene cluster.

## Δ6 Desaturase

Δ6 desaturase catalyzes the addition of a double bond that the Δ6 position (Nakamura and Nara, 2004). This enzyme was cloned originally from Synechocystis (Reddy et al., 1993) and subsequently from Borage officinalis (Sayanova et al., 1997), C. elegans (Napier et al., 1998), mice (Cho et al., 1999b), rats (Aki et al., 1999), and humans (Cho et al., 1999b). The deduced amino acid sequence suggests that Δ6 desaturase includes a N-terminal cytochrome b5 domain, which is different from that of SCD. Δ6 desaturases in animals include two membrane anchored domains and three histidine box motifs which are the unique feature of membrane-bound desaturases.

This enzyme is a rate-limiting desaturase for the synthesis of PUFA (Lee et al., 2016). Δ6 desaturase in zebrafish also exhibits Δ5 desaturase activity (Hastings et al., 2001) (Chapter 3). However, mammalian Δ6 desaturase does not exhibit Δ5 desaturase activity. Δ6 desaturase has been reported to catalyze the conversion of 16:0 to 16:1n-10 and that this activity competes with conversion of α-linolenic acid and linoleic acid to longer chain PUFA. This activity is important for skin of animals, including humans, because 16:1n-10 accounts for 25% of skin total fatty acids (Guillou et al., 2003; Park et al., 2016; Picardo et al., 2009). In addition, Δ6 desaturase of primates catalyzes the Δ8 desaturation of eicosadienoic acid (20:2n-6) and eicosatrienoic acid (ETE, 20:3n-3) to dihomogammalinolenic acid (DGLA, 20:3n-6) and ETA (20:4n-3), respectively in Saccharomyces cerevisiae transformed cells as well as MCF-7 breast cancer cells (Park et al., 2009a; Park et al., 2012). *Fads2* deficient mice showed altered reproduction, skin, and intestine, which are characteristic of essential fatty acid deficiency. However, supplementation of *Fads2* null mice with arachidonic acid (ARA) showed partial amelioration of these changes. In addition, *Fads2* transgenic mice also showed tissue-dependent variation in LCPUFA contents. The proportions of docosahexaenoic acid (DHA) and ARA were lower while oleic acid and linoleic were increased in *Fads2* deficient mice (Stoffel et al., 2008; Stroud et al., 2009).

The product of the FADS2 gene also possesses $\Delta 4$ desaturase activity in MCF-7 cells, which catalyzed the conversion of 22:4n-6 to DPAn-6, and 22:5n-3 to DHA instead of $\Delta 6$ desaturation (Park et al., 2015). This is the first report using molecular biological evidence to show the specific gene containing $\Delta 4$ desaturation acitivity in mammals, suggesting that DPAn-6 and DHA are synthesized by $\Delta 4$ desaturase and more studies associated with $\Delta 4$ desaturation in mammals are needed.

### $\Delta 5$ Desaturase

$\Delta 5$ desaturase catalyzes the addition of the 5th carbon–carbon bond from the carboxylic end of PUFA and is encoded by the *FADS1* gene. Because the predicted structure of $\Delta 5$ desaturase is similar to $\Delta 6$ desaturase and that the FADSs1 and FAD2 genes lie closely together in mammals, although in opposite orientations, it has been suggested that these gene arose by duplication (Lattka et al., 2012). Since fish express a FADS2 product with both $\Delta 6$ and $\Delta 5$ desaturase activities (Lattka et al., 2012), it is possible that the FADS2 gene arose first.

$\Delta 5$ desaturases catalyze the conversion of DGLA (20:3n-6) and ETA (eicosatetraenoic acid, 20:4n-3) to ARA (20:4n-6) and EPA (20:5n-3), respectively (see Chapter 2). $\Delta 5$ desaturase was first found in a fungus, *Mortierella alpine* (Knutzon et al., 1998; Michaelson et al., 1998), and it has since been reported in a wide range of organisms (Castro et al., 2012; Cho et al., 1999a). $\Delta 5$ desaturase has been shown to catalyze the synthesis of 5,11,14–20:3 and 5,11,14,17–20:4 in MCF-7 breast cancer cells that lack $\Delta 6$ activity. Since these fatty acid lack the 8–9 double bond present in ARA and EPA, these novel PUFA may potentially act as substrates for unusual lipid mediators, which could contribute to the tumorigenic process (Park et al., 2011). Deletion of the *Fads1* gene in mice prevented synthesis of ARA (Fan et al., 2012). Furthermore, *Fads1* knockout mice died within 12 weeks of birth, although ARA supplementation extended their life span. The reason for this dramatic reduction in life span is not known, but it may reflect the inability to synthesize the normal range of eicosanoids required for physiological processes essential to life, including immunity and growth. If so, this suggests that ARA synthesized endogenously may be at least as important as that consumed preformed from the diet as a substrate for eicosanoid synthesis.

# BIOTECHNOLOGY AND DESATURASES IN ANIMALS

Mammals do not express $\Delta 12$ and $\Delta 15$ desaturases and hence it is not possible to alter the ratio of omega-3 and omega-6 PUFA by changing the actives of these enzymes. However, by introducing *fat-1* from *C. elegans* into mice enabled these animals to maintain a higher proportion of $\omega$-3 fatty acids [ALA, EPA, $\omega$ -3 DPA (22:5n-3), and DHA] with lower proportions of the $\omega$-6 fatty acids LA, ARA, and $\omega$-6 DPA (22:5n-6) (Kang et al., 2004). Transgenic pigs containing *hfat-1* gene, a humanized *fat-1*, were also engineered to yield high levels of $\omega$-3 fatty acids, such as ALA, EPA, $\omega$ -3 DPA (22:5n-3), and DHA, increasing from approximately 2% to approximately 8% (Lai et al., 2006), suggesting that those animals may be nutritionally beneficial to produce pork rich in $\omega$-3 fatty acids.

Moreover, *fat-1* gene transfer inhibited neuronal apoptosis in rat cortical neurons (Ge et al., 2002). The incidence of colitis-associated colon cancer and prostate cancer was also found to be lower in transgenic *fat-1* mice compared to wild type animals. Together these findings suggest that the ratio of omega-3 and omega-6 fatty acids is an important factor in neuronal function and the development of some cancers (Jia et al., 2008; Lu et al., 2008). Therefore, one possible application is gene therapy using desaturases in the treatment of neurodegenerative diseases and some cancers.

## References

Aki, T., Shimada, Y., Inagaki, K., Higashimoto, H., Kawamoto, S., Shigeta, S., Ono, K., Suzuki, O., 1999. Molecular cloning and functional characterization of rat $\Delta 6$ fatty acid desaturase. Biochem. Biophys. Res. Commun. 255, 575–579.

Borgeson, C.E., de Renobales, M., Blomquist, G.J., 1990. Characterization of the delta 12 desaturase in the American cockroach, *Periplaneta americana*: the nature of the substrate. Biochim. Biophys. Acta 1047, 135–140.

Brenna, J.T., Kothapalli, K.S., Park, W.J., 2010. Alternative transcripts of fatty acid desaturase (FADS) genes. Prostaglandins Leukot. Essent. Fatty Acids 82, 281–285.

Calder, P.C., 2013. Omega-3 polyunsaturated fatty acids and inflammatory processes: nutrition or pharmacology? Br. J. Clin. Pharmacol. 75, 645–662.

Castro, L.F., Monroig, O., Leaver, M.J., Wilson, J., Cunha, I., Tocher, D.R., 2012. Functional desaturase Fads1 (Delta5) and Fads2 (Delta6) orthologues evolved before the origin of jawed vertebrates. PLoS One 7, e31950.

Cho, H.P., Nakamura, M., Clarke, S.D., 1999a. Cloning, expression, and fatty acid regulation of the human delta-5 desaturase. J. Biol. Chem. 274, 37335–37339.

Cho, H.P., Nakamura, M.T., Clarke, S.D., 1999b. Cloning, expression, and nutritional regulation of the mammalian delta-6 desaturase. J. Biol. Chem. 274, 471–477.

Cripps, C., Borgeson, C., Blomquist, G.J., de Renobales, M., 1990. The delta 12-desaturase from the house cricket, *Acheta domesticus* (Orthoptera: Gryllidae): characterization and form of the substrate. Arch. Biochem. Biophys. 278, 46–51.

Eigenheer, A.L., Young, S., Blomquist, G.J., Borgeson, C.E., Tillman, J.A., Tittiger, C., 2002. Isolation and molecular characterization of *Musca domestica* delta-9 desaturase sequences. Insect. Mol. Biol. 11, 533–542.

Fan, Y.Y., Monk, J.M., Hou, T.Y., Callway, E., Vincent, L., Weeks, B., Yang, P., Chapkin, R.S., 2012. Characterization of an arachidonic acid-deficient (Fads1 knockout) mouse model. J. Lipid Res. 53, 1287–1295.

Fasano, E., Serini, S., Cittadini, A., Calviello, G., 2015. Long-chain n-3 PUFA against breast and prostate cancer: which are the appropriate doses for intervention studies in animals and humans? Crit. Rev. Food Sci. Nutr. 57 (11), 0.

Flowers, M.T., Ntambi, J.M., 2008. Role of stearoyl-coenzyme A desaturase in regulating lipid metabolism. Curr. Opin. Lipidol. 19, 248–256.

Ge, Y., Wang, X., Chen, Z., Landman, N., Lo, E.H., Kang, J.X., 2002. Gene transfer of the *Caenorhabditis elegans* n-3 fatty acid desaturase inhibits neuronal apoptosis. J. Neurochem. 82, 1360–1366.

Guillou, H., Rioux, V., Catheline, D., Thibault, J.N., Bouriel, M., Jan, S., D'Andrea, S., Legrand, P., 2003. Conversion of hexadecanoic acid to hexadecenoic acid by rat delta 6-desaturase. J. Lipid Res. 44, 450–454.

Guillou, H., Zadravec, D., Martin, P.G., Jacobsson, A., 2010. The key roles of elongases and desaturases in mammalian fatty acid metabolism: insights from transgenic mice. Prog. Lipid Res. 49, 186–199.

Hastings, N., Agaba, M., Tocher, D.R., Leaver, M.J., Dick, J.R., Sargent, J.R., Teale, A.J., 2001. A vertebrate fatty acid desaturase with delta 5 and delta 6 activities. Proc. Natl. Acad. Sci. U.S.A. 98, 14304–14309.

Heinemann, F.S., Ozols, J., 2003. Stearoyl-CoA desaturase, a short-lived protein of endoplasmic reticulum with multiple control mechanisms. Prostaglandins Leukot. Essent. Fatty Acids 68, 123–133.

Jia, Q., Lupton, J.R., Smith, R., Weeks, B.R., Callaway, E., Davidson, L.A., Kim, W., Fan, Y.Y., Yang, P., Newman, R.A., Kang, J.X., McMurray, D.N., Chapkin, R.S., 2008. Reduced colitis-associated colon cancer in Fat-1 (n-3 fatty acid desaturase) transgenic mice. Cancer Res. 68, 3985–3991.

Kaestner, K.H., Ntambi, J.M., Kelly, Jr., T.J., Lane, M.D., 1989. Differentiation-induced gene expression in 3T3-L1 preadipocytes: a second differentially expressed gene encoding stearoyl-CoA desaturase. J. Biol. Chem. 264, 14755–14761.

Kang, J.X., Wang, J., Wu, L., Kang, Z.B., 2004. Transgenic mice: fat-1 mice convert n-6 to n-3 fatty acids. Nature 427, 504.

Knutzon, D.S., Thurmond, J.M., Huang, Y.S., Chaudhary, S., Bobik, Jr., E.G., Chan, G.M., Kirchner, S.J., Mukerji, P., 1998. Identification of delta5-desaturase from *Mortierella alpina* by heterologous expression in Bakers' yeast and canola. J. Biol. Chem. 273, 29360–29366.

Lai, L., Kang, J.X., Li, R., Wang, J., Witt, W.T., Yong, H.Y., Hao, Y., Wax, D.M., Murphy, C.N., Rieke, A., Samuel, M., Linville, M.L., Korte, S.W., Evans, R.W., Starzl, T.E., Prather, R.S., Dai, Y., 2006. Generation of cloned transgenic pigs rich in omega-3 fatty acids. Nat. Biotechnol. 24, 435–436.

Lee, J.M., Lee, H., Kang, S., Park, W.J., 2016. Fatty acid desaturases, polyunsaturated fatty acid regulation, and biotechnological advances. Nutrients 8 (1), 8.

Lu, Y., Nie, D., Witt, W.T., Chen, Q., Shen, M., Xie, H., Lai, L., Dai, Y., Zhang, J., 2008. Expression of the fat-1 gene diminishes prostate cancer growth in vivo through enhancing apoptosis and inhibiting GSK-3 beta phosphorylation. Mol. Cancer Ther. 7, 3203–3211.

Lattka, E., Klopp, N., Demmelmair, H., Klingler, M., Heinrich, J., Koletzko, B., 2012. Genetic variations in polyunsaturated fatty acid metabolism: implications for child health? Ann. Nutr. Metab. 60 (Suppl. 3), 8–17.

Marquardt, A., Stohr, H., White, K., Weber, B.H., 2000. cDNA cloning, genomic structure, and chromosomal localization of three members of the human fatty acid desaturase family. Genomics 66, 175–183.

Meesapyodsuk, D., Reed, D.W., Savile, C.K., Buist, P.H., Ambrose, S.J., Covello, P.S., 2000. Characterization of the regiochemistry and cryptoregiochemistry of a *Caenorhabditis elegans* fatty acid desaturase (FAT-1) expressed in *Saccharomyces cerevisiae*. Biochemistry 39, 11948–11954.

Michaelson, L.V., Lazarus, C.M., Griffiths, G., Napier, J.A., Stobart, A.K., 1998. Isolation of a delta5-fatty acid desaturase gene from *Mortierella alpina*. J. Biol. Chem. 273, 19055–19059.

Miyazaki, M., Bruggink, S.M., Ntambi, J.M., 2006. Identification of mouse palmitoyl-coenzyme A delta9-desaturase. J. Lipid Res. 47, 700–704.

Miyazaki, M., Dobrzyn, A., Elias, P.M., Ntambi, J.M., 2005. Stearoyl-CoA desaturase-2 gene expression is required for lipid synthesis during early skin and liver development. Proc. Natl. Acad. Sci. U.S.A. 102, 12501–12506.

Miyazaki, M., Jacobson, M.J., Man, W.C., Cohen, P., Asilmaz, E., Friedman, J.M., Ntambi, J.M., 2003. Identification and characterization of murine SCD4, a novel heart-specific stearoyl-CoA desaturase isoform regulated by leptin and dietary factors. J. Biol. Chem. 278, 33904–33911.

Nakamura, M.T., Nara, T.Y., 2004. Structure, function, and dietary regulation of delta6, delta5, and delta9 desaturases. Annu. Rev. Nutr. 24, 345–376.

Napier, J.A., Hey, S.J., Lacey, D.J., Shewry, P.R., 1998. Identification of a *Caenorhabditis elegans* delta6-fatty-acid-desaturase by heterologous expression in *Saccharomyces cerevisiae*. Biochem. J. 330 (Pt 2), 611–614.

Ntambi, J.M., 1995. The regulation of stearoyl-CoA desaturase (SCD). Prog. Lipid Res. 34, 139–150.

Ntambi, J.M., 1999. Regulation of stearoyl-CoA desaturase by polyunsaturated fatty acids and cholesterol. J. Lipid Res. 40, 1549–1558.

Ntambi, J.M., Buhrow, S.A., Kaestner, K.H., Christy, R.J., Sibley, E., Kelly, Jr., T.J., Lane, M.D., 1988. Differentiation-induced gene expression in 3T3-L1 preadipocytes: characterization of a differentially expressed gene encoding stearoyl-CoA desaturase. J. Biol. Chem. 263, 17291–17300.

Ntambi, J.M., Miyazaki, M., Stoehr, J.P., Lan, H., Kendziorski, C.M., Yandell, B.S., Song, Y., Cohen, P., Friedman, J.M., Attie, A.D., 2002. Loss of stearoyl-CoA desaturase-1 function protects mice against adiposity. Proc. Natl. Acad. Sci. U.S.A. 99, 11482–11486.

Oura, T., Kajiwara, S., 2004. *Saccharomyces kluyveri* FAD3 encodes an omega3 fatty acid desaturase. Microbiology 150, 1983–1990.

Park, H.G., Kothapalli, K.S., Park, W.J., DeAllie, C., Liu, L., Liang, A., Lawrence, P., Brenna, J.T., 2016. Palmitic acid (16:0) competes with omega-6 linoleic and omega-3 a-linolenic acids for FADS2 mediated delta6-desaturation. Biochim. Biophys. Acta 1861, 91–97.

Park, H.G., Park, W.J., Kothapalli, K.S., Brenna, J.T., 2015. The fatty acid desaturase 2 (FADS2) gene product catalyzes delta4 desaturation to yield n-3 docosahexaenoic acid and n-6 docosapentaenoic acid in human cells. FASEB J. 29, 3911–3919.

Park, W.J., Kothapalli, K.S., Lawrence, P., Brenna, J.T., 2011. FADS2 function loss at the cancer hotspot 11q13 locus diverts lipid signaling precursor synthesis to unusual eicosanoid fatty acids. PLoS One 6, e28186.

Park, W.J., Kothapalli, K.S., Lawrence, P., Tyburczy, C., Brenna, J.T., 2009a. An alternate pathway to long-chain polyunsaturates: the FADS2 gene product delta8-desaturates 20:2n-6 and 20:3n-3. J. Lipid Res. 50, 1195–1202.

Park, W.J., Kothapalli, K.S., Reardon, H.T., Kim, L.Y., Brenna, J.T., 2009b. Novel fatty acid desaturase 3 (FADS3) transcripts generated by alternative splicing. Gene 446, 28–34.

Park, W.J., Kothapalli, K.S., Reardon, H.T., Lawrence, P., Qian, S.B., Brenna, J.T., 2012. A novel FADS1 isoform potentiates FADS2-mediated production of eicosanoid precursor fatty acids. J. Lipid Res. 53, 1502–1512.

Park, W.J., Reardon, H.T., Tyburczy, C., Kothapalli, K.S., Brenna, J.T., 2010. Alternative splicing generates a novel FADS2 alternative transcript in baboons. Mol. Biol. Rep. 37, 2403–2406.

Pedrono, F., Blanchard, H., Kloareg, M., D'Andrea, S., Daval, S., Rioux, V., Legrand, P., 2010. The fatty acid desaturase 3 gene encodes for different FADS3 protein isoforms in mammalian tissues. J. Lipid Res. 51, 472–479.

Picardo, M., Ottaviani, M., Camera, E., Mastrofrancesco, A., 2009. Sebaceous gland lipids. Dermatoendocrinol 1, 68–71.

Reddy, A.S., Nuccio, M.L., Gross, L.M., Thomas, T.L., 1993. Isolation of a delta 6-desaturase gene from the cyanobacterium *Synechocystis* sp. strain PCC 6803 by gain-of-function expression in *Anabaena* sp. strain PCC 7120. Plant Mol. Biol. 22, 293–300.

Rioux, V., Pedrono, F., Blanchard, H., Duby, C., Boulier-Monthean, N., Bernard, L., Beauchamp, E., Catheline, D., Legrand, P., 2013. Trans-vaccenate is delta13-desaturated by FADS3 in rodents. J. Lipid Res. 54, 3438–3452.

Roche, H.M., 1999. Unsaturated fatty acids. Proc. Nutr. Soc. 58, 397–401.

Sakamoto, T., Los, D.A., Higashi, S., Wada, H., Nishida, I., Ohmori, M., Murata, N., 1994. Cloning of omega 3 desaturase from cyanobacteria and its use in altering the degree of membrane-lipid unsaturation. Plant Mol. Biol. 26, 249–263.

Sampath, H., Ntambi, J.M., 2014. Role of stearoyl-CoA desaturase-1 in skin integrity and whole body energy balance. J. Biol. Chem. 289, 2482–2488.

Sayanova, O., Smith, M.A., Lapinskas, P., Stobart, A.K., Dobson, G., Christie, W.W., Shewry, P.R., Napier, J.A., 1997. Expression of a borage desaturase cDNA containing an *N*-terminal cytochrome b5 domain results in the accumulation of high levels of delta6-desaturated fatty acids in transgenic tobacco. Proc. Natl. Acad. Sci. U.S.A. 94, 4211–4216.

Sinner, D.I., Kim, G.J., Henderson, G.C., Igal, R.A., 2012. StearoylCoA desaturase-5: a novel regulator of neuronal cell proliferation and differentiation. PLoS One 7, e39787.

Stoffel, W., Holz, B., Jenke, B., Binczek, E., Gunter, R.H., Kiss, C., Karakesisoglou, I., Thevis, M., Weber, A.A., Arnhold, S., Addicks, K., 2008. Delta6-desaturase (FADS2) deficiency unveils the role of omega3- and omega6-polyunsaturated fatty acids. EMBO J. 27, 2281–2292.

Strittmatter, P., Spatz, L., Corcoran, D., Rogers, M.J., Setlow, B., Redline, R., 1974. Purification and properties of rat liver microsomal stearyl coenzyme A desaturase. Proc. Natl. Acad. Sci. U.S.A. 71, 4565–4569.

Stroud, C.K., Nara, T.Y., Roqueta-Rivera, M., Radlowski, E.C., Lawrence, P., Zhang, Y., Cho, B.H., Segre, M., Hess, R.A., Brenna, J.T., Haschek, W.M., Nakamura, M.T., 2009. Disruption of FADS2 gene in mice impairs male reproduction and causes dermal and intestinal ulceration. J. Lipid Res. 50, 1870–1880.

Superko, H.R., Superko, A.R., Lundberg, G.P., Margolis, B., Garrett, B.C., Nasir, K., Agatston, A.S., 2014. Omega-3 fatty acid blood levels clinical significance update. Curr. Cardiovasc. Risk Rep. 8, 407.

Venegas-Caleron, M., Sayanova, O., Napier, J.A., 2010. An alternative to fish oils: metabolic engineering of oil-seed crops to produce omega-3 long chain polyunsaturated fatty acids. Prog. Lipid Res. 49, 108–119.

Wang, J., Yu, L., Schmidt, R.E., Su, C., Huang, X., Gould, K., Cao, G., 2005. Characterization of HSCD5, a novel human stearoyl-CoA desaturase unique to primates. Biochem. Biophys. Res. Commun. 332, 735–742.

Wang, M., Chen, H., Gu, Z., Zhang, H., Chen, W., Chen, Y.Q., 2013. omega3 fatty acid desaturases from microorganisms: structure, function, evolution, and biotechnological use. Appl. Microbiol. Biotechnol. 97, 10255–10262.

Watts, J.L., Browse, J., 2000. A palmitoyl-CoA-specific delta9 fatty acid desaturase from *Caenorhabditis elegans*. Biochem. Biophys. Res. Commun. 272, 263–269.

Watts, J.L., Browse, J., 2002. Genetic dissection of polyunsaturated fatty acid synthesis in *Caenorhabditis elegans*. Proc. Natl. Acad. Sci. U.S.A. 99, 5854–5859.

Wolfe, L.S., 1982. Eicosanoids: prostaglandins, thromboxanes, leukotrienes, and other derivatives of carbon-20 unsaturated fatty acids. J. Neurochem. 38, 1–14.

Zhang, L., Ge, L., Parimoo, S., Stenn, K., Prouty, S.M., 1999. Human stearoyl-CoA desaturase: alternative transcripts generated from a single gene by usage of tandem polyadenylation sites. Biochem. J. 340 (Pt 1), 255–264.

Zheng, J.S., Huang, T., Yang, J., Fu, Y.Q., Li, D., 2012. Marine N-3 polyunsaturated fatty acids are inversely associated with risk of type 2 diabetes in Asians: a systematic review and meta-analysis. PLoS One 7, e44525.

Zheng, Y., Prouty, S.M., Harmon, A., Sundberg, J.P., Stenn, K.S., Parimoo, S., 2001. Scd3-a novel gene of the stearoyl-CoA desaturase family with restricted expression in skin. Genomics 71, 182–191.

Zhou, X.R., Green, A.G., Singh, S.P., 2011. *Caenorhabditis elegans* delta12-desaturase FAT-2 is a bifunctional desaturase able to desaturate a diverse range of fatty acid substrates at the delta12 and delta15 positions. J. Biol. Chem. 286, 43644–43650.

# Further Readings

Berg, M., Soreide, K., 2011. Prevention: will an aspirin a day keep the colorectal cancer away? Nat. Rev. Clin. Oncol. 8, 130–131.

Brookes, K.J., Chen, W., Xu, X., Taylor, E., Asherson, P., 2006. Association of fatty acid desaturase genes with attention-deficit/hyperactivity disorder. Biol. Psychiatry 60, 1053–1061.

Caspi, A., Williams, B., Kim-Cohen, J., Craig, I.W., Milne, B.J., Poulton, R., Schalkwyk, L.C., Taylor, A., Werts, H., Moffitt, T.E., 2007. Moderation of breastfeeding effects on the IQ by genetic variation in fatty acid metabolism. Proc. Natl. Acad. Sci. U.S.A. 104, 18860–18865.

Cohn, S.M., Schloemann, S., Tessner, T., Seibert, K., Stenson, W.F., 1997. Crypt stem cell survival in the mouse intestinal epithelium is regulated by prostaglandins synthesized through cyclooxygenase-1. J. Clin. Invest. 99, 1367–1379.

Dennis, E.A., Norris, P.C., 2015. Eicosanoid storm in infection and inflammation. Nat. Rev. Immunol. 15, 511–523.

Elbein, S.C., Kern, P.A., Rasouli, N., Yao-Borengasser, A., Sharma, N.K., Das, S.K., 2011. Global gene expression profiles of subcutaneous adipose and muscle from glucose-tolerant, insulin-resistant individuals matched for BMI. Diabetes 60, 1019–1029.

Ford, J.H., 2010. Saturated fatty acid metabolism is key link between cell division, cancer, and senescence in cellular and whole organism aging. Age 32, 231–237.

Ford, J.H., Tavendale, R., 2010. Analysis of fatty acids in early mid-life in fertile women: implications for reproductive decline and other chronic health problems. Am. J. Hum. Biol. 22, 134–136.

Kim, O.Y., Lim, H.H., Yang, L.I., Chae, J.S., Lee, J.H., 2011. Fatty acid desaturase (FADS) gene polymorphisms and insulin resistance in association with serum phospholipid polyunsaturated fatty acid composition in healthy Korean men: cross-sectional study. Nutr. Metab. 8, 24.

Kroger, J., Zietemann, V., Enzenbach, C., Weikert, C., Jansen, E.H., Doring, F., Joost, H.G., Boeing, H., Schulze, M.B., 2011. Erythrocyte membrane phospholipid fatty acids, desaturase activity, and dietary fatty acids in relation to risk of type 2 diabetes in the European Prospective Investigation into Cancer and Nutrition (EPIC)-Potsdam Study. Am. J. Clin. Nutr. 93, 127–142.

Li, Y., Monroig, O., Zhang, L., Wang, S., Zheng, X., Dick, J.R., You, C., Tocher, D.R., 2010. Vertebrate fatty acyl desaturase with delta4 activity. Proc. Natl. Acad. Sci. U.S.A. 107, 16840–16845.

Liew, C.F., Groves, C.J., Wiltshire, S., Zeggini, E., Frayling, T.M., Owen, K.R., Walker, M., Hitman, G.A., Levy, J.C., O'Rahilly, S., Hattersley, A.T., Johnston, D.G., McCarthy, M.I., 2004. Analysis of the contribution to type 2 diabetes susceptibility of sequence variation in the gene encoding stearoyl-CoA desaturase, a key regulator of lipid and carbohydrate metabolism. Diabetologia 47, 2168–2175.

Ma, X.H., Hu, S.J., Ni, H., Zhao, Y.C., Tian, Z., Liu, J.L., Ren, G., Liang, X.H., Yu, H., Wan, P., Yang, Z.M., 2006. Serial analysis of gene expression in mouse uterus at the implantation site. J. Biol. Chem. 281, 9351–9360.

Malerba, G., Schaeffer, L., Xumerle, L., Klopp, N., Trabetti, E., Biscuola, M., Cavallari, U., Galavotti, R., Martinelli, N., Guarini, P., Girelli, D., Olivieri, O., Corrocher, R., Heinrich, J., Pignatti, P.F., Illig, T., 2008. SNPs of the FADS gene cluster are associated with polyunsaturated fatty acids in a cohort of patients with cardiovascular disease. Lipids 43, 289–299.

Martinelli, N., Girelli, D., Malerba, G., Guarini, P., Illig, T., Trabetti, E., Sandri, M., Friso, S., Pizzolo, F., Schaeffer, L., Heinrich, J., Pignatti, P.F., Corrocher, R., Olivieri, O., 2008. FADS genotypes and desaturase activity estimated by the ratio of arachidonic acid to linoleic acid are associated with inflammation and coronary artery disease. Am. J. Clin. Nutr. 88, 941–949.

Peyou-Ndi, M.M., Watts, J.L., Browse, J., 2000. Identification and characterization of an animal delta(12) fatty acid desaturase gene by heterologous expression in Saccharomyces cerevisiae. Arch. Biochem. Biophys. 376, 399–408.

Qiu, X., Hong, H., MacKenzie, S.L., 2001. Identification of a Delta 4 fatty acid desaturase from Thraustochytrium sp. involved in the biosynthesis of docosahexanoic acid by heterologous expression in Saccharomyces cerevisiae and Brassica juncea. J. Biol. Chem. 276, 31561–31566.

Reardon, H.T., Zhang, J., Kothapalli, K.S., Kim, A.J., Park, W.J., Brenna, J.T., 2012. Insertion-deletions in a FADS2 intron 1 conserved regulatory locus control expression of fatty acid desaturases 1 and 2 and modulate response to simvastatin. Prostaglandins Leukot. Essent. Fatty Acids 87, 25–33.

Sergeant, S., Hugenschmidt, C.E., Rudock, M.E., Ziegler, J.T., Ivester, P., Ainsworth, H.C., Vaidya, D., Case, L.D., Langefeld, C.D., Freedman, B.I., Bowden, D.W., Mathias, R.A., Chilton, F.H., 2012. Differences in arachidonic acid levels and fatty acid desaturase (FADS) gene variants in African Americans and European Americans with diabetes or the metabolic syndrome. Br. J. Nutr. 107, 547–555.

Serhan, C.N., Chiang, N., Van Dyke, T.E., 2008. Resolving inflammation: dual anti-inflammatory and pro-resolution lipid mediators. Nat. Rev. Immunol. 8, 349–361.

Simopoulos, A.P., 1999. Essential fatty acids in health and chronic disease. Am. J. Clin. Nutr. 70, 560S–569S.

Simopoulos, A.P., 2008. The importance of the omega-6/omega-3 fatty acid ratio in cardiovascular disease and other chronic diseases. Exp. Biol. Med. 233, 674–688.

Spychalla, J.P., Kinney, A.J., Browse, J., 1997. Identification of an animal omega-3 fatty acid desaturase by heterologous expression in Arabidopsis. Proc. Natl. Acad. Sci. U.S.A. 94, 1142–1147.

Tanaka, T., Shen, J., Abecasis, G.R., Kisialiou, A., Ordovas, J.M., Guralnik, J.M., Singleton, A., Bandinelli, S., Cherubini, A., Arnett, D., Tsai, M.Y., Ferrucci, L., 2009. Genome-wide association study of plasma polyunsaturated fatty acids in the InCHIANTI Study. PLoS Genet. 5, e1000338.

Tonon, T., Harvey, D., Larson, T.R., Graham, I.A., 2003. Identification of a very long chain polyunsaturated fatty acid delta4-desaturase from the microalga Pavlova lutheri. FEBS Lett. 553, 440–444.

Voss, A., Reinhart, M., Sankarappa, S., Sprecher, H., 1991. The metabolism of 7,10,13,16,19-docosapentaenoic acid to 4,7,10,13,16,19-docosahexaenoic acid in rat liver is independent of a 4-desaturase. J. Biol. Chem. 266, 19995–20000.

Wang, Y.D., Peng, K.C., Wu, J.L., Chen, J.Y., 2014. Transgenic expression of salmon delta-5 and delta-6 desaturase in zebrafish muscle inhibits the growth of Vibrio alginolyticus and affects fish immunomodulatory activity. Fish Shellfish Immunol. 39, 223–230.

Warensjo, E., Ingelsson, E., Lundmark, P., Lannfelt, L., Syvanen, A.C., Vessby, B., Riserus, U., 2007. Polymorphisms in the SCD1 gene: associations with body fat distribution and insulin sensitivity. Obesity 15, 1732–1740.

Xie, L., Innis, S.M., 2008. Genetic variants of the FADS1 FADS2 gene cluster are associated with altered (n-6) and (n-3) essential fatty acids in plasma and erythrocyte phospholipids in women during pregnancy and in breast milk during lactation. J. Nutr. 138, 2222–2228.

# 6

# Biochemistry and Regulation of Elongases 2 and 5 in Mammals

*Graham C. Burdge*

**University of Southampton, Southampton, United Kingdom**

## INTRODUCTION

Mammals are able to synthesise *de novo* fatty acids up to C16 from malonyl-CoA in a cycle of four reactions that extend the length of the carbon chain by two carbons. These reactions are catalyzed by the cytoplasmaic homodimeric multifunctional enzyme fatty acid synthase. Fatty acid synthesis *de novo* starts with condensation of acetyl-CoA and malonyl-CoA form a four carbon beta-ketoacyl-CoA, which is then converted by beta-ketoreductase activity to a beta-hydroxyacyl-CoA, dehydrated to form an enoyl-CoA and finally reduced to a fatty acid CoA by enoyl reductase activity. In the next cycle, butyryl-CoA, instead of acetyl-CoA, is condensed with malonyl-CoA. However, mammals can synthesize 18 carbon saturated and monounsaturated fatty acids *de novo,* and convert 18 carbon essential polyunsaturated fatty acids (PUFA) to longer chain, more unsaturated species. Synthesis of fatty acids longer than 16 carbons requires the activity of fatty acid elongase complexes.

Polyunsaturated Fatty Acid Metabolism. http://dx.doi.org/10.1016/B978-0-12-811230-4.00006-5

# FATTY ACID ELONGATION

Fatty acid elongase complexes are distinct in terms of location and structure from fatty acid synthase. Fatty acid elongase complexes are composed of single enzymes (Bernert and Sprecher, 1979) rather than a multifunctional enzyme, which is located in the cytoplasmic face of the endoplasmic reticulum with the hydroxyacyl-CoA dehydratase imbedded within the membrane (Osei et al., 1989) rather than being a cytosolic enzyme. However, the individual enzymes within the fatty acid elongase complex may be associated physically (Kohlwein et al., 2001) and catalyze extension of long chain fatty acids by the same series of reactions as the fatty acid synthase complex (Fig. 6.1).

# THE FUNCTIONS OF THE CONDENSING ENZYME

The first reaction that is catalyzed by the condensing enzyme is the rate-limiting step in the pathway (Bernert and Sprecher, 1977) and also confers substrate specificity. In humans and mice the condensation reaction

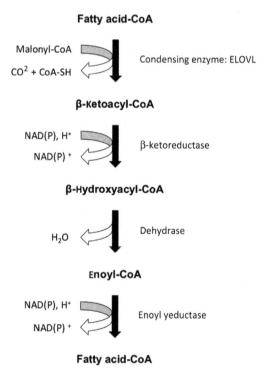

FIGURE 6.1  The long-chain fatty acid elongation pathway in mammals. The first reaction is of 2C from malonyl-CoA by a condensation reaction catalyzed by an elongase (ELOVL) to form β-ketoacyl-CoA. This is converted by β-ketoreductase activity to a β-hydroxyacyl-CoA, dehydrated to form an enoyl-CoA and finally reduced to a fatty acid-CoA by enoyl reductase.

is catalyzed by a family of enzymes known as ELOngation of Very Long chain fatty acids (ELOVL) (Denic and Weissman, 2007; Moon et al., 2001), which have distinct tissue distributions (Table 6.1) (Matsuzaka et al., 2002; Moon et al., 2001; Tvrdik et al., 1997, 2000; Zhang et al., 2001). ELOVL exhibit up to 30% sequence homology between mammals, yeast and nematode worms (Leonard et al., 2004) and the catalytic domain responsible for formation of 3-keoacyl-CoA is fully conserved between humans, rats and mice (Denic and Weissman, 2007).

ELOVL proteins are characterized by 5–7 transmembrane domains (Monne et al., 1999; Tvrdik et al., 2000) and either lysine or arginine at the carboxyl terminus, which may function as endoplasmic reticulum retrieval signals (Beaudoin et al., 2000; Michelsen et al., 2005; Tvrdik et al., 2000; Zhang et al., 2001). Elongases exhibit distinct substrate specificities (Table 6.1) of which two, *ELOVL2* and *ELOVL5*, catalyze the conversion of PUFA to their longer chain metabolites. *ELOVL2* and *ELOVL5* share 56% sequence identity, which suggest these enzymes may have arisen by gene duplication (Leonard et al., 2002) in a jawed vertebrate ancestor (Monroig et al., 2016). The main region of heterogeneity between the two genes is located in a sequence between the sixth and seventh membrane-spanning domains (Gregory et al., 2013). However, expression of human *ELOVL2* in yeast showed that it is unable to elongate 18 carbon fatty acids, but preferentially catalyzes the elongation of C20–C22 fatty acids (Leonard et al., 2002) to produce 24:4n-6 and 24:5n-3, which are substrates for the second reaction catalyzed by Δ6 deasturase in the PUFA synthesis pathway (Leonard et al., 2002; Wang et al., 2008). There is some evidence from kinetic studies that *ELOVL2* may regulate flux through the PUFA synthesis pathway downstream of 22:5n-3 and 22:4n-6 (Gregory et al., 2011). In contrast, the rat homologue of *ELOVL5*, *rElO1*,

**TABLE 6.1** Fatty Acid Substrate Specificity of Mammalian Elongases

| Elongase | Fatty acid substrate | References |
|---|---|---|
| ELOVL1 | C18:0–C26:0 and C20:1n—9 and C22:1n—9 acyl-CoAs | Ohno et al. (2010) |
| ELOVL2 | C20 and C22 PUFA | Leonard et al. (2002) |
| ELOVL3 | C16–C22 saturated and monounsaturated fatty acids | Ohno et al. (2010) |
| ELOVL4 | >C26 saturated and monounsaturated fatty acids | Vasireddy et al. (2007) |
| ELOVL5 | 16:1n-7, 18:1n-9, 18:3n-6, 18:4n-3, 20:4n-6, 20:5n-3 | Leonard et al. (2000) |
| ELOVL6 | C12–C16 saturated and monounsaturated fatty acids | Matsuzaka et al. (2002) |
| ELOVL7 | C16–C22 saturated fatty acids | Naganuma et al. (2011); Tamura et al. (2009) |

has been shown to elongate monounsaturated fatty acids and PUFA with chain lengths between 16 and 20 carbons (Inagaki et al., 2002). The chain length specificity of *ELOVL5* for C18 PUFA has been confirmed in *Elovl5* -/- mice (Moon et al., 2009). Microsomes from wild type or *Elovl5* -/- mice were incubated with radiolabeled malonyl-CoA and a range of fatty acids. Incorporation of radiolabel into 18:3n-6 and 18:4n-3 was substantially lower in *Elovl5* -/- mice than controls, while labeling of 20:4n-6, 20:5n-3, 22:4n-6, and 22:5n-6 was increased in mice lacking *ELOVL5*, indicating that elongation of C20 PUFA is independent of *ELOVL5*. In addition, the proportion of 18:n-3 was higher and that of 18:4n-3 and 22:6n-3 lower in the liver of null mice, compared to controls (Moon et al., 2009). Such differences in substrate preference have been shown to be due to the presence of a cysteine residue at position of 217 in *ELOVL2*, while *ELOVL5* contains a tryptophan residue at the equivalent position (Gregory et al., 2013). Consistent with their role in PUFA synthesis, *ELOVL2* and *ELOVL5* are highly expressed in liver and testis (Leonard et al., 2002; Tvrdik et al., 2000; Wang et al., 2005) and to a lesser extent in lung and white adipose tissue (Tvrdik et al., 2000; Wang et al., 2005). *ELOVL2* null mice have lower concentrations of 22:6n-3 and 22:5n-6, accompanied by increased levels of the substrates 22:5n-3 and 22:4n-6 and their precursors (Pauter et al., 2014), indicating that *ELOVL2* is important for synthesis of very long china PUFA. However, 22:5n-3n-3 and 22:6n-3 were not completely absent from liver phospholipids or triglyceride, or serum total lipids. For example, the proportion of 22:6n-3 in liver triglyceride was only reduced by approximately 30%, although the reduction in liver phospholipids and serum lipids was greater (90%) (Pauter et al., 2014). Others have reported incomplete ablation of 22:6n-3 in liver and testis from *ELOVL2* null mice (Zadravec et al., 2011). Furthermore, despite active conversion of 18:3n-3 to 22:5n-3, *ELOVL2* is not expressed in human, mouse, or rat primary vascular smooth muscle cells, or in whole rat or mouse aortae (Irvine et al., 2015). Together these findings suggest that *ELOVL2* may not be the sole *ELOVL* responsible for elongation of >C20 PUFA in mammals. If so, this suggests that the assertion that *ELOVL2* is crucial for 22:6n-3 synthesis (Gregory et al., 2013) may be over-stated. *ELOVL2* has also been shown to be important for the synthesis of 28:5n-6 and 30:5n-6 in testis, which are required for sperm formation. *ELOVL2* null mice exhibit complete arrest of spermatogenesis accompanied by a dose-related reduction in 28:5n-6 and 30:5n-3 in testis such that concentrations of these fatty acids were wild type > heterozygous > homozygous such that these fatty acids were essentially absent from *Elovl2* -/- mice (Zadravec et al., 2011). *ELOVL2* is also involved in the regulation of fatty acid synthesis *de novo*. *ELOVL2* null mice exhibit increased expression of the sterol response element binding protein (SREBP)-1 in liver accompanied by upregulation of the target genes stearoyl-CoA desaturase-1 and fatty acid synthase (Pauter et al., 2014). However, there was no accumulation of triglyceride in the liver. Over expression of *ELOVL2* in preadipocytes induced increased levels of diacylglycerol acyltransferae-2 and fatty acid binding protein-4, but there was no change in the expression of fatty acid synthase (Kobayashi et al., 2007). Therefore, it appears that in addition to catalyzing PUFA chain elongation, *ELOVL2* is also involved in the regulation of fatty acid synthesis *de novo* although the precise mechanism is not known.

In addition to its role in PUFA synthesis *ELOVL5* is also involved in the regulation of genes which control cellular energy balance. Mice with diet-induced obesity and impaired glucose homeostasis exhibit reduced *ELOVL5* activity (Tripathy et al., 2010). Restoration of hepatic *ELOVL5* levels using a adenovirus construct improved glucose homeostasis (Tripathy et al., 2010). This effect has been shown to involve increased phosphorylation of Akts-[473] and PP2Acat-Y[307] and decreased nuclear content of Fox01 and lower expression of phosphoenolpyruvate carboxykinase and glucose-6-phosphatase (Tripathy et al., 2010). Further analysis showed these changes in the regulation of gluconeogenesis to be mediated by the *Elovl5*-catalyzed synthesis of 18:1n-7 via altered activity of the mTORC2-Akt-Fox01 pathway (Tripathy and Jump, 2013). In addition, increased *ELOVL5* expression in obese mice prevented fatty liver (Tripathy et al., 2013).

## REGULATION OF ELOVL2 AND 5 EXPRESSION

The regulation of *ELOVL2* has yet to be characterized in detail. However, SREBP-1 has been implicated in the regulation of *ELOVL2*. Adenovirus over-expression of SREBP-1c in primary rat hepatocytes increased the expression of *ELOVL2* (Wang et al., 2006). However, treatment of these cells with the LXRα agonist T1317, glucose or insulin did not alter *ELOVL2* expression. This inconsistency in the findings has not been explained. However, *ELOVL2* expression has been shown to be independent of whether an animal is in the fed or fasted state, and is not modulated by dietary fatty acid intake (Wang et al., 2005). These findings suggest that *ELOVL2* activity may be largely independent of nutritional inputs.

Gene expression studies in SREBP-1 transgenic mice (Horton et al., 2003) and assessment of promoter activity (Qin et al., 2009) have shown that *ELOVL5* is regulated directly by SREBP-1c. Human the *ELOVL5* gene contains two SREBP-1 response elements, one located 10 kb upstream of the transcription start site and the other within intron 1, which are conserved in humans and mice (Shikama et al., 2015). These response elements were shown to regulate *ELOVL5* transcription in luciferase reporter constructs and in SREBP-1 binding assays, and SREBP-1-induced transcription was abolished by mutation of these binding domains (Shikama et al., 2015).

A third SREBP-1 response element in the promoter region is only found in the human ELOVL5 (Qin et al., 2009) and was not active in luciferase reporter assays (Shikama et al., 2015). Unlike *ELOVL2* null mice, *Elovl5* −/− mice develop hepatic steatosis due to upregulation of SREBP-1c, acetyl-CoA carboxylase, fatty acid synthase and *ELOVL6* (Moon et al., 2009). This appears to be a consequence of reduction in the concentrations of 20:4n-6 and 22:6n-3, which exert a repressive effect on SREBP-1c activity (Takeuchi et al., 2010; Yahagi et al., 1999). This suggests that *ELOVL5* is important in the negative regulation of lipogenesis. In addition, liver X receptor-alpha (LXRα) is required for basal and ligand-mediated activation of *ELOVL5* transcription, but may also regulate *ELOVL5* expression via transcriptional control of SREBP-1c (Qin et al., 2009). Thus *ELOVL5* is regulated by two transcription factors that involved centrally in the control of lipid metabolism. PPARα has also been shown to regulate *ELOVL5* transcription (Wang et al., 2005).

The effects of polymorphisms in *ELOVL2* and *5* on their function are discussed in Chapter 11. The epigenetic regulation of *ELOVL2* and *5* transcription has not been described in detail and dietary fatty acids modify the DNA methylation status of specific CpG dinuleotides in the 5′-regulatory region in humans (Hoile et al., 2014). Several recent studies have reported reported that *ELOVL2* DNA methylation is positively associated with chronological age in humans (Garagnani et al., 2012; Steegenga et al., 2014) and explained up to 85% of variation in age between individuals (Zbiec-Piekarska et al., 2015a). One possible application of these findings is establishment of age in forensic analysis (Bekaert et al., 2015; Zbiec-Piekarska et al., 2015b). Furthermore, it is possible that *ELOVL2* may become progressively transcriptionally repressed with increasing age due to higher DNA methylation. If so, because of its role in spermatogenesis (Zadravec et al., 2011), progressive increase in DNA methylation of *ELOVL2* could contribute to decreasing male fertility with age.

# References

Beaudoin, F., Michaelson, L.V., Lewis, M.J., Shewry, P.R., Sayanova, O., Napier, J.A., 2000. Production of C20 polyunsaturated fatty acids (PUFAs) by pathway engineering: identification of a PUFA elongase component from *Caenorhabditis elegans*. Biochem. Soc. Trans. 28, 661–663.

Bekaert, B., Kamalandua, A., Zapico, S.C., Van de Voorde, W., Decorte, R., 2015. Improved age determination of blood and teeth samples using a selected set of DNA methylation markers. Epigenetics 10, 922–930.

Bernert, Jr., J.T., Sprecher, H., 1977. An analysis of partial reactions in the overall chain elongation of saturated and unsaturated fatty acids by rat liver microsomes. J. Biol. Chem. 252, 6736–6744.

Bernert, Jr., J.T., Sprecher, H., 1979. Solubilization and partial purification of an enzyme involved in rat liver microsomal fatty acid chain elongation: beta-hydroxyacyl-CoA dehydrase. J. Biol. Chem. 254, 11584–11590.

Denic, V., Weissman, J.S., 2007. A molecular caliper mechanism for determining very long-chain fatty acid length. Cell 130, 663–677.

Garagnani, P., Bacalini, M.G., Pirazzini, C., Gori, D., Giuliani, C., Mari, D., Di Blasio, A.M., Gentilini, D., Vitale, G., Collino, S., Rezzi, S., Castellani, G., Capri, M., Salvioli, S., Franceschi, C., 2012. Methylation of *ELOVL2* gene as a new epigenetic marker of age. Aging Cell 11, 1132–1134.

Gregory, M.K., Cleland, L.G., James, M.J., 2013. Molecular basis for differential elongation of omega-3 docosapentaenoic acid by the rat Elovl5 and Elovl2. J. Lipid Res. 54, 2851–2857.

Gregory, M.K., Gibson, R.A., Cook-Johnson, R.J., Cleland, L.G., James, M.J., 2011. Elongase reactions as control points in long-chain polyunsaturated fatty acid synthesis. PLoS One 6 (12).

Hoile, S.P., Clarke-Harris, R., Huang, R.C., Calder, P.C., Mori, T.A., Beilin, L.J., Lillycrop, K.A., Burdge, G.C., 2014. Supplementation with n-3 long-chain polyunsaturated fatty acids or olive oil in men and women with renal disease induces differential changes in the DNA methylation of FADS2 and ELOVL5 in peripheral blood mononuclear cells. PLoS One 9, e109896.

Horton, J.D., Shah, N.A., Warrington, J.A., Anderson, N.N., Park, S.W., Brown, M.S., Goldstein, J.L., 2003. Combined analysis of oligonucleotide microarray data from transgenic and knockout mice identifies direct SREBP target genes. Proc. Natl. Acad. Sci. U.S.A. 100, 12027–12032.

Inagaki, K., Aki, T., Fukuda, Y., Kawamoto, S., Shigeta, S., Ono, K., Suzuki, O., 2002. Identification and expression of a rat fatty acid elongase involved in the biosynthesis of C18 fatty acids. Biosci. Biotechnol. Biochem. 66, 613–621.

Irvine, N.A., Lillycrop, K.A., Fielding, B., Torrens, C., Hanson, M.A., Burdge, G.C., 2015. Polyunsaturated fatty acid biosynthesis is involved in phenylephrine-mediated calcium release in vascular smooth muscle cells. Prostagl. Leuk. Essent. Fatty Acids 101, 31–39.

Kobayashi, T., Zadravec, D., Jacobsson, A., 2007. ELOVL2 overexpression enhances triacylglycerol synthesis in 3T3-L1 and F442A cells. FEBS Lett. 581, 3157–3163.

Kohlwein, S.D., Eder, S., Oh, C.S., Martin, C.E., Gable, K., Bacikova, D., Dunn, T., 2001. Tsc13p is required for fatty acid elongation and localizes to a novel structure at the nuclear-vacuolar interface in *Saccharomyces cerevisiae*. Mol. Cell Biol. 21, 109–125.

Leonard, A.E., Bobik, E.G., Dorado, J., Kroeger, P.E., Chuang, L.T., Thurmond, J.M., Parker-Barnes, J.M., Das, T., Huang, Y.S., Mukerji, P., 2000. Cloning of a human cDNA encoding a novel enzyme involved in the elongation of long-chain polyunsaturated fatty acids. Biochem. J. 350, 765–770.

Leonard, A.E., Kelder, B., Bobik, E.G., Chuang, L.T., Lewis, C.J., Kopchick, J.J., Mukerji, P., Huang, Y.S., 2002. Identification and expression of mammalian long-chain PUFA elongation enzymes. Lipids 37, 733–740.

Leonard, A.E., Pereira, S.L., Sprecher, H., Huang, Y.S., 2004. Elongation of long-chain fatty acids. Prog. Lipid Res. 43, 36–54.

Matsuzaka, T., Shimano, H., Yahagi, N., Yoshikawa, T., Amemiya-Kudo, M., Hasty, A.H., Okazaki, H., Tamura, Y., Iizuka, Y., Ohashi, K., Osuga, J., Takahashi, A., Yato, S., Sone, H., Ishibashi, S., Yamada, N., 2002. Cloning and characterization of a mammalian fatty acyl-CoA elongase as a lipogenic enzyme regulated by SREBPs. J. Lipid Res. 43, 911–920.

Michelsen, K., Yuan, H., Schwappach, B., 2005. Hide and run. Arginine-based endoplasmic-reticulum-sorting motifs in the assembly of heteromultimeric membrane proteins. EMBO Rep. 6, 717–722.

Monne, M., Gafvelin, G., Nilsson, R., von Heijne, G., 1999. N-tail translocation in a eukaryotic polytopic membrane protein: synergy between neighboring transmembrane segments. Eur. J. Biochem. 263, 264–269.

Monroig, O., Lopes-Marques, M., Navarro, J.C., Hontoria, F., Ruivo, R., Santos, M.M., Venkatesh, B., Tocher, D.R., Castro, L.F., 2016. Evolutionary functional elaboration of the *Elovl2/5* gene family in chordates. Sci. Rep. 6, 20510.

Moon, Y.A., Hammer, R.E., Horton, J.D., 2009. Deletion of ELOVL5 leads to fatty liver through activation of SREBP-1c in mice. J. Lipid Res. 50, 412–423.

Moon, Y.A., Shah, N.A., Mohapatra, S., Warrington, J.A., Horton, J.D., 2001. Identification of a mammalian long chain fatty acyl elongase regulated by sterol regulatory element-binding proteins. J. Biol. Chem. 276, 45358–45366.

Naganuma, T., Sato, Y., Sassa, T., Ohno, Y., Kihara, A., 2011. Biochemical characterization of the very long-chain fatty acid elongase ELOVL7. FEBS Lett. 585, 3337–3341.

Ohno, Y., Suto, S., Yamanaka, M., Mizutani, Y., Mitsutake, S., Igarashi, Y., Sassa, T., Kihara, A., 2010. ELOVL1 production of C24 acyl-CoAs is linked to C24 sphingolipid synthesis. Proc. Natl. Acad. Sci. U.S.A. 107, 18439–18444.

Osei, P., Suneja, S.K., Laguna, J.C., Nagi, M.N., Cook, L., Prasad, M.R., Cinti, D.L., 1989. Topography of rat hepatic microsomal enzymatic components of the fatty acid chain elongation system. J. Biol. Chem. 264, 6844–6849.

Pauter, A.M., Olsson, P., Asadi, A., Herslof, B., Csikasz, R.I., Zadravec, D., Jacobsson, A., 2014. Elovl2 ablation demonstrates that systemic DHA is endogenously produced and is essential for lipid homeostasis in mice. J. Lipid Res. 55, 718–728.

Qin, Y., Dalen, K.T., Gustafsson, J.A., Nebb, H.I., 2009. Regulation of hepatic fatty acid elongase 5 by LXR alpha-SREBP-1c. Biochim. Biophys. Acta 1791, 140–147.

Shikama, A., Shinozaki, H., Takeuchi, Y., Matsuzaka, T., Aita, Y., Murayama, T., Sawada, Y., Piao, X.Y., Toya, N., Oya, Y., Takarada, A., Masuda, Y., Makiko, N., Kubota, M., Izumida, Y., Nakagawa, Y., Iwasaki, H., Kobayashi, K., Yatoh, S., Suzuki, H., Yagyu, H., Kawakami, Y., Yamada, N., Shimano, H., Yahagi, N., 2015. Identification of human ELOVL5 enhancer regions controlled by SREBP. Biochem. Biophys. Res. Commun. 465, 857–863.

Steegenga, W.T., Boekschoten, M.V., Lute, C., Hooiveld, G.J., de Groot, P.J., Morris, T.J., Teschendorff, A.E., Butcher, L.M., Beck, S., Muller, M., 2014. Genome-wide age-related changes in DNA methylation and gene expression in human PBMCs. Age 36, 1523–1540.

Takeuchi, Y., Yahagi, N., Izumida, Y., Nishi, M., Kubota, M., Teraoka, Y., Yamamoto, T., Matsuzaka, T., Nakagawa, Y., Sekiya, M., Iizuka, Y., Ohashi, K., Osuga, J., Gotoda, T., Ishibashi, S., Itaka, K., Kataoka, K., Nagai, R., Yamada, N., Kadowaki, T., Shimano, H., 2010. Polyunsaturated fatty acids selectively suppress sterol regulatory element-binding protein-1 through proteolytic processing and autoloop regulatory circuit. J. Biol. Chem. 285, 11681–11691.

Tamura, K., Makino, A., Hullin-Matsuda, F., Kobayashi, T., Furihata, M., Chung, S., Ashida, S., Miki, T., Fujioka, T., Shuin, T., Nakamura, Y., Nakagawa, H., 2009. Novel lipogenic enzyme ELOVL7 is involved in prostate cancer growth through saturated long-chain fatty acid metabolism. Cancer Res. 69, 8133–8140.

Tripathy, S., Jump, D.B., 2013. Elovl5 regulates the mTORC2-Akt-FOXO1 pathway by controlling hepatic cis-vaccenic acid synthesis in diet-induced obese mice. J. Lipid Res. 54, 71–84.

Tripathy, S., Stevens, R.D., Bain, J.R., Newgard, C.B., Jump, D.B., 2013. Elevated hepatic fatty acid elongase-5 (Elovl5) attenuates fatty liver in high fat diet induced obese mice. FASEB J. 27.

Tripathy, S., Torres-Gonzalez, M., Jump, D.B., 2010. Elevated hepatic fatty acid elongase-5 activity corrects dietary fat-induced hyperglycemia in obese C57BL/6J mice. J. Lipid Res. 51, 2642–2654.

Tvrdik, P., Asadi, A., Kozak, L.P., Nedergaard, J., Cannon, B., Jacobsson, A., 1997. Cig30, a mouse member of a novel membrane protein gene family, is involved in the recruitment of brown adipose tissue. J. Biol. Chem. 272, 31738–31746.

Tvrdik, P., Westerberg, R., Silve, S., Asadi, A., Jakobsson, A., Cannon, B., Loison, G., Jacobsson, A., 2000. Role of a new mammalian gene family in the biosynthesis of very long chain fatty acids and sphingolipids. J. Cell Biol. 149, 707–718.

Vasireddy, V., Uchida, Y., Salem, N., Kim, S.Y., Mandal, M.N.A., Reddy, G.B., Bodepudi, R., Alderson, N.L., Brown, J.C., Hama, H., Dlugosz, A., Elias, P.M., Holleran, W.M.,

Ayyagari, R., 2007. Loss of functional ELOVL4 depletes very long-chain fatty acids (≥C28) and the unique omega-O-acylceramides in skin leading to neonatal death. Human Mol. Genet. 16, 471–482.

Wang, Y., Botolin, D., Christian, B., Busik, J., Xu, J., Jump, D.B., 2005. Tissue-specific, nutritional, and developmental regulation of rat fatty acid elongases. J. Lipid Res. 46, 706–715.

Wang, Y., Botolin, D., Xu, J.H., Christian, B., Mitchell, E., Jayaprakasam, B., Nair, M., Peters, J.M., Busik, J., Olson, L.K., Jump, D.B., 2006. Regulation of hepatic fatty acid elongase and desaturase expression in diabetes and obesity. J. Lipid Res. 47, 2028–2041.

Wang, Y., Torres-Gonzalez, M., Tripathy, S., Botolin, D., Christian, B., Jump, D.B., 2008. Elevated hepatic fatty acid elongase-5 activity affects multiple pathways controlling hepatic lipid and carbohydrate composition. J. Lipid Res. 49, 1538–1552.

Yahagi, N., Shimano, H., Hasty, A.H., Amemiya-Kudo, M., Okazaki, H., Tamura, Y., Iizuka, Y., Shionoiri, F., Ohashi, K., Osuga, J., Harada, K., Gotoda, T., Nagai, R., Ishibashi, S., Yamada, N., 1999. A crucial role of sterol regulatory element-binding protein-1 in the regulation of lipogenic gene expression by polyunsaturated fatty acids. J. Biol. Chem. 274, 35840–35844.

Zadravec, D., Tvrdik, P., Guillou, H., Haslam, R., Kobayashi, T., Napier, J.A., Capecchi, M.R., Jacobsson, A., 2011. ELOVL2 controls the level of n-6 28:5 and 30:5 fatty acids in testis, a prerequisite for male fertility and sperm maturation in mice. J. Lipid Res. 52, 245–255.

Zbiec-Piekarska, R., Spolnicka, M., Kupiec, T., Makowska, Z., Spas, A., Parys-Proszek, A., Kucharczyk, K., Ploski, R., Branicki, W., 2015a. Examination of DNA methylation status of the ELOVL2 marker may be useful for human age prediction in forensic science. Forensic Sci. Int. Genet. 14, 161–167.

Zbiec-Piekarska, R., Spolnicka, M., Kupiec, T., Parys-Proszek, A., Makowska, Z., Paleczka, A., Kucharczyk, K., Ploski, R., Branicki, W., 2015b. Development of a forensically useful age prediction method based on DNA methylation analysis. Forensic Sci. Int. Genet. 17, 173–179.

Zhang, K., Kniazeva, M., Han, M., Li, W., Yu, Z., Yang, Z., Li, Y., Metzker, M.L., Allikmets, R., Zack, D.J., Kakuk, L.E., Lagali, P.S., Wong, P.W., MacDonald, I.M., Sieving, P.A., Figueroa, D.J., Austin, C.P., Gould, R.J., Ayyagari, R., Petrukhin, K., 2001. A 5-bp deletion in ELOVL4 is associated with two related forms of autosomal dominant macular dystrophy. Nat. Genet. 27, 89–93.

# Polyunsaturated Fatty Acids: Metabolism and Nutritional Requirements in Pregnancy and Infancy

*Beverly S. Muhlhausler\*,\*\*, Carmel T. Collins\*\*,*
*Jacqueline F. Gould\*\*, Karen P. Best\*\*,*
*Gabriela E. Leghi\**

\*FOODplus Research Centre, The University of Adelaide, Adelaide, SA, Australia; \*\*Healthy Mothers, Babies and Children Theme, South Australian Health and Medical Research Institute (SAHMRI), Adelaide, SA, Australia

## OUTLINE

Polyunsaturated Fatty Acid Metabolism. http://dx.doi.org/10.1016/B978-0-12-811230-4.00007-7

# INTRODUCTION

The fatty acid composition of the maternal diet during pregnancy and lactation plays a critical role in the supply of essential fatty acids to the fetus and neonate to support their development and metabolism. The omega-3 long chain polyunsaturated fatty acids (n-3 LCPUFAs) play a particularly important role in both maternal and infant outcomes (Innis, 2005; Makrides and Gibson, 2000, 2002). The most significant effect of the n-3 LCPUFA on pregnancy outcomes are their ability to extend gestation by ~2 days, and thereby reduce the incidence of preterm birth, especially very preterm birth, by up to 40% (Carlson et al., 2013; Makrides et al., 2010). There is limited evidence, however, of beneficial effects of maternal n-3 LCPUFA on other maternal outcomes, including the incidence of gestational diabetes or postnatal depression.

In terms of the infant, the n-3 LCPUFAs, in particular docosahexaenoic acid (DHA, 22:6n-3), play a major role in supporting the optimal development of the brain, nervous, and visual systems (Makrides and Gibson, 2002), and n-3 LCPUFA deficiency is associated with impaired neurological outcomes, cognitive performance, and visual acuity in animal models (Neuringer et al., 1986). However, while deficits in n-3 LCPUFA are clearly associated with poorer outcomes, there is little evidence to suggest that increasing n-3 LCPUFA supply above required levels results in further improvements in neurodevelopmental outcomes or visual acuity (Gould et al., 2013). In addition to their role in brain development and function, the n-3 LCPUFAs also have other key physiological properties, including immunomodulation, antiadipogenic/antilipogenic, and insulin sensitizing effects, and this has led to suggestions that an increased supply of these fatty acids during fetal life and/or in early infancy could potentially decrease the risk of allergic and metabolic diseases in the infants (Klemens et al., 2011). However, despite some encouraging evidence from animal studies, and biological plausibility, there is insufficient evidence from human randomized controlled trials (RCTs) to support the suggestion that increasing the supply of n-3 LCPUFA to an individual in whom levels are already sufficient either before birth or in early infancy has any additional benefits on these outcomes.

The purpose of this chapter is to provide an overview of our current understanding of the physiological effects of the n-3 LPCUFA, DHA, and eicosapentaenoic acid (EPA, 20:5n-3), during pregnancy and lactation, with a particular emphasis on metabolism and nutritional requirements.

# PART 1: n-3 LCPUFA IN PREGNANCY

The fetus has a limited capacity to synthesize EPA and DHA, and therefore relies almost exclusively on EPA/DHA transferred across the placenta from the maternal circulation. DHA is preferentially transported across the placenta in the last trimester of gestation (Innis, 2005) and during this period there is rapid accumulation of DHA in fetal tissues, particularly in the retinal membrane synapses and neural cortex tissues (Gibson and Makrides, 1998). Between 35 and 40 weeks of gestation, approximately 42%–50% of the total DHA supply to the fetus accumulates in the adipose tissue, while 23% is taken up by the fetal brain. The final 5 weeks of gestation represent the period of highest rates of accretion of all the major n-3 and omega-6 (n-6) PUFA/LCPUFA, and the rate of accretion increases progressively for DHA, arachidonic acid (AA), and linoleic acid (LA) (Kuipers et al., 2012). The estimated fetal accretion rate during the last trimester of pregnancy is 4 mg/kg/day for α-linolenic acid (ALA) and 43 mg/kg/day for DHA (Lapillonne and Jensen, 2009). Failure of an individual to accumulate sufficient DHA during this period, due to either an insufficient supply from the mother and/or preterm delivery, has long-term implications for developmental outcomes. Growth-restricted and preterm infants are at highest risk of early DHA deficits.

# ROLE OF n-3 LCPUFA IN MATERNAL AND INFANT OUTCOMES

## Maternal/Pregnancy Outcomes

Despite some early evidence from small studies that n-3 LCPUFA supplementation could have beneficial effects on postnatal depression and gestational diabetes, the results of subsequent, and appropriately powered, studies have been largely disappointing. This is highlighted by the results of the largest RCT of maternal n-3 LCPUFA supplementation, chiefly as DHA, during pregnancy, the DHA to optimize mother infant outcome (DOMInO) trial (Makrides et al., 2010). In this trial, 2399 women were randomized to either a high dose n-3 LCPUFA supplement (800 mg DHA and 100 mg EPA/day) or placebo from <21 weeks gestation until delivery of their infant. However, while the intervention was associated with a significant increase in n-3 LCPUFA concentrations in the umbilical cord blood at delivery, there was no difference in the incidence of post-partum depression between groups (Makrides et al., 2010). This same trial found no effect of n-3 LCPUFA supplementation on the incidence of either gestational diabetes or preeclampsia (Zhou et al., 2012). Thus, there is currently no compelling evidence that increasing the supply of DHA during pregnancy reduces the risk of either gestational diabetes or postpartum depression; however, further research is required to determine the effects of providing a higher dose of EPA.

The most consistent evidence to date in support of a role for n-3 LCP-UFA in pregnancy relates to its role in prolonging gestation and reducing the incidence of early preterm birth (<34 weeks gestation). The potential role of n-3 LCPUFA in the length of gestation originated from the results of observational studies in the early 1980s, in which women from fish-eating communities were observed to have longer durations of gestation and infants with higher birthweights than women who consumed little or no fish (Olsen et al., 1986; Olsen and Joensen, 1985). The effect of maternal n-3 LCPUFA was subsequently tested and confirmed in RCTs (Olsen et al., 1992). The earlier trials almost without exception used very high doses of n-3 LCPUFA and mostly involved women thought to be at higher risk of preterm birth. The role of DHA in extending gestation length and increasing infant birth weight was supported by a Cochrane Systematic Review, published in 2006 (Makrides et al., 2006). From 2006 onward, systematic reviews have also included women with low-risk pregnancies and continued to support the role of n-3 LCPUFA, in particular DHA, in increasing the duration of gestation and reducing the risk of early preterm birth. The effectiveness of n-3 LCPUFA has been validated by the high level of concordance between the two largest randomized trials of omega-3 supplementation in pregnancy, the DOMInO and Kansas University DHA outcome study (KUDOS) trials (Carlson et al., 2013; Makrides et al., 2010). The DOMInO trial showed that maternal n-3 LCPUFA supplementation in the second half of pregnancy increased the mean duration of gestation by ~2 days, resulting in a 10% reduction in the incidence of preterm birth overall (<37 weeks) and a ~40% reduction in the incidence of early preterm birth (<34 weeks), with the KUDOS trial producing similar results (Carlson et al., 2013; Makrides et al., 2010). The mechanisms through which n-3 LCPUFA mediate this effect remain unclear; however, it has been suggested that the antiinflammatory properties of the n-3 LCPUFA derivatives may play a significant role, though this is an area that requires further research. It is also important to note, however, that maternal n-3 LCPUFA supplementation was also associated with an increase in the number of women who required interventions due to extended gestation beyond term (Makrides et al., 2010), suggesting that not all women may benefit from supplementation, and this also requires further study. Further research in this area is underway to investigate the optimal timing of supplementation to achieve the best outcomes.

## n-3 LCPUFA DURING PREGNANCY AND INFANT/ CHILD OUTCOMES

### Neurodevelopmental Outcomes

Studies into the long-term effects of perinatal n-3 LCPUFA exposure on subsequent health outcomes in an individual were initially focused largely on cognitive/neurodevelopmental outcomes, due to the established role of these fatty acids in the brain and nervous system. There is evidence that membrane lipid composition of the fetal brain is sensitive to changes in DHA supply from the mother, and this led to suggestions that increasing the supply of n-3 LCPUFA during critical periods of brain development could result in lasting improvements in brain function/cognitive performance. However, despite evidence that overt n-3 LCPUFA deficiency during development leads to impaired development of the brain and visual systems (Abedin et al., 1999; Levant et al., 2004; Moriguchi et al., 2000; Pawlosky et al., 1997), there is limited evidence that supplementing the mother with n-3 LCPUFA during pregnancy in populations in which there is no overt n-3 LCPUFA deficiency has any benefit on subsequent cognitive performance of the child. This is highlighted in the most comprehensive systematic review/meta-analysis of all available trials in this area, which includes data from 11 RCTs involving 5272 participants (Gould et al., 2013). Outcomes assessed included general cognitive functioning, language, problem solving, behavior, temperament, and motor development (Gould et al., 2013). This view is supported by the results of the DOMInO study, which found no effect of high-dose maternal n-3 LCPUFA supplementation in the second half of pregnancy on neurodevelopmental outcomes in the children at 18 months of age (Makrides et al., 2010). Since DHA predominately accumulates in the hippocampus and frontal cortex of the brain, it has been suggested that more pronounced effects of prenatal n-3 LCPUFA supplementation would be seen if specific aspects of cognition, that is, higher order executive functions, which are regulated by these brain regions, were measured. However, a follow-up of 158 children born to women in the DOMInO trial found no effect of maternal n-3 LCPUFA supplementation in pregnancy on higher order cognitive tasks, including attention, working memory, and inhibitory control at 2 years of age (Gould et al., 2014). Further follow-up of the larger cohort of DOMInO trial children at 4 years showed no effects of prenatal supplements on cognition, language, or executive functioning assessed by a psychologist and no evidence of benefit to parent-rated behavior or executive functioning (Makrides et al., 2014). Other prenatal DHA trials published subsequent to the meta-analyses (Gould et al., 2013) have similarly shown no benefit of supplements on child cognition, motor or language scores at 3 months (Keenan et al., 2016), cognition or motor at 18 months (Mulder et al., 2014), cognition or behavior at 5 years (Ramakrishnan et al., 2016), attention at 8.5 years (Catena et al., 2016), or cognition, motor, or language at 12 years (Meldrum et al., 2015). Thus, there is limited evidence that providing additional DHA in utero results in enhanced cognitive performance in children whose mothers are not overtly n-3 LCPUFA-deficient. However, a noteworthy limitation of these trials is that children who are born preterm, and are at risk of poorer developmental outcomes, are typically excluded from assessment.

Since DHA is the major constituent of the fetal brain, the majority of studies to date have focused chiefly on DHA, rather than EPA. However, while DHA is the major fatty acid in the brain, EPA is also present and is an important component of neuronal membranes and there are studies suggesting that the uptake of EPA by neurons of the brain and retina is increased substantially during the neuronal growth spurt (Clandinin et al., 1980; Janssen and Kiliaan, 2014). Despite this, there is limited information on the role of EPA alone on brain development.

## Visual Development

Interest in the potential for n-3 LCPUFA supplementation to improve the visual development of the child originated from the early studies of Neuringer et al. (1986) in nonhuman primates, that elegantly demonstrated that n-3 LCPUFA deficiency in the maternal diet during pregnancy and lactation led to permanent deficits in visual function of the offspring, which could not be corrected by supplying them with n-3 LCPUFA after birth. While it is therefore clear that there is a requirement for an adequate supply of n-3 LCPUFA, in particular DHA, for optimal development of the visual system, there is limited evidence that increasing the fetal n-3 LCPUFA supply above the required level leads to further improvements in visual function. Due to the significant variety of visual assessments and ages at assessment between studies, it has not been possible to combine studies with visual outcomes in a meta-analysis in any meaningful way, however a systematic evaluation of these studies reported that six of the eight assessments in five trials reported no difference between the supplemented and control groups in visual outcomes

(Gould et al., 2013). The largest of these studies, which used the most robust measure of visual function, found no effect of prenatal DHA supplements on visual evoked potential at 4 months (Smithers et al., 2011). A more recently published trial also found no effect of DHA on visual acuity at 2 or 12 months (Mulder et al., 2014). There is therefore no conclusive evidence that maternal n-3 LCPUFA supplementation has any significant benefits for infant visual function, at least at a population level. However, again, the majority of trials only include infants born full-term in their outcome assessments.

## Body Fat Mass and Metabolic Health

Findings from in vitro and experimental animal studies that n-3 LCPUFA, in particular DHA, had the capacity to inhibit both the hyperplastic and hypertrophic expansion of fat depots, led to the hypothesis that exposing individuals to an increased supply of n-3 LCPUFA during the period of fat cell development (before birth and in early infancy) would reduce body fat mass in the child later in life (Al-Hasani and Joost, 2005; Hauner et al., 2009). However, this has only been tested in relatively few RCT, and the results to date have been largely negative. A systematic review of humans studies that had addressed this question identified five studies and four trials that had measured body max index (BMI) and/or body fat mass in children born to women involved in RCTs of maternal n-3 LCPUFA supplementation during pregnancy and/or lactation (Muhlhausler et al., 2010). This review identified a number of important limitations in the studies conducted to date, the most important being the high rates of attrition (>40%). Not surprisingly, the studies to date had produced mixed results, with one study reporting the maternal n-3 LCPUFA supplementation was associated with a reduction in BMI z-score in infants at 21 months, one reporting an increase in BMI z-score at 2.5 years and others finding no significant effects. Overall, however, the conclusion of this systematic review was that there was a lack of robust evidence to support a role of maternal n-3 LCPUFA supplementation in affecting body composition in the child. Since then, a follow-up of growth, body composition, and insulin sensitivity in >1500 children born to women in the DOMInO trial at 3 and 5 years of age, found no significant effect of maternal n-3 LCPUFA supplementation on BMI z-score or body fat mass in the children (Muhlhausler et al., 2016). This study was well-powered and achieved a >90% follow-up rate, and therefore provides the strongest evidence to date that exposure to a higher n-3 LCPUFA supply in utero does not affect body fat mass in childhood, either positively or negatively. Similarly, there are no studies to date in humans to suggest benefits of maternal n-3 LCPUFA supplementation on insulin sensitivity/risk of type 2 diabetes in the offspring.

## Allergy/Immune Outcomes

The extensive immunomodulatory properties of n-3 LCPUFA are well recognized (Calder and Yaqoob, 2009; Everts et al., 2000), however, the evidence involving n-3 LCPUFA supplementation in established disease has been unconvincing (Mihrshahi et al., 2004). A Cochrane review and meta-analysis conducted in 2002 concluded that there was little evidence to recommend that people with asthma supplement or modify their diet to increase intake of omega-3 fatty acids in order to improve their symptoms (Woods et al., 2002). However, the period of fetal immune system development is more susceptible to influence and there are plausible mechanisms by which diets high in n-3 LCPUFA may modulate the development of immunoglobulin E (IgE)-mediated allergic disease. More recent evidence has emerged, which suggests that maternal n-3 LCPUFA supplementation during pregnancy may have long-term benefits for immune function and the risk of allergic disease in the offspring. Diets high in n-3 LCPUFA, particularly DHA, increase cell membrane n-3 LCPUFA and reduce prostaglandin-E synthesis and proinflammatory cytokine responses known to be associated with allergies (Calder, 2013). There is increasing evidence demonstrating modification of immune function and influence on perinatal immune programming at a number of different stages in this complex process (Beck et al., 2000; Denburg et al., 2005; Gottrand, 2008; Prescott and Dunstan, 2007). However, the exact mechanisms through which n-3 LCPUFA act to alter neonatal immune responses remain unclear.

Despite the biological rationale for increased exposure to n-3 LCPUFA during immune system development to favor a reduction in the risk of allergic disease, the data from clinical trials has not been entirely consistent (Best et al., 2016a). While some RCTs have demonstrated that maternal n-3 LCPUFA supplementation is associated with improved immune/allergy outcomes, including a lower rate of sensitization to food allergens and a lower prevalence of allergic disease symptoms in the first year of life, there are limited RCTs in which the infants are followed up at older ages. Our DOMInO RCT that assessed children at 1, 3, and 6 years of age, suggests that early beneficial effects on sensitization observed in the first 12 months of life are not evident at 3 or 6 years (Best et al., 2016b; Palmer et al., 2012, 2013). Two recent reviews in this area, including a Cochrane Systematic Review and meta-analysis/systematic review of RCT and observational studies concluded that while there was evidence to support a beneficial effect of maternal n-3 LCPUFA supplementation on some aspects of allergic disease, further studies were required to enable a definite conclusion to be drawn as to whether maternal n-3 LCPUFA supplementation can reduce the risk of allergic disease in the child (Best et al., 2016a; Gunaratne et al., 2015).

While the separate contributions of EPA and DHA to the development of the immune system, and subsequent risk of allergic disease in the infants, remains unclear, EPA has been shown to have an important role in immune system function in adults (Chapkin et al., 2009). Recent studies have suggested that the balance of omega-6 and omega-3 fats in the diet may be important for this process of immune development—with the AA-derived mediators tending to shift the T-cell populations towards T-helper Cells Type 2 Th2 (allergy mediators) and the EPA-derived mediators shifting the populations in the opposite direction (toward a less allergic phenotype).

## DETERMINANTS OF FETAL DHA/EPA STATUS

While maternal DHA/EPA status is considered to be the single most important contributor to fetal DHA status, a number of studies have demonstrated that maternal n-3 LCPUFA intake cannot fully account for cord blood DHA/EPA levels. Previous studies have suggested that maternal DHA concentrations account for ~20%–30% of the variation in fetal DHA concentrations. This suggests that factors beyond maternal dietary DHA intake also play an important role in determining fetal DHA (and EPA) concentrations. Fig. 7.1 summarizes our current understanding of the determinants of n-3 LCPUFA concentrations in the fetal circulation, each of which is discussed here.

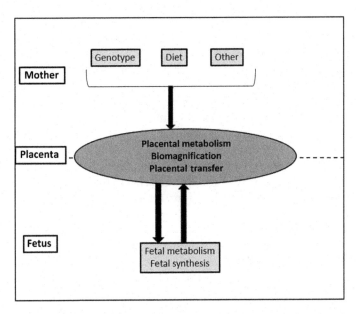

FIGURE 7.1 Summary of maternal, placental, and fetal factors contributing to fetal Omega-3 long chain polyunsaturated fatty acids supply.

### Maternal Circulating Omega-3 Long Chain Polyunsaturated Fatty Acids Concentrations

During pregnancy and lactation, there is a higher metabolic requirement for n-3 LCPUFA than in nonpregnant women, because the mother must meet her own needs as well as meeting the requirements of the developing infant. The n-3 LCPUFA status of the mother (i.e., levels in the maternal circulation), which can be derived from either the maternal diet (preformed or synthesized from precursor fatty acids) or mobilized from adipose tissue stores, is the major determinant of fetal DHA/EPA supply. There is a significant correlation between maternal DHA concentrations and concentrations in cord blood ($r = 0.48$), but this is weaker for EPA than for DHA (Elias and Innis, 2001).

While maternal dietary intake of n-3 LCPUFA, and to a lesser extent intake of the precursor short-chain n-3 PUFA (ALA), there are a number of other factors that have been shown to influence maternal n-3 LCPUFA status, and therefore fetal supply, during pregnancy. These include maternal genotype (particularly for the fatty acid desaturases, FADS genes), and other factors that can influence fatty acid metabolism (particular disease states, gastric malabsorption, smoking) (Muhlhausler et al., 2014). Single nucleotide polymorphisms (SNPs) in the genes encoding the desaturase enzymes responsible for the conversion of short-chain n-3 and n-6 PUFA to their long-chain derivatives (FADS1 and 2) can modulate the relationship between dietary fat intake and circulating n-3 LCPUFA concentrations. However, in the largest study to date to assess the contribution of SNP variants to fatty acid status of red blood cells (RBC) in pregnant women, found that the genetically explained variability of RBC DHA amounts was only 0.51% (Koletzko et al., 2011), and this is therefore unlikely to make a significant contribution to fetal supply.

Furthermore, in the last trimester of pregnancy, the increased lipolytic activity in maternal adipose tissue results in release of fatty acids from maternal tissue stores. Since the fatty acid composition of the adipose tissue is a marker of long-term fatty acid intake, the fatty acid composition of the maternal diet before pregnancy, independent of dietary composition during pregnancy, can contribute to differences in maternal n-3 LCPUFA status in late pregnancy, and, therefore, fetal concentrations.

### Placental Metabolism

The placenta has a high-energy demand and it is likely that at least some of the fatty acids from the maternal circulation are used by the placenta as an energy source (Herrera and Amusquivar, 2000; Larque et al., 2011). The placenta is also likely to utilize some fatty acids in preference to others, which contributes to the lack of a linear relationship between maternal and fetal DHA/EPA concentrations.

## *Placental n-3 LCPUFA Transfer*

Aside from maternal n-3 LCPUFA status, placental n-3 LCPUFA transfer capacity is perhaps the most important, but also least well understood, determinant of fetal n-3 LCPUFA supply. While the fundamental aspects of placental fatty acid metabolism and transport have been described, there are few studies that have examined the extent to which these vary between individuals and whether and to what it extent it is modified by other dietary/environmental factors (Innis, 2005; Larque et al., 2011). It has been suggested that variations between individual women in the efficiency of placental n-3 LCPUFA transport may explain, at least in part, why maternal n-3 LCPUFA supplementation results in improved outcomes in some children, but not in others; however, this remains to be proven.

In general, the rate of transplacental transfer is related to the surface area available for exchange, the concentration gradient between the maternal and fetal circulations and the number of transporter proteins inserted per unit area of the trophoblast membrane. Placental transfer is also selective for certain fatty acids over others, with the LCPUFA being transferred in preference to saturated and monounsaturated fatty acids, and DHA being transferred in preference to LA. This is best illustrated by tracer studies, in which different $^{13}$C labeled fatty acids ($^{13}$C-DHA, $^{13}$C-LA, $^{13}$C-oleic acid, and $^{13}$C-palmitic acid) were administered to women 4 or 12 h prior to caesarean section and then measured in the placenta, and which identified a significantly higher $^{13}$C-DHA incorporation into placenta lipids as compared with the other fatty acids (Larque et al., 2011).

Delta 5 and 6 desaturase activities are not detectable in the human placenta (Kuhn and Crawford, 1986), suggesting that the placenta does not have the capacity for DHA synthesis. Therefore, the ability of the placenta to extract DHA/EPA from the maternal circulation and deliver them to the fetus is a critical determinant of fetal DHA/EPA delivery. N-3 LCPUFA circulate in maternal plasma bound to lipoprotein triglycerides, and in a minor proportion in the form of free fatty acids (FFA). Despite the lack of a direct placental transfer of triglycerides, diffusion of fatty acids to the fetus is ensured by means of lipoprotein receptors, lipoprotein lipase activity, and intracellular lipase activities in the placenta (Fig. 7.2; Herrera and Amusquivar, 2000). Maternal plasma FFA are also an important source of fetal n-3 LPCUFA, and their placental uptake occurs via a selective process of facilitated membrane translocation involving a plasma membrane fatty acid-binding protein. This mechanism, together with a selective cellular metabolism, determine the actual rate of placental transfer and its selectivity, and results in an enrichment of certain LCPUFA, including DHA, in the fetal circulation as compared to maternal. The concentrations of DHA in the cord blood increase across gestation, and several studies have reported positive relationships between cord blood DHA and gestational

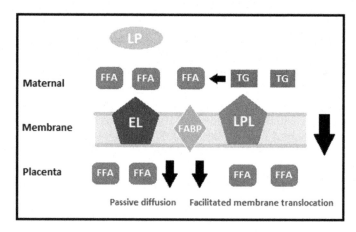

FIGURE 7.2   **Summary of mechanisms involved in transplacental transfer of fatty acids from maternal to fetal circulation.** *LP*, Lipoprotein; *FFA*, free fatty acids; *TG*, triglycerides; *EL*, endothelial lipase; *FABP*, fatty acid-binding protein; *LPL*, lipoprotein lipase.

age at delivery (Innis, 2005; Larque et al., 2011; Muhlhausler et al., 2014). This appears to be due to the increased transfer of DHA across the placenta as gestation progresses, in order to meet the increasing fetal demands.

In addition, variations in placental and fetal metabolism can influence the rate and extent to which fatty acids are converted or metabolized during or after transport. Our group recently conducted a study, using the cord blood DHA concentrations measured in 1500 women in the DOMInO RCT, which attempted to identify the key predictors of fetal DHA in late gestation. We found no relationship between cord blood DHA and infant birth weight in this study, which, given the tight relationship between fetal and placental weight, suggests that the size of the placenta is not a key determinant of DHA transfer (Muhlhausler et al., 2014). One possibility is that differences between individuals in the expression of fatty acid transport proteins (FATP) within the placenta could influence the extent of placental DHA transfer. In support of this, Larque et al. (2006) demonstrated that placental FATP1 and FATP4 gene expressions were directly correlated with the percentage of DHA in the placental and maternal plasma phospholipids. Furthermore, FATP4 expression was significantly correlated with the DHA in the PL of fetal blood (Larque et al., 2006). There is evidence of aberrant expression of fatty acid binding proteins in pregnancies complicated by gestational diabetes and intrauterine growth restriction (Gil-Sanchez et al., 2011), but it is unclear to what extent this can vary in healthy pregnancies. There is at least some evidence from human studies that placental expression of fatty acid binding proteins (FABPs) can be influenced by the fatty acid content of the maternal diet (Larque et al., 2006), but very few studies have investigated the environmental and/or genetic factors, which regulate placental fatty acid transfer.

## Biomagnification

The concept of biomagnification is a well-described phenomenon that describes the observation, first reported by Crawford et al. (1976), that DHA concentrations are higher in the fetal circulation compared to the maternal circulation. This is an important physiological process, since it ensures that the fetus is protected from maternal n-3 LCPUFA deficiency, at least until it reaches critical levels. The source of this biomagnification is not entirely understood, but is thought to be related to preferential transfer of n-3 LCPUFA from maternal to fetal circulation in late gestation. This, in turn, appears to be related to the fact that FABPs within the placenta preferentially bind n-3 LCPUFA, in particular DHA, compared to other fatty acids (Gil-Sanchez et al., 2011).

More recent studies have shown that the process of biomagnification does not follow a linear relationship with maternal n-3 LCPUFA concentrations, but that there is an upper limit of maternal DHA concentrations beyond which there are no further increases in fetal DHA status, implying that placental DHA transfer becomes saturated at very high maternal n-3 LCPUFA levels concentrations. This means that in pregnant women with a very high DHA status, fetal/cord blood DHA concentrations are no longer higher than those in the maternal circulation, and may even be lower (a process known as bioattenuation). The reason for this is unclear, but does suggest that n-3 LCPUFA excess, as well as deficiency, is maybe suboptimal for fetal development. Recent studies have suggested that bioattenuation occurs at maternal plasma/erythrocyte DHA concentrations in excess of >8/100 g (8%). In one study, it was reported that biomagnification occurs up to 8 g% DHA in maternal RBC and maternal RBC-DHA equilibrium is reached at ~6 g%, which corresponds with an infant RBC-DHA of 7 g% at delivery (Luxwolda et al., 2012).

## Fetal Synthesis

In addition to the supply of n-3 LCPUFA from the mother, the fetus also has the capacity for DHA synthesis, particularly in the last trimester of gestation. This has been demonstrated in studies of preterm and very preterm infants, which have established that humans have the capacity to endogenously synthesize long-chain PUFA (DHA, EPA, and AA) from short-chain precursors (LA and ALA) (Carnielli et al., 1996). The capacity of the human fetus to synthesize LCPUFA has been confirmed by studies that have measured fetal/neonatal synthesis directly using tracer techniques (Mayes et al., 2006; Pawlosky et al., 2006). While this process is likely to make a relatively minor contribution to fetal DHA/EPA supply, it nevertheless may have a role. In addition, the FADS genotype of the fetus has been shown to be related to the cord blood fatty acid concentrations independent of the maternal genotype (Lattka et al., 2013), suggesting that variations in the fetal, as well as maternal, capacity for DHA synthesis is an important determinant of fetal DHA supply.

### Fetal Metabolism

The majority of DHA/EPA supplied to the fetus is incorporated into fetal tissues, and it does not appear that any significant proportion of the n-3 LCPUFA is used as an energy source in late gestation. DHA is accumulated at a particularly rapid rate into the fetal brain in late gestation, and fetal DHA requirements in the final trimester are estimated ~43 mg/kg/day (Lapillonne and Jensen, 2009).

## Current Recommendations of DHA and EPA Intakes in Pregnant Women From National and International Health Agencies

The European Commission charged the research project PERILIP, jointly with the Early Nutrition Programming Project, to develop recommendations on dietary fat intake in pregnancy and lactation. This involved a comprehensive review of the existing literature and a meeting involving international experts in the field, including representatives of international scientific associations, in order to discuss and decide this. They concluded that pregnant and lactating women should aim to achieve an average dietary intake of at least 200 mg DHA/day (Koletzko et al., 2007). This can be achieved either through fish-oil supplements or consumption of at least two oily fish meals per week. The authors also indicated that it was important for pregnant and lactating women to note that the intake of the DHA precursor, ALA, was far less effective with regard to DHA deposition in fetal brain than preformed DHA. While 200 mg DHA/day is considered the minimum acceptable intake, the international experts who authored the consensus statement noted that maternal supplements of up to 1 g/day DHA or 2.7 g/day n-3 long-chain PUFA had been used in RCTs without significant adverse effects.

It is also important to note that n-3 LCPUFA requirements in pregnancy and whether a woman will require n-3 LCPUFA supplementation to achieve optimal maternal and neonatal outcomes is likely to vary substantially between individuals depending on their background diet, prepregnancy habitual dietary intake and a range of other factors. Perhaps the most critical of these is the level of n-6 PUFA in the background diet. This has become particularly significant due to the massive increase in the levels of n-6 PUFA in modern western diets, compared to those of traditional diets, and the resulting imbalance between the level of n-6 PUFA and n-3 PUFA (ratio of ~10–15:1 vs. 2:1) (Calder, 2012). This imbalance has important physiological/biological implications, because the n-6 PUFA compete with n-3 LPCUFA for incorporation into cells and tissues, which is required in order for these fatty acids to mediate their biological effects (Lands, 2008; Lands et al., 1992). Consequently, higher intakes of n-6 PUFA in the background diet mean that higher levels of n-3 LCPUFA are needed to achieve the same n-3 LCPUFA level in tissues (n-3 LCPUFA status) and, therefore, the associated health benefits.

# ROLE OF n-3 LCPUFA IN EARLY INFANCY

During lactation, the breast-fed infant relies exclusively on their mother's breast milk to obtain the n-3 LCPUFA supplies that are essential for optimal growth and development. A full-term breastfeed newborn can accrete upon to 1.9 g of DHA/day in the first 6 months, and DHA deficiency during this time can have profound and long-lasting effects on infant growth and neurodevelopmental outcomes (Cunnane et al., 2000). Both maternal dietary intake and PUFA reserves in adipose tissue influence the fat composition of the breast milk. As a result of storage of fatty acids consumed during pregnancy in maternal adipose stores, which are subsequently mobilized during breast feeding, maternal DHA and EPA intake during the last trimester of gestation is directly associated to their concentrations in the breast milk in the early postpartum period (Makrides et al., 2010). There is little influence of maternal ALA levels on breast milk DHA/EPA, because there is limited conversion of ALA into DHA by the mammary gland. In addition, it does not appear that the mammary gland has any mechanisms through which to maintain individual fatty acids constant with varying maternal intakes, as is the case for some other key nutrients (Innis, 2007).

The early infant formulas contained no added DHA. However, it was quickly determined that DHA was present at relatively high levels in breast milk, which suggested that it was important for infant development. Critically, the levels of DHA in the infant brain were found to be higher in breast fed compared to formula fed infants (Makrides et al., 1994). Some early studies also showed that the formula fed infants had less optimal indexes of visual function and neurodevelopmental status in comparison with infants who were breast fed (Makrides and Gibson, 2000). The finding of significant DHA, and to a lesser extent EPA, concentrations in breast milk ultimately led to recommendations that infant formulas contain these fats.

There have since been a large number of studies that have investigated the effect of providing infants with higher levels of DHA/EPA, by either supplementing the mother (to increase breast milk levels) or adding additional n-3 LCPUFA to infant formulas, on a range of outcomes in the infant/child. However, while it is clear that an overt deficiency of DHA/EPA during infancy is detrimental, there is limited evidence that providing additional DHA/EPA above the required levels provides additional benefits. There is currently no conclusive evidence that providing term infants with additional DHA/EPA during early infancy has substantial benefits for their long-term neurodevelopmental or visual function (Simmer et al., 2011). Similarly, there is no evidence that providing additional DHA/EPA to term infants has any significant effect on their growth or growth quality (Makrides et al., 2005) or on allergy/immune outcomes (Gunaratne et al., 2015).

## RELATIONSHIP BETWEEN MATERNAL INTAKES OF DHA/EPA IN PREGNANCY, BREAST MILK FATTY ACID CONCENTRATIONS, AND INFANT DHA/EPA STATUS

Human breast milk contains a high amount of fat, and this makes up ~50% of its energy. The vast majority (~98%) of these fats are in the form of triglycerides, each of which contain three fatty acids whose composition depends on the composition of the maternal diet (Innis, 2007). Human milk fatty acids are derived from endogenous synthesis in the mammary gland and uptake from maternal plasma. Both mammary gland fatty acid synthesis and the fatty acids available for uptake from the maternal plasma are influenced by maternal nutrition (Makrides et al., 1995). The importance of maternal fatty acids intakes for breast milk composition is highlighted by a large number of studies that have compared the fatty acid composition of breast milk from women in different countries and in response to maternal n-3 LCPUFA supplementation (Innis, 2007). These have consistently demonstrated that concentrations of most fatty acids in human milk, including the n-3 LCPUFA, vary widely in accordance with the fatty acid composition of the maternal diet (Table 7.1).

A seminal study conducted by Makrides et al. (1996) established that there was a linear relationship between maternal DHA intake and breast milk DHA concentrations across a wide range of DHA concentrations. In this trial, breastfeeding women were randomized to take identical capsules providing either 0 (placebo), 0.2, 0.4, 0.9, or 1.3 g DHA/day from day 5 to 12 weeks postpartum. The breast milk DHA concentrations of the women in this study ranged from 0.2% to 1.7% of total fatty acids and increased in direct correlation with the level of supplementation ($r^2 = 0.89$, $P < .01$). Maternal plasma ($r^2 = 0.71$, $P < .01$) and erythrocyte ($r^2 = 0.77$, $P < .01$) phospholipid DHA levels increased and were also strongly associated with dietary dose of DHA. The results of this study demonstrated that the level of maternal DHA supplementation in the maternal diet has a strong, specific, and dose-dependent effect on breast milk DHA ($r^2 = 0.89$, $P < .01$). It

**TABLE 7.1** Breast milk composition in countries consuming modern western diets versus a traditional diet

| Fatty acid | Modern diet | | | Traditional diet |
| | United States ($n$ = 81) | Canada ($n$ = 149) | United States ($n$ = 35) | Bolivia ($n$ = 35) |
| --- | --- | --- | --- | --- |
| 18:2n-6 | 12.7 (8.1–21.7) | 13.6 (7.2–26.2) | 18.1 (12.8–24.1) | 9.31 (3.86–14.8) |
| 20:3n-6 | NR | 0.31 (0.14–0.60) | 0.33 (0.23–0.43) | 0.44 (0.29–0.59) |
| 20:4n-6 | 0.47 (0.25–0.70) | 0.36 (0.10–0.64) | 0.56 (0.43–0.69) | 0.96 (0.44–1.48) |
| 18:3n-3 | 0.95 (0.54–1.70) | 1.60 (0.68–3.18) | 1.39 (0.61–2.17) | 1.64 (0.60–2.68) |
| 20:5n-3 | 0.02 (0.00–0.22) | 0.08 (0.02–0.45) | 0.06 (0.02–0.10) | 1.17 (0.04–0.30) |
| 22:5n-3 | 0.10 (0.05–0.41) | 0.12 (0.04–0.52) | 0.14 (0.11–0.17) | 0.36 (0.53–1.06) |
| 22:6n-3 | 0.23 (0.07–1.20) | 0.21 (0.03–1.13) | 0.13 (0.04–0.21) | 0.62 (0.31–0.93) |

*Source: Data taken from: Innis, S.M., 2007. Human milk: maternal dietary lipids and infant development. Proc. Nutr. Soc. 66, 397–404.*

also identified a positive relationship between the concentrations of DHA in maternal plasma and RBC phospholipids and the levels of DHA in the breast milk. Very similar results have been reported in a number of other studies, including a study by Jensen et al. (2000). This study demonstrated that the concentration of DHA and EPA in breast milk was directly related to maternal status (EPA: $r^2 = 0.30$, $P < .004$; DHA: $r^2 = 0.66$, $P < .001$). This same study identified a significant positive relationship between breast milk DHA and infant DHA status across a wide range of breast milk DHA concentrations. Thus, the supply of DHA and EPA to the breast fed infant is almost exclusively dependent on the n-3 LCPUFA status of the mother, which in turn is closely linked to maternal diet.

# CHANGES IN FATTY ACID COMPOSITION OF HUMAN MILK ACROSS LACTATION

The fatty acid composition of human breast milk varies across lactation and is acutely sensitive to the current maternal diet, as well as to mobilization of fatty acids from maternal fat stores (Makrides et al., 1995). Breast milk fatty acid concentrations change rapidly in response to shifts in the fatty acid composition of the maternal diet. This is clearly illustrated in a study conducted in the 1950s by Insull et al. (1959), in which lactating women were provided with 40% of dietary energy from lard, corn oil, or linseed oil. This study showed that the fatty acid composition of the breast milk changed to reflect the composition of the fat type to which the women were assigned within 2–3 days of the women changing from their usual diet. Since this very early study, studies from around the world have confirmed that the fatty acid composition of the maternal diet is the single most important factor contributing to the variability in human milk fatty acids. In addition, this relationship is almost completely accounted for by the intakes of preformed n-3 LCPUFA, and increases in maternal intake of the precursor n-3 PUFA, ALA, does not result in increases in breast milk DHA concentrations (Francois et al., 2003).

Even in the absence of changes in diet composition, however, the content of n-3 LCPUFA in the breast milk changes across pregnancy and lactation. Longitudinal studies have demonstrated that maternal DHA status decreases progressively across the second and third trimesters of pregnancy and the first 6 months of lactation (Markhus et al., 2015), after which there is a slow recovery to prepregnancy levels by the end of the first year postpartum (van Houwelingen et al., 1999). Importantly, maternal seafood consumption and/or intake of EPA/DHA supplements during pregnancy is positively correlated with her n-3 LCPUFA status during lactation (Markhus et al., 2015), suggesting that supplementation during pregnancy may also be beneficial for the development of breast-fed infants. For women with a low n-3 LCPUFA status at delivery, however, supplementation during the lactation period may be required to achieve adequate breast milk DHA concentrations.

# CURRENT RECOMMENDATIONS OF DHA AND EPA INTAKES IN LACTATING WOMEN FROM NATIONAL AND INTERNATIONAL HEALTH AGENCIES

Current recommendations are that lactating women should consume at least 200 mg of DHA/day. This was based on clinical trials that showed that supplementation of lactating women with 200 mg DHA/day increased human milk DHA content by about 0.2% fatty acids to a level considered desirable for infant outcomes (Koletzko et al., 2007). More recent extrapolation from dietary and supplementation studies has suggested that intakes of ~300 mg DHA/day are associated with a DHA content of 0.3%–0.35% of total fatty acids, and that this level is likely to be required to achieve optimal outcomes for infants (Mulder et al., 2014).

As with the level of n-3 LCPUFA supplementation required to achieve optimal maternal and infant outcomes, the optimal level of n-3 LCPUFA (and of DHA and EPA, separately) in the diet of breast-feeding women to support the development of their infant remains unclear. Again, however, this is likely to vary considerably between women, and to be heavily dependent on the content of other fatty acids in the diet, in particular the n-6 PUFA. Determining the optimal level of n-3 LCPUFA in breast milk and the level of n-3 LCPUFA intake/supplementation required to achieve this is a critical area for future research.

## Preterm Infants: A Special Case

It is important to note that the earlier discussion on the impact of n-3 LCPUFA supplementation in the neonatal period, describes the effects for the ~90% of infants who are born at term and without significant perinatal complications. However, those infants born preterm, particularly those born before 34 weeks gestation, miss out on the period of rapid DHA/EPA accumulation in the final trimester of gestation, and are therefore born with lower n-3 LCPUFA stores in comparison to infants born at or close to term (Lapillonne and Jensen, 2009). In addition, the various enzymatic systems responsible for the digestion and metabolism of fatty acids are less mature in preterm compared to term infants. Consequently, preterm infants, particularly those born very preterm, have increased n-3 LCPUFA requirements than term born infants in the early postnatal period, and the amount of DHA in breast milk and standard infant formulas is generally thought to be insufficient to meet their requirements (Lapillonne et al., 2013; Lapillonne and Jensen, 2009). The visual development of formula fed preterm infants has shown improvements with DHA given at amount similar to that in breast milk (DHA 0.2%–0.3% of total fatty acids), which led to changes in clinical practice to ensure DHA was present in preterm infant formula (Schulzke et al., 2011). However, no consistent benefit on cognitive development has been shown (Smithers et al., 2008a). Of more current relevance are the two trials that were inlcusive of breast milk feeding and used DHA doses at the estimated in utero accretion rate ( 60 mg/kg/day) (Henriksen et al., 2008; Makrides et al., 2009). Both trials reported improvements in visual acuity (Smithers et al., 2008b), problem-solving ability (Henriksen et al., 2008), and neurodevelopment (Makrides et al., 2009) up to 18 months corrected age, with the latter outcomes more evident in girls and infants <1250 g (Makrides et al., 2009). However, by middle childhood (7–8 years of age) differences between groups were no longer evident (Almaas et al., 2015; Collins et al., 2015). It is possible that the dose tested was not high enough or administered for sufficient length of time, but the most likely explanation is that the effect of family and the environment exert a greater influency on cognitive outcomes and the early benefits of specfic interventions in the neonatal period diminish during the early school years.

Perhaps even more so than with pregnant and lactating women, there is no clear consensus as to the precise n-3 LCPUFA requirements of this infant population, and it is likely that this will vary considerably depending on the gestational age at birth and other comorbidities. This remains an important area for further research.

# References

Abedin, L., Lien, E.L., Vingrys, A.J., Sinclair, A.J., 1999. The effects of dietary alpha-linolenic acid compared with docosahexaenoic acid on brain, retina, liver and heart in the guinea pig. Lipids 34, 475–482.

Al-Hasani, H., Joost, H.-G., 2005. Nutrition-/diet-induced changes in gene expression in white adipose tissue. Best Pract. Res. Clin. Endocrinol. Metab. 19, 589–603.

Almaas, A.N., Tamnes, C.K., Nakstad, B., Henriksen, C., Walhovd, K.B., Fjell, A.M., Due-Tonnessen, P., Drevon, C.A., Iversen, P.O., 2015. Long-chain polyunsaturated fatty acids and cognition in VLBW infants at 8 years: an RCT. Pediatrics 135, 972–980.

Beck, M., Zelczak, G., Lentze, M.J., 2000. Abnormal fatty acid composition in umbilical cord blood of infants at high risk of atopic disease. Acta Paediatr. 89, 279–284.

Best, K.P., Gold, M., Kennedy, D., Martin, J., Makrides, M., 2016a. Omega-3 long-chain PUFA intake during pregnancy and allergic disease outcomes in the offspring: a systematic review and meta-analysis of observational studies and randomized controlled trials. Am. J. Clin. Nutr. 103, 128–143.

Best, K.P., Sullivan, T., Palmer, D., Gold, M., Kennedy, D.J., Martin, J., Makrides, M., 2016b. Prenatal fish oil supplementation and allergy: 6-year follow-up of a randomized controlled trial. Pediatrics, 137.

Calder, P.C., 2012. Mechanisms of action of (n-3) fatty acids. J. Nutr. 142, 592s–599s.

Calder, P.C., 2013. n-3 Fatty acids, inflammation and immunity: new mechanisms to explain old actions. Proc. Nutr. Soc. 72, 326–336.

Calder, P.C., Yaqoob, P., 2009. Omega-3 polyunsaturated fatty acids and human health outcomes. Biofactors 35, 266–272.

Carlson, S.E., Colombo, J., Gajewski, B.J., Gustafson, K.M., Mundy, D., Yeast, J., Georgieff, M.K., Markley, L.A., Kerling, E.H., Shaddy, D.J., 2013. DHA supplementation and pregnancy outcomes. Am. J. Clin. Nutr. 97, 808–815.

Carnielli, V.P., Wattimena, D.J., Luijendijk, I.H., Boerlage, A., Degenhart, H.J., Sauer, P.J., 1996. The very low birth weight premature infant is capable of synthesizing arachidonic and docosahexaenoic acids from linoleic and linolenic acids. Pediatr. Res. 40, 169–174.

Catena, A., Munoz-Machicao, J.A., Torres-Espinola, F.J., Martinez-Zaldivar, C., Diaz-Piedra, C., Gil, A., Haile, G., Gyorei, E., Molloy, A.M., Decsi, T., Koletzko, B., Campoy, C., 2016. Folate and long-chain polyunsaturated fatty acid supplementation during pregnancy has long-term effects on the attention system of 8.5-year-old offspring: a randomized controlled trial. Am. J. Clin. Nutr. 103, 115–127.

Chapkin, R.S., Kim, W., Lupton, J.R., McMurray, D.N., 2009. Dietary docosahexaenoic and eicosapentaenoic acid: emerging mediators of inflammation. Prostagl. Leuk. Essent. Fatty Acids 81, 187–191.

Clandinin, M.T., Chappell, J.E., Leong, S., Heim, T., Swyer, P.R., Chance, G.W., 1980. Intrauterine fatty acid accretion rates in human brain: implications for fatty acid requirements. Early Hum. Dev. 4, 121–129.

Collins, C.T., Gibson, R.A., Anderson, P.J., McPhee, A.J., Sullivan, T.R., Gould, J.F., Ryan, P., Doyle, L.W., Davis, P.G., McMichael, J.E., French, N.P., Colditz, P.B., Simmer, K., Morris, S.A., Makrides, M., 2015. Neurodevelopmental outcomes at 7 years' corrected age in preterm infants who were fed high-dose docosahexaenoic acid to term equivalent: a follow-up of a randomised controlled trial. BMJ Open 5, e007314.

Crawford, M.A., Hassam, A.G., Williams, G., 1976. Essential fatty acids and fetal brain growth. Lancet 1, 452–453.

Cunnane, S.C., Francescutti, V., Brenna, J.T., Crawford, M.A., 2000. Breast-fed infants achieve a higher rate of brain and whole body docosahexaenoate accumulation than formula-fed infants not consuming dietary docosahexaenoate. Lipids 35, 105–111.

Denburg, J.A., Hatfield, H.M., Cyr, M.M., Hayes, L., Holt, P.G., Sehmi, R., Dunstan, J.A., Prescott, S.L., 2005. Fish oil supplementation in pregnancy modifies neonatal progenitors at birth in infants at risk of atopy. Pediatr. Res. 57, 276–281.

Elias, S.L., Innis, S.M., 2001. Infant plasma trans, n-6, and n-3 fatty acids and conjugated linoleic acids are related to maternal plasma fatty acids, length of gestation, and birth weight and length. Am. J. Clin. Nutr. 73, 807–814.

Everts, B., Wahrborg, P., Hedner, T., 2000. COX-2-Specific inhibitors: the emergence of a new class of analgesic and anti-inflammatory drugs. Clin. Rheumatol. 19, 331–343.

Francois, C.A., Connor, S.L., Bolewicz, L.C., Connor, W.E., 2003. Supplementing lactating women with flaxseed oil does not increase docosahexaenoic acid in their milk. Am. J. Clin. Nutr. 77, 226–233.

Gibson, R.A., Makrides, M., 1998. The role of long chain polyunsaturated fatty acids (LCPUFA) in neonatal nutrition. Acta Paediatr. 87, 1017–1022.

Gil-Sanchez, A., Demmelmair, H., Parrilla, J.J., Koletzko, B., Larque, E., 2011. Mechanisms involved in the selective transfer of long chain polyunsaturated fatty acids to the fetus. Front. Genet. 2, 57.

Gottrand, F., 2008. Long-chain polyunsaturated fatty acids influence the immune system of infants. J. Nutr. 138, 1807s–1812s.

Gould, J.F., Makrides, M., Colombo, J., Smithers, L.G., 2014. Randomized controlled trial of maternal omega-3 long-chain PUFA supplementation during pregnancy and early childhood development of attention, working memory, and inhibitory control. Am. J. Clin. Nutr. 99, 851–859.

Gould, J.F., Smithers, L.G., Makrides, M., 2013. The effect of maternal omega-3 (n-3) LCPUFA supplementation during pregnancy on early childhood cognitive and visual development: a systematic review and meta-analysis of randomized controlled trials. Am. J. Clin. Nutr. 97, 531–544.

Gunaratne, A.W., Makrides, M., Collins, C.T., 2015. Maternal prenatal and/or postnatal n-3 long chain polyunsaturated fatty acids (LCPUFA) supplementation for preventing allergies in early childhood. Cochrane Database Syst. Rev., Cd010085.

Hauner, H., Vollhardt, C., Schneider, K.T., Zimmermann, A., Schuster, T., Amann-Gassner, U., 2009. The impact of nutritional fatty acids during pregnancy and lactation on early human adipose tissue development. Rationale and design of the INFAT study. Ann. Nutr. Metab. 54, 97–103.

Henriksen, C., Haugholt, K., Lindgren, M., Aurvag, A.K., Ronnestad, A., Gronn, M., Solberg, R., Moen, A., Nakstad, B., Berge, R.K., Smith, L., Iversen, P.O., Drevon, C.A., 2008. Improved cognitive development among preterm infants attributable to early supplementation of human milk with docosahexaenoic acid and arachidonic acid. Pediatrics 121, 1137–1145.

Herrera, E., Amusquivar, E., 2000. Lipid metabolism in the fetus and the newborn. Diabetes Metab. Res. Rev. 16, 202–210.

Innis, S.M., 2005. Essential fatty acid transfer and fetal development. Placenta 26, S70–S75.

Innis, S.M., 2007. Human milk: maternal dietary lipids and infant development. Proc. Nutr. Soc. 66, 397–404.

Insull, Jr., W., Hirsch, J., James, T., Ahrens, Jr., E.H., 1959. The fatty acids of human milk. II. Alterations produced by manipulation of caloric balance and exchange of dietary fats. J. Clin. Invest. 38, 443–450.

Janssen, C.I., Kiliaan, A.J., 2014. Long-chain polyunsaturated fatty acids (LCPUFA) from genesis to senescence: the influence of LCPUFA on neural development, aging, and neurodegeneration. Prog. Lipid Res. 53, 1–17.

Jensen, C.L., Maude, M., Anderson, R.E., Heird, W.C., 2000. Effect of docosahexaenoic acid supplementation of lactating women on the fatty acid composition of breast milk lipids and maternal and infant plasma phospholipids. Am. J. Clin. Nutr. 71, 292s–299s.

Keenan, K., Hipwel, l.A., McAloon, R., Hoffmann, A., Mohanty, A., Magee, K.P., 2016. The effect of prenatal docosahexaenoic acid supplementation on infant outcomes in African American women living in low-income environments: a randomized, controlled trial. Psychoneuroendocrinology 71, 170–175.

Klemens, C.M., Berman, D.R., Mozurkewich, E.L., 2011. The effect of perinatal omega-3 fatty acid supplementation on inflammatory markers and allergic diseases: a systematic review. BJOG 118, 916–925.

Koletzko, B., Cetin, I., Brenna, J.T., Group, P.L.I.W., 2007. Dietary fat intakes for pregnant and lactating women. Br. J. Nutr. 98, 873–877.

Koletzko, B., Lattka, E., Zeilinger, S., Illig, T., Steer, C., 2011. Genetic variants of the fatty acid desaturase gene cluster predict amounts of red blood cell docosahexaenoic and other polyunsaturated fatty acids in pregnant women: findings from the Avon Longitudinal Study of Parents and Children. Am. J. Clin. Nutr. 93, 211–219.

Kuhn, D.C., Crawford, M., 1986. Placental essential fatty acid transport and prostaglandin synthesis. Prog. Lipid Res. 25, 345–353.

Kuipers, R.S., Luxwolda, M.F., Offringa, P.J., Boersma, E.R., Dijck-Brouwer, D.A., Muskiet, F.A., 2012. Fetal intrauterine whole body linoleic, arachidonic and docosahexaenoic acid contents and accretion rates. Prostagl. Leuk. Essent. Fatty Acids 86, 13–20.

Lands, B., 2008. A critique of paradoxes in current advice on dietary lipids. Prog. Lipid Res. 47, 77–106.

Lands, W.E., Libelt, B., Morris, A., Kramer, N.C., Prewitt, T.E., Bowen, P., Schmeisser, D., Davidson, M.H., Burns, J.H., 1992. Maintenance of lower proportions of (n-6) eicosanoid precursors in phospholipids of human plasma in response to added dietary (n-3) fatty acids. Biochim. Biophys. Acta 1180, 147–162.

Lapillonne, A., Groh-Wargo, S., Gonzalez, C.H., Uauy, R., 2013. Lipid needs of preterm infants: updated recommendations. J. Pediatr. 162, S37–S47.

Lapillonne, A., Jensen, C.L., 2009. Reevaluation of the DHA requirement for the premature infant. Prostagl. Leuk. Essent. Fatty Acids 81, 143–150.

Larque, E., Demmelmair, H., Gil-Sanchez, A., Prieto-Sanchez, M.T., Blanco, J.E., Pagan, A., Faber, F.L., Zamora, S., Parrilla, J.J., Koletzko, B., 2011. Placental transfer of fatty acids and fetal implications. Am. J. Clin. Nutr. 94, 1908S–1913S.

Larque, E., Krauss-Etschmann, S., Campoy, C., Hartl, D., Linde, J., Klingler, M., Demmelmair, H., Cano, A., Gil, A., Bondy, B., Koletzko, B., 2006. Docosahexaenoic acid supply in pregnancy affects placental expression of fatty acid transport proteins. Am. J. Clin. Nutr. 84, 853–861.

Lattka, E., Koletzko, B., Zeilinger, S., Hibbeln, J.R., Klopp, N., Ring, S.M., Steer, C.D., 2013. Umbilical cord PUFA are determined by maternal and child fatty acid desaturase (FADS) genetic variants in the Avon Longitudinal Study of Parents and Children (ALSPAC). Br. J. Nutr. 109, 1196–1210.

Levant, B., Radel, J.D., Carlson, S.E., 2004. Decreased brain docosahexaenoic acid during development alters dopamine-related behaviors in adult rats that are differentially affected by dietary remediation. Behav. Brain Res. 152, 49–57.

Luxwolda, M.F., Kuipers, R.S., Sango, W.S., Kwesigabo, G., Dijck-Brouwer, D.A., Muskiet, F.A., 2012. A maternal erythrocyte DHA content of approximately 6 g% is the DHA status at which intrauterine DHA biomagnifications turns into bioattenuation and postnatal infant DHA equilibrium is reached. Eur. J. Nutr. 51, 665–675.

Makrides, M., Duley, L., Olsen, S.F., 2006. Marine oil, and other prostaglandin precursor, supplementation for pregnancy uncomplicated by pre-eclampsia or intrauterine growth restriction. Cochrane Database Syst. Rev. 3, CD003402.

Makrides, M., Gibson, R.A., 2000. Long-chain polyunsaturated fatty acid requirements during pregnancy and lactation. Am. J. Clin. Nutr. 71, 307S–311S.

Makrides, M., Gibson, R.A., 2002. The role of fats in the lifecycle stages: pregnancy and the first year of life. Med. J. Aust. 176, S111–S112.

Makrides, M., Gibson, R.A., McPhee, A.J., Collins, C.T., Davis, P.G., Doyle, L.W., Simmer, K., Colditz, P.B., Morris, S., Smithers, L.G., Willson, K., Ryan, P., 2009. Neurodevelopmental outcomes of preterm infants fed high-dose docosahexaenoic acid: a randomized controlled trial. JAMA 301, 175–182.

Makrides, M., Gibson, R.A., McPhee, A.J., Yelland, L., Quinlivan, J., Ryan, P., 2010. Effect of DHA supplementation during pregnancy on maternal depression and neurodevelopment of young children: a randomized controlled trial. JAMA 304, 1675–1683.

Makrides, M., Gibson, R.A., Udell, T., Ried, K., 2005. Supplementation of infant formula with long-chain polyunsaturated fatty acids does not influence the growth of term infants. Am. J. Clin. Nutr. 81, 1094–1101.

Makrides, M., Gould, J.F., Gawlik, N.R., Yelland, L.T., 2014. Four-year follow-up of children born to women in a randomized trial of prenatal DHA supplementation. JAMA 311, 1802–1804.

Makrides, M., Neumann, M.A., Byard, R.W., Simmer, K., Gibson, R.A., 1994. Fatty acid composition of brain, retina, and erythrocytes in breast- and formula-fed infants. Am. J. Clin. Nutr. 60, 189–194.

Makrides, M., Neumann, M.A., Gibson, R.A., 1996. Effect of maternal docosahexaenoic acid (DHA) supplementation on breast milk composition. Eur. J. Clin. Nutr. 50, 352–357.

Makrides, M., Simmer, K., Neumann, M., Gibson, R., 1995. Changes in the polyunsaturated fatty acids of breast milk from mothers of full-term infants over 30 week of lactation. Am. J. Clin. Nutr. 61, 1231–1233.

Markhus, M.W., Rasinger, J.D., Malde, M.K., Froyland, L., Skotheim, S., Braarud, H.C., Stormark, K.M., Graff, I.E., 2015. Docosahexaenoic acid status in pregnancy determines the maternal docosahexaenoic acid status 3-, 6- and 12 months postpartum: results from a longitudinal observational study. PLoS One 10, e0136409.

Mayes, C., Burdge, G.C., Bingham, A., Murphy, J.L., Tubman, R., Wootton, S.A., 2006. Variation in [U-13C] alpha linolenic acid absorption, beta-oxidation and conversion to docosahexaenoic acid in the pre-term infant fed a DHA-enriched formula. Pediatr. Res. 59, 271–275.

Meldrum, S., Dunstan, J.A., Foster, J.K., Simmer, K., Prescott, S.L., 2015. Maternal fish oil supplementation in pregnancy: a 12 year follow-up of a randomised controlled trial. Nutrients 7, 2061–2067.

Mihrshahi, S., Peat, J.K., Webb, K., Oddy, W., Marks, G.B., Mellis, C.M., 2004. Effect of omega-3 fatty acid concentrations in plasma on symptoms of asthma at 18 months of age. Pediatr. Allergy Immunol. 15, 517–522.

Moriguchi, T., Greiner, R.S., Salem, N.J., 2000. Behavioral deficits associated with dietary induction of decreased brain docosahexaenoic acid concentration. J. Neurochem. 75, 2563–2573.

Muhlhausler, B.S., Gibson, R.A., Makrides, M., 2010. Effect of long-chain polyunsaturated fatty acid supplementation during pregnancy or lactation on infant and child body composition: a systematic review. Am. J. Clin. Nutr. 92, 857–863.

Muhlhausler, B.S., Gibson, R.A., Yelland, L.N., Makrides, M., 2014. Heterogeneity in cord blood DHA concentration: towards an explanation. Prostagl. Leuk. Essent. Fatty Acids 91, 135–140.

Muhlhausler, B.S., Yelland, L.N., McDermott, R., Tapsell, L., McPhee, A., Gibson, R.A., Makrides, M., 2016. DHA supplementation during pregnancy does not reduce BMI or body fat mass in children: follow-up of the DHA to optimize mother infant outcome randomized controlled trial. Am. J. Clin. Nutr. 103, 1489–1496.

Mulder, K.A., King, D.J., Innis, S.M., 2014. Omega-3 fatty acid deficiency in infants before birth identified using a randomized trial of maternal DHA supplementation in pregnancy. PLoS One 9, e83764.

Neuringer, M., Connor, W.E., Lin, D.S., Barstad, L., Luck, S., 1986. Biochemical and functional effects of prenatal and postnatal omega 3 fatty acid deficiency on retina and brain in rhesus monkeys. Proc. Natl. Acad. Sci. U.S.A. 83, 4021–4025.

Olsen, S.F., Hansen, H.S., Sorensen, T.I., Jensen, B., Secher, N.J., Sommer, S., Knudsen, L.B., 1986. Intake of marine fat, rich in (n-3)-polyunsaturated fatty acids, may increase birthweight by prolonging gestation. Lancet 2, 367–369.

Olsen, S.F., Joensen, H.D., 1985. High liveborn birth weights in the Faroes: a comparison between birth weights in the Faroes and in Denmark. J. Epidemiol. Commun. Health 39, 27–32.

Olsen, S.F., Sorensen, J.D., Secher, N.J., Hedegaard, M., Henriksen, T.B., Hansen, H.S., Grant, A., 1992. Randomised controlled trial of effect of fish-oil supplementation on pregnancy duration. Lancet 339, 1003–1007.

Palmer, D.J., Sullivan, T., Gold, M.S., Prescott, S.L., Heddle, R., Gibson, R.A., Makrides, M., 2012. Effect of n-3 long chain polyunsaturated fatty acid supplementation in pregnancy on infants' allergies in first year of life: randomised controlled trial. BMJ 344, e184.

Palmer, D.J., Sullivan, T., Gold, M.S., Prescott, S.L., Heddle, R., Gibson, R.A., Makrides, M., 2013. Randomized controlled trial of fish oil supplementation in pregnancy on childhood allergies. Allergy 68, 1370–1376.

Pawlosky, R.J., Denkins, Y., Ward, G., Salem, N.J., 1997. Retinal and brain accretion of long-chain polyunsaturated fatty acids in developing felines: the effects of corn oil-based maternal diets. Am. J. Clin. Nutr. 65, 465–472.

Pawlosky, R.J., Lin, Y.H., Llanos, A., Mena, P., Uauy, R., Salem, Jr., N., 2006. Compartmental analyses of plasma 13C- and 2H-labeled n-6 fatty acids arising from oral administrations of 13C-U-18:2n-6 and 2H5-20:3n-6 in newborn infants. Pediatr. Res. 60, 327–333.

Prescott, S.L., Dunstan, J.A., 2007. Prenatal fatty acid status and immune development: the pathways and the evidence. Lipids 42, 801–810.

Ramakrishnan, U., Gonzalez-Casanova, I., Schnaas, L., DiGirolamo, A., Quezada, A.D., Pallo, B.C., Hao, W., Neufeld, L.M., Rivera, J.A., Stein, A.D., Martorell, R., 2016. Prenatal supplementation with DHA improves attention at 5 years of age: a randomized controlled trial. Am. J. Clin. Nutr. 104, 1075–1082.

Schulzke, S.M., Patole, S.K., Simmer, K., 2011. Long-chain polyunsaturated fatty acid supplementation in preterm infants. Cochrane Database Syst. Rev. 2, CD000375.

Simmer, K., Patole, S.K., Rao, S.C., 2011. Long-chain polyunsaturated fatty acid supplementation in infants born at term. Cochrane Database Syst. Rev. 12, CD000376.

Smithers, L.G., Gibson, R.A., Makrides, M., 2011. Maternal supplementation with docosahexaenoic acid during pregnancy does not affect early visual development in the infant: a randomized controlled trial. Am. J. Clin. Nutr. 93, 1293–1299.

Smithers, L.G., Gibson, R.A., McPhee, A., Makrides, M., 2008a. Effect of long-chain polyunsaturated fatty acid supplementation of preterm infants on disease risk and neurodevelopment: a systematic review of randomized controlled trials. Am. J. Clin. Nutr. 87, 912–920.

Smithers, L.G., Gibson, R.A., McPhee, A., Makrides, M., 2008b. Higher dose of docosahexaenoic acid in the neonatal period improves visual acuity of preterm infants: results of a randomized controlled trial. Am. J. Clin. Nutr. 88, 1049–1056.

van Houwelingen, A.C., Ham, E.C., Hornstra, G., 1999. The female docosahexaenoic acid status related to the number of completed pregnancies. Lipids 34 (Suppl.), S229.

Woods, R.K., Thien, F.C., Abramson, M.J., 2002. Dietary marine fatty acids (fish oil) for asthma in adults and children. Cochrane Database Syst. Rev. 3, Cd001283.

Zhou, S.J., Yelland, L., McPhee, A.J., Quinlivan, J., Gibson, R.A., Makrides, M., 2012. Fish-oil supplementation in pregnancy does not reduce the risk of gestational diabetes or preeclampsia. Am. J. Clin. Nutr. 95, 1378–1384.

# Metabolism of Polyunsaturated Fatty Acids by Cells of the Immune System

*Philip C. Calder*

University of Southampton, Southampton, United Kingdom

OUTLINE

Polyunsaturated Fatty Acid Metabolism. http://dx.doi.org/10.1016/B978-0-12-811230-4.00008-9

# INTRODUCTION

The immune system is vital to host defense against pathogenic organisms like bacteria and viruses. The immune system also plays a central role in assuring tolerance to harmless microorganisms, to innocuous components of the foods than an individual eats, and to the tissues of the host. Any breakdown in these functions of defence or tolerance can lead to disease. The number of individual structures that the immune system regularly encounters is vast and its responses fall into four general types: barrier function; recognition and deciphering; actions leading to elimination or tolerance; memory. To enable such a sophisticated array of responses to so many possible structures from the external and internal environments the immune system comprises many different cell types, each with their own specific role in the immune response. Many of these cell types exist as a large number of different subtypes, which display, often subtle, differences in function. Furthermore, there are many different cellular interactions that occur as part of the immune response and there are many different chemical mediators and effector molecules produced and secreted. Most cells of the immune system have a discrete array of interactions and secretions linked to the specific function of that cell type. The secreted products include proteins like antibacterial peptides, cytokines, chemokines, adhesion molecules, proteases and antibodies; amino acid derivatives like histamine and nitric oxide; lipid-derived mediators like prostaglandins (PGs), leukotrienes (LTs), and platelet activating factor; and oxygen-derived mediators like superoxide and hydrogen peroxide. These are produced as a result of increased enzyme activity so that substrates can be made available and then metabolized or as a result of upregulated transcription (mRNA synthesis) and translation (protein synthesis). Such enhanced metabolism requires energy. The immune response also involves significant cellular proliferation; again this will involve much biosynthesis requiring substrate (e.g., to form new cell membranes) and energy. Fatty acids in general, and polyunsaturated fatty acids (PUFAs) in particular, have key roles in supporting and regulating the immune response: fatty acids are important energy sources for cells of the immune system, they are vital structural and functional components of immune cell membranes, they are regulators of the immune response or structural components of regulators, and they are substrates for the synthesis of some of the most important immunoregulatory lipid mediators. This chapter provides an overview of the metabolism of PUFAs by cells of the immune system, as summarized in Fig. 8.1.

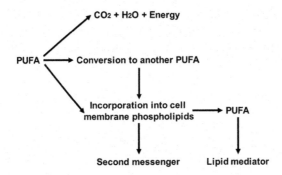

FIGURE 8.1 Summary of the possible pathways of metabolism of polyunsaturated fatty acids (PUFAs) in cells of the immune system.

## USE OF (POLYUNSATURATED) FATTY ACIDS AS FUELS BY CELLS OF THE IMMUNE SYSTEM

Up until the early 1980s it was generally considered that glucose was the sole fuel substrate for cells of the immune system and that much of their metabolism was glycolytic and nonoxidative (Calder, 1995). However, Ardawi and Newsholme (1984) reported that rat lymph node lymphocytes oxidized oleic acid added to their incubation medium, although oxidation accounted for only a small proportion of oleic acid utilisation. Nevertheless, fatty acid oxidation was calculated to contribute to more than 30% of oxygen consumption of isolated lymphocytes (Ardawi and Newsholme, 1984). Mitogenic stimulation of lymphocytes did not increase the rate of oxidation of oleic acid in short-term incubations (Ardawi and Newsholme, 1984). In similar in vitro experiments, Newsholme and Newsholme (1989) showed that murine peritoneal macrophages oxidized a small proportion of oleic acid from their incubation medium and that oleic acid oxidation could account for more than 20% of oxygen consumption in such incubations. Spolarics et al. (1991) calculated that for cultured rat Kupffer cells (these are specialized macrophages found in the liver) the oxidation of palmitic acid contributed about 30% of cellular ATP generated. These studies did not examine the ability of lymphocytes or macrophages to oxidize PUFAs. However, Yaqoob et al. (1994) identified that rat lymph node lymphocytes oxidized some of the linoleic acid (LA; 18:2n-6) and arachidonic acid (AA; 20:4n-6) provided to them in short-term incubations, although the rates of oxidation were less than seen for palmitic and oleic acids. Yaqoob et al. (1994) confirmed that mitogenic stimulation of lymphocytes did not significantly enhance the rate of oxidation of any of the fatty acids tested. In recent years there has been a significant interest in the role of fatty acid oxidation in determining immune cell phenotype. This has been studied for both macrophages and T lymphocytes (also called T-cells). This work has been summarized and discussed in several recent reviews (Chiaranunt et al., 2015; Lochner et al., 2015; Mills and O'Neill, 2016; Namgaladze and Brüne, 2016; Van den Bossche et al., 2017; Wang and Green, 2012).

Macrophages are classified into two phenotypes, M1 or classically activated macrophages and M2 or alternatively activated macrophages. M1 macrophages are generated in response to stimulation with bacterial or viral products (e.g., lipopolysaccharide) through toll-like receptors in combination with the presence of the cytokine interferon (IFN)-gamma. M2 macrophages are generated in response to the cytokines interleukin (IL)-4 and IL-13. In general, M1 macrophages are important for dealing with microbial infections, and they are considered to be more inflammatory, being potent producers of proinflammatory cytokines and nitric oxide. In contrast, M2 macrophages have roles in the regulation of inflammation (i.e., they are antiinflammatory), tissue repair and return to homeostasis. Interestingly, M1 and M2 macrophages differ in their metabolic signatures. The metabolism of M1 macrophages is dominated by a high rate of glycolysis producing energy largely nonoxidatively. M1 macrophages also show high rates of activity of the pentose phosphate pathway and of fatty acid biosynthesis and impaired activity of the Krebs cycle. In contrast, the metabolism of M2 macrophages is dominated by fatty acid oxidation and oxidative phosphorylation with high Krebs cycle activity. The data suggest that when fatty acid oxidation becomes the dominant energy-generating pathway in the macrophage, an antiinflammatory phenotype results. Consistent with this conclusion, exposure of macrophages to the fatty acid oxidation inhibitor Etomoxir prevented polarisation to the M2 phenotype (Vats et al., 2006), although this finding is not reported by all studies.

The finding that enhanced fatty acid oxidation is intimately linked with a less inflammatory macrophage phenotype should not lead to the conclusion that simply exposing macrophages to fatty acids will be antiinflammatory. This is evident from in vitro studies with lauric and palmitic acids, which each induce a range of proinflammatory macrophage responses (Lee et al., 2001; Maloney et al., 2009). Both these fatty acids activate inflammation through toll-like receptor 4, while palmitic acid promotes generation of proinflammatory signaling molecules like ceramides and diacylglycerols and promotes endoplasmic reticulum stress, which is proinflammatory. Interestingly, some PUFAs, especially the long chain n-3 PUFA docosahexaenoic acid (DHA; 22:6n-6), are able to oppose the actions of lauric and palmitic acids resulting in less inflammation (Hwang et al., 2016; Lee et al., 2001). This removal of saturated fatty acid driven inflammation may allow macrophages to adopt a less inflammatory, fatty acid oxidation dominant, more M2-like phenotype.

Interestingly, T-cell activation and differentiation to the different functional T-cell phenotypes show strong parallels with the macrophage M1 versus M2 pathway. The metabolism of naive T-cells is dominated by fatty acid oxidation. Following T-cell stimulation their metabolism switches to one dominated by glycolysis, the pentose-phosphate pathway and metabolism of the amino acid glutamine. This metabolic profile is consistent with the large amount of biosynthesis that is required at this time prior to cell division. As activated T-lymphocytes begin to proliferate, they are driven toward different functional subsets depending on the availability of various extracellular signals including key cytokines like IFN-gamma, IL-2, and IL-4. These different T-cell subsets determine the nature of the immune response that is generated. For example, CD4$^+$ T-cells differentiate into either T helper 1, T helper 2, or T helper 17 subsets that mediate various aspects of the immune response depending on the initial activators or into induced regulatory T-cells that suppress uncontrolled immune responses. CD8$^+$ T-cells differentiate into cytotoxic T lymphocytes that kill host cells infected with pathogens. After the pathogens are cleared, most differentiated T-cells undergo apoptosis. The antigen-specific T-cells that remain (these are memory T-cells) are responsible for enhanced immunity after re-exposure to the pathogen. Of these various T-cell subsets, the induced regulatory T-cells and memory T-cells rely on fatty acid oxidation as their energy source, whereas cytotoxic T-lymphocytes and most effector T-cells sustain high glycolytic activity.

Most studies investigating fatty acid oxidation by cells of the immune system have used palmitic or oleic acids. There are few studies investigating the oxidation of PUFAs by cells of the immune system and how that might affect cell phenotype and functionality, despite the large number of studies of the effects of PUFAs, especially n-3 PUFAs, on immune and inflammatory cell responses. It is possible that some of the reported impact of different fatty acids on immune and inflammatory cell responses could be due a switch in metabolism that exposure to those fatty acids induces.

## POLYUNSATURATED FATTY ACID BIOSYNTHESIS BY CELLS OF THE IMMUNE SYSTEM

The ability of cells of the immune system to convert one PUFA to another has been studied in cell-culture systems usually involving exposure of the cells to a fairly high concentration of a single added PUFA, which in many studies was radioisotopically labeled. Chapkin et al. (1988a) demonstrated that murine peritoneal macrophages could not convert LA to AA but instead produced the LA elongation product 20:2n-6. They concluded that the high amounts of AA in the macrophage membrane must be supplied from outside of the cell. In a follow-up study the same authors showed that di-homo-gamma linolenic acid (DGLA; 20:3n-6) could be converted to AA and to other n-6 PUFAs, including 22:4n-6, by murine peritoneal macrophages, suggesting some Δ5-desaturase and elongase activity (Chapkin et al., 1988b). Taking these two studies together suggests that peritoneal macrophages lack Δ6-desaturase activity. It was also reported that gamma-linolenic acid (GLA; 18:3n-6) could be converted to DGLA by these cells (Chapkin and Coble, 1991), indicative of elongase activity. Likewise, Chilton-Lopez et al. (1996) demonstrated that GLA could be converted to DGLA by human blood neutrophils in vitro. In another study Chapkin and Miller (1990) reported that murine peritoneal macrophages could elongate AA to adrenic acid (22:4n-6) and eicosapentaenoic acid (EPA; 20:5n-3) to docosapentaenoic acid (DPA; 22:5n-3), with the latter conversion being more active. No conversion of EPA to DHA was observed in that study. Calder et al. (1990) found that incubating murine peritoneal macrophages with EPA resulted in higher proportions of both EPA and DPA in cellular phospholipids, consistent with the findings of Chapkin and Miller (1990). Incubation with DHA resulted in higher proportions of both DHA and EPA in cellular phospholipids (Calder et al., 1990), suggesting that the process of retroconversion occurs in murine macrophages. Incubation with alpha-linolenic acid (ALA; 18:3n-3) did not result in increased EPA in the phospholipids, consistent with the absence of Δ6-desaturase activity in these cells. In similar experiments, rat lymph node lymphocytes were incubated with different fatty acids. Incubation with ALA resulted in higher proportions of EPA, DPA, and even DHA in these cells (Calder et al., 1994), suggesting they have an intact PUFA metabolism pathway. Consistent with this, incubation with EPA increased both DPA and DHA as well as EPA. Incubation with DHA resulted in higher proportions of EPA and DPA as well as DHA, demonstrating that lymphocytes also retroconvert.

## ACTIVATION OF LYMPHOCYTES INDUCES CHANGES IN THEIR POLYUNSATURATED FATTY ACID CONTENT AND ACTIVATES POLYUNSATURATED FATTY ACID BIOSYNTHESIS

Anel et al. (1990a) determined the changes in the fatty acid composition of human peripheral blood mononuclear cells (PBMCs; these are a mix of about 85% lymphocytes, in which T-cells dominate, and about 15% monocytes) and of purified T-cells upon mitogenic activation. After 24 h of stimulation the proportions of oleic acid, DPA and DHA increased while the proportions of LA and AA decreased. The changes in fatty acid composition were associated with changes in membrane order (i.e., the physical nature of the membrane). It is not clear if the changes in fatty acid composition reported reflect alterations in PUFA biosynthetic pathways, in phospholipid metabolism, or in PUFA turnover in phospholipids.

In follow-up experiments, Anel et al. (1990b) studied the incorporation and metabolism of fatty acids by mitogen-activated human PBMCs. Addition of stearic acid, oleic acid, LA, ALA, or AA to the culture medium resulted in appearance of the elongation products of these fatty acids in the mitogen-activated PBMCs. Incubation with stearic acid resulted in an increased proportion of oleic acid in the cells indicating the presence of Δ9-desaturase activity. PBMCs were incubated with radioactively labeled stearic acid, LA or ALA in order to track the metabolism of these fatty acids. Unstimulated cells showed little Δ9-desaturase and no Δ6-desaturase activity. Mitogenic stimulation of the cells resulted in appearance of labeled fatty acids consistent with increased activity of Δ9-, Δ6-, and Δ5-desaturases. LA was converted to DGLA and AA, while ALA was converted to EPA and DPA, but not to DHA. These findings demonstrate that within the PBMC population some cell types, most likely T-cells, have an intact Δ6-desaturase, elongase, and Δ5-desaturase pathway for conversion of the two essential PUFAs to longer-chain, more unsaturated derivatives. The findings of the later study by Calder et al. (1994) described in the previous section are consistent with this conclusion.

Calder et al. (1994) also studied the effect of mitogenic activation on immune cell fatty acid composition using a preparation of rat lymph node cells which are almost entirely T-cells. Consistent with the report of Anel et al. (1990b) the fatty acid composition of the cell phospholipids was not altered in culture if the cells were not stimulated. However, 24 h of mitogenic stimulation resulted in decreased proportions of stearic acid and LA and an increased proportion of oleic acid compared with the fresh cells. More marked changes occurred with 48 h of stimulation: the proportions of palmitic, stearic, linoleic, and arachidonic acids were reduced, and the proportion of oleic acid was markedly increased. The proportions of EPA and DHA were not significantly affected.

# INCORPORATION OF EXOGENOUS POLYUNSATURATED FATTY ACIDS INTO CELLS OF THE IMMUNE SYSTEM

The fatty acid composition of immune tissues, such as the spleen, thymus, or lymph nodes, of immune cells from those tissues or from the blood, and of immune cell phospholipids has been reported many times. It is generally assumed that these measurements largely reflect the fatty acids in the membranes of the cells. This is believed to be of functional significance for two reasons (Fig. 8.2). First, the fatty acid

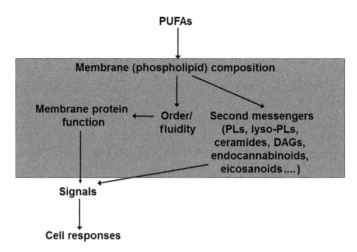

FIGURE 8.2  Summary of the mechanisms by which incorporation of polyunsaturated fatty acids (PUFAs) into immune cell membrane phospholipids can influence the functional responses of those cells.
*DAG*, diacylglycerol; *PL*, phospholipid.

components of membrane phospholipids influence the physical nature of the membrane (often called membrane order or membrane fluidity), which affects the environment and activity of membrane proteins. Second, membrane phospholipids are precursors for the biosynthesis of many important signaling molecules, including lysophospholipids, diacylglycerols, ceramides, endocannabinoids, and eicosanoids, the potency of which can be influenced the fatty acid makeup of the parent phospholipid. However, it is important to keep in mind that there are many different types of membrane (e.g., plasma, nuclear, mitochondrial, endoplasmic reticulum) and that different membranes adopt different fatty acid compositions. Furthermore, different membranes contain different relative amounts of different types of phospholipids (e.g., phosphatidylcholine, phosphatidyethanolamine, phosphatidylserine) and these have different fatty acid compositions. Thus, when the fatty acid composition of the entire immune cell or even of isolated immune cell phospholipids is measured this reflects quite a mixture of membranes and phospholipids.

The phospholipids of immune cells (such as lymph node or splenic lymphocytes or peritoneal macrophages) taken from rodents maintained on normal laboratory chow typically contain 5%–10% of fatty acids as LA and 15%–20% of fatty acids as AA and contain very little ALA, EPA, DPA, or DHA (Anel et al., 1990a; Calder et al., 1990, 1994). Incubating such cells with specific fatty acids in culture results in quite marked changes in fatty acid composition, typically with significant increases in the proportion of the fatty acid added to the culture medium as well as some further metabolism of that fatty acid (see earlier sections). Calder et al. (1994) incubated rat lymph node lymphocytes with radioactively labeled AA. Incorporation was modest into unstimulated cells over 64 h but was increased 6-fold by mitogenic stimulation. Most of the AA was incorporated into cellular phospholipids and within the phospholipids the incorporation was phosphatidylethanolamine > phosphatidylcholine > phosphatidylserine + phosphatidylinositol.

Modifying the fatty acid composition of the diet of experimental animals leads to altered fatty acid composition of the immune cells, typically with increased appearance of the fatty acid(s) in which the diet has become enriched. Thus, it is possible to enrich immune cells in AA by feeding experimental animals a diet containing AA (Peterson et al., 1998; Whelan et al., 1993) or in EPA and DHA by feeding a diet containing these fatty acids (Brouard and Pascaud, 1990; Chapkin et al., 1988c, 1992; Hardardóttir and Kinsella, 1992; Kew et al., 2003a; Lokesh et al., 1986; Peterson et al., 1998; Surette et al., 1995; Wallace et al., 2000, 2001; Yaqoob et al., 1995). Incorporation of EPA and DHA into rodent immune cells is accompanied by a decrease in the content of AA (Kew et al., 2003a; Peterson et al., 1998; Wallace et al., 2000, 2001; Yaqoob et al., 1995). Fritsche (2007) collated much data from rodent feeding studies with n-3 PUFAs and demonstrated that incorporation of both EPA and DHA into rodent lymphocytes and macrophages, and the corresponding reduction in the content of AA, occurs in a dose-dependent manner. However, there are marked subtleties to the changes in the PUFA composition of immune cells that are not revealed by measuring fatty acids in bulk phospholipids. This is clearly revealed by the study of Kew et al. (2003a) where EPA was shown to be incorporated into specific molecular species of phosphatidylcholine within spleen cells and to modify the amount of specific AA-containing phosphatidylcholine molecular species. For example including EPA in the mouse diet reduced the amount of stearic acid-AA phosphatidylcholine in spleen cells but did not affect the amounts of palmitic acid-AA or oleic acid-AA phosphatidylcholine. Furthermore, AA containing phosphatidylinositol molecular species in spleen cells were unaffected by inclusion of EPA or DHA in the diet.

The bulk phospholipids of immune cells (e.g., neutrophils, lymphocytes, monocytes) from the blood of humans consuming typical Western diets also contain about 10% of fatty acids as LA and about 15%–20% as AA, with about 0.5%–1% EPA and about 2%–3% DHA (Browning et al., 2012; Caughey et al., 1996; Endres et al., 1989; Faber et al., 2011; Gibney and Hunter, 1993; Healy et al., 2000; Kew et al., 2003b, 2004; Lee et al., 1985; Miles et al., 2004; Rees et al., 2006; Sperling et al., 1993; Thies et al., 2001; Yaqoob et al., 2000), although there are differences between phospholipid classes in terms of the content of these fatty acids (Sperling et al., 1993).

The finding that fatty acid composition of human immune cells can be modified by altering oral intakes of certain fatty acids has been well described. For example, increased AA intake resulted in a higher proportion of AA in PBMCs (Thies et al., 2001), while increased intake of GLA increased the DGLA content of PBMCs (Miles et al., 2004; Thies et al., 2001; Yaqoob et al., 2000) and neutrophils (Johnson et al., 1997). However, GLA did not increase AA in either PBMCs or neutrophils. Increasing the intake of ALA resulted in increased content of EPA in PBMCs (Caughey et al., 1996; Kelley et al., 1993; Kew et al., 2003b; Zhao et al., 2004) and neutrophils (Healy et al., 2000; Mantzioris et al., 1994). Likewise, increasing the intake of SDA resulted in increased content of EPA in PBMCs (Miles et al., 2004) and neutrophils (Surette et al., 2004). However, increasing intake of ALA or SDA did not increase the DHA content of human PBMCs or neutrophils.

Increasing intake of EPA and DHA from fish oil or from purer n-3 preparations resulted in increased proportions of the respective n-3 PUFAs in blood immune cells in humans (Browning et al., 2012; Caughey et al., 1996; Endres et al., 1989; Faber et al., 2011; Fisher et al., 1990; Gibney and Hunter, 1993; Healy et al., 2000; Kelley et al., 1999; Kew et al., 2003b, 2004; Lee et al., 1985; Luostarinen and Saldeen, 1996; Miles et al., 2004; Mølvig et al., 1991; Rees et al., 2006; Sperling et al., 1993; Schmidt et al., 1996; Thies et al., 2001; Yaqoob et al., 2000). Typically, the increase in content of EPA and DHA occurs at the expense of n-6 PUFAs, especially AA. Time-course studies have indicated that the incorporation of EPA and DHA into human immune cells reaches its peak within a few weeks of commencing increased intake (Browning et al., 2012; Faber et al., 2011; Healy et al., 2000; Kew et al., 2004; Miles et al., 2004; Rees et al., 2006; Thies et al., 2001; Yaqoob et al., 2000) (Fig. 8.3) and studies using multiple doses of n-3 PUFAs show that the incorporation of these fatty acids occurs in a manner that is highly correlated with the amount of the fatty acid consumed (Browning et al., 2012; Healy et al., 2000; Rees et al., 2006). In Fig. 8.4 data from a number of studies have been pooled together to reveal the nature of the dose-response incorporation of EPA and DHA into human PBMCs. It is clear that EPA incorporation displays a good linear dose-response relationship (Fig. 8.4), at least over the range of EPA intakes used. Reasons for the variation in response seen among different studies include duration of supplementation, the starting EPA content of the cells, the amount of DHA used, the background n-6 PUFA intake, differences between subjects studied (e.g., age), and differences in laboratory processing and analysis of samples. Although pooling data from a number of studies also reveals a linear dose response relationship for DHA incorporation, this relationship is less clear when lower doses of DHA have been used. The reason for this is most likely that when EPA and DHA are used in a combination in which EPA predominates, EPA is

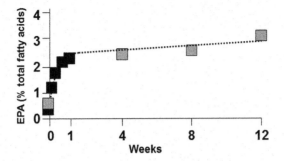

FIGURE 8.3   **Time-dependent changes in the eicosapentaenoic acid (EPA) content of human mononuclear cells when EPA intake is increased.** Healthy human volunteers consumed fish oil providing 2.1 g EPA and 1.1 g DHA per day for 1 week (Faber et al., 2011) or for 12 weeks (Yaqoob et al., 2000). Blood was sampled at several time points in each study and mononuclear cells prepared. Fatty acid composition of the cells was determined by gas chromatography. Mean values for EPA as a % of total fatty acids are shown. *Source:* Black squares *represent data from Faber, J., Berkhout, M., Vos, A.P., Sijben, J.W., Calder, P.C., Garssen, J., van Helvoort, A., 2011. Supplementation with a fish oil-enriched, high-protein medical food leads to rapid incorporation of EPA into white blood cells and modulates immune responses within one week in healthy men and women. J. Nutr. 141, 964–970 and* gray squares *represent data from Yaqoob, P., Pala, H.S., Cortina-Borja, M., Newsholme, E.A., Calder, P.C., 2000. Encapsulated fish oil enriched in α-tocopherol alters plasma phospholipid and mononuclear cell fatty acid compositions but not mononuclear cell functions. Eur. J. Clin. Invest. 30, 260–274.*

readily incorporated into the cells so increasing its abundance and this limits the extent to which DHA can be incorporated.

The results of the studies described here clearly demonstrate effects of increased oral intake of GLA, AA, ALA, SDA, EPA, and DHA on the fatty acid composition of immune cells in both laboratory animals and humans. Where the increased intake of a particular fatty acid results in increased content of one of its metabolic products (e.g., increased content of AA after GLA administration or increased content of EPA after ALA or SDA administration) this probably mostly reflects production of the metabolic product in the liver rather than in the immune cells themselves.

## POLYUNSATURATED FATTY ACIDS AS SUBSTRATES FOR THE SYNTHESIS OF BIOACTIVE LIPID MEDIATORS BY CELLS OF THE IMMUNE SYSTEM

Cells of the immune system contain several enzymes capable of oxidatively modifying PUFAs to produce bioactive lipid mediators. These include the cyclooxygenase (COX), lipoxygenase (LOX), and cytochrome P450 enzymes. PUFAs may also be subject to nonenzymatic oxidative modification to produce lipid mediators. Although it is likely that all

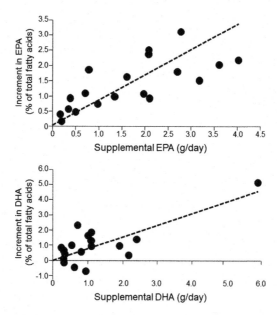

**FIGURE 8.4** **Collated data demonstrating the dose-dependent incorporation of eicosa-pentaenoic acid (EPA) and docosahexaenoic acid (DHA) into human immune cells.** Data are shown from studies providing supplemental EPA and/or DHA to human volunteers and which reported the fatty acid composition of the total lipid or of the bulk phospholipid of blood mononuclear cells or monocytes. Data are the increment in EPA or DHA content from study entry as % of total fatty acids. Data for EPA are for 19 doses used in 13 studies; for DHA, 18 doses used in 12 studies. Data from Rees et al. (2006) are for the young subjects only. Data from Browning et al. (2012) were recalculated as daily EPA and DHA intakes. Studies were of 1 week to 12 months duration. The linear correlation shown for each graph is the line of best fit through the data points but forced to go through the origin. *Source for EPA data: From Browning, L.M., Walker, C.G., Mander, A.P., West, A.L., Madden, J., Gambell, J.M., Young, S., Wang, L., Jebb, S.A., Calder, P.C., 2012. Incorporation of eicosapentaenoic and docosahexaenoic acids into lipid pools when given as supplements providing doses equivalent to typical intakes of oily fish. Am. J. Clin. Nutr. 96, 748–758; Caughey, G.E., Mantzioris, E., Gibson, R.A., Cleland, L.G., James, M.J., 1996. The effect on human tumor necrosis factor a and interleukin 1b production of diets en-riched in n-3 fatty acids from vegetable oil or fish oil. Am. J. Clin. Nutr. 63, 116–122; Endres, S., Ghorbani, R., Kelley, V.E., Georgilis, K., Lonnemann, G., van der Meer, J.M.W., Cannon, J.G., Rog-ers, T.S., Klempner, M.S., Weber, P.C., Schaeffer, E.J. Wolff, S.M., Dinarello, C.A., 1989. The effect of dietary supplementation with n-3polyunsaturated fatty acids on the synthesis of interleukin-1 and tumornecrosis factor by mononuclear cells. N. Engl. J. Med. 320, 265–271; Faber, J., Berkhout, M., Vos, A.P., Sijben, J.W., Calder, P.C., Garssen, J., van Helvoort, A., 2011. Supplementation with a fish oil-enriched; high-protein medical food leads to rapid incorporation of EPA into white blood cells and modulates immune responses within one week in healthy men and women. J. Nutr. 141, 964–970; Fisher, M., Levine, P.H., Weiner, B.H., Johnson, M.H., Doyle, E.M., Ellis, P.A., Hoogasian, J.J., 1990. Dietary n-3 fatty acid supplementation reduces superoxide production and chemiluminescence in a monocyte-enriched preparation of leukocytes. Am. J. Clin. Nutr. 51, 804–808; Kew, S., Banerjee, T., Minihane, A.M., Finnegan, Y.E., Muggli, R., Albers, R., Williams, C.M., Calder, P.C., 2003b. Lack of effect of foods enriched with plant- or marine-derived n-3 fatty acids on human immune function. Am. J. Clin. Nutr. 77, 1287–1295; Lee, T.H., Hoover, R.L., Williams, J.D., Sperling, R.I., Ravalese, J., Spur, B.W., Robinson, D.R., Corey, E.J. Lewis, R.A., Austen, K.F., 1985. Effects of dietary enrichment*

*(Continued)*

*with eicosapentaenoic acid and docosahexaenoic acid on in vitro neutrophil and monocyte leukotriene generation and neutrophil function. N. Engl. J. Med. 312, 1217–1224; Miles, E.A., Banerjee, T., Dooper, M.W.B.W., M'Rabet, L., Graus, Y.M.F., Calder, P.C., 2004. The influence of different combinations of γ-linolenic acid, stearidonic acid and EPA on immune function in healthy young male subjects. Br. J. Nutr. 91, 893–903; Mølvig, J., Pociot, F., Worsaae, H., Wogensen, L.D., Baek, L., Christensen, P., Mandrup-Poulsen, T., Andersen, K., Madsen, P., Dyerberg, J., 1991. Dietary supplementation with omega 3 polyunsaturated fatty acids decreases mononuclear cell proliferation and interleukin 1 beta content but not monokine secretion in healthy and insulin dependent diabetic individuals. Scand. J. Immunol. 34, 399–410; Rees, D., Miles, E.A., Banerjee, T., Wells, S.J., Roynette, C.E., Wahle, K.W.J.W., Calder, P.C., 2006. Dose-related effects of eicosapentaenoic acid on innate immune function in healthy humans: a comparison of young and older men. Am. J. Clin. Nutr. 83, 331–342; Schmidt, E.B., Varming, K., Møller, J.M., Bü low Pedersen, I., Madsen, P., Dyerberg, J., 1996. No effect of a very low dose of n-3 fatty acids on monocyte function in healthy humans. Scand. J. Clin. Lab. Invest. 56, 87–92; Thies, F., Nebe-von-Caron, G., Powell, J.R., Yaqoob, P., Newsholme, E.A., Calder, P.C., 2001. Dietary supplementation with γ-linolenic acid or fish oil decreases T lymphocyte proliferation in healthy older humans. J. Nutr. 131, 1918–1927; Yaqoob, P., Pala, H.S., Cortina-Borja, M., Newsholme, E.A., Calder, P.C., 2000. Encapsulated fish oil enriched in α-tocopherol alters plasma phospholipid and mononuclear cell fatty acid compositions but not mononuclear cell functions. Eur. J. Clin. Invest. 30, 260–274. Source for DHA data: From Browning, L.M., Walker, C.G., Mander, A.P., West, A.L., Madden, J., Gambell, J.M., Young, S., Wang, L., Jebb, S.A., Calder, P.C., 2012. Incorporation of eicosapentaenoic and docosahexaenoic acids into lipid pools when given as supplements providing doses equivalent to typical intakes of oily fish. Am. J. Clin. Nutr. 96, 748–758; Caughey, G.E., Mantzioris, E., Gibson, R.A., Cleland, L.G., James, M.J., 1996. The effect on human tumor necrosis factor a and interleukin 1b production of diets enriched in n-3 fatty acids from vegetable oil or fish oil. Am. J. Clin. Nutr. 63, 116–122; Endres, S., Ghorbani, R., Kelley, V.E., Georgilis, K., Lonnemann, G., van der Meer, J.M.W., Cannon, J.G., Rogers, T.S., Klempner, M.S., Weber, P.C., Schaeffer, E.J. Wolff, S.M., Dinarello, C.A., 1989. The effect of dietary supplementation with n-3 polyunsaturated fatty acids on the synthesis of interleukin-1 and tumornecrosis factor by mononuclear cells. N. Engl. J. Med. 320, 265–271; Faber, J., Berkhout, M., Vos, A.P., Sijben, J.W., Calder, P.C., Garssen, J., van Helvoort, A., 2011. Supplementation with a fish oil-enriched, high-protein medical food leads to rapid incorporation of EPA into white blood cells and modulates immune responses within one week in healthy men and women. J. Nutr. 141, 964–970; Fisher, M., Levine, P.H., Weiner, B.H., Johnson, M.H., Doyle, E.M., Ellis, P.A., Hoogasian, J.J. 1990. Dietary n-3 fatty acid supplementation reduces superoxide production and chemiluminescence in a monocyte-enriched preparation of leukocytes. Am. J. Clin. Nutr. 51, 804–808; Kelley, D.S., Taylor, P.C., Nelson, G.J., Schmidt, P.C., Ferretti, A., Erickson, K.L., Yu, R., Chandra, R.K., 1999. Docosahexaenoic acid ingestion inhibits natural killer cell activity and production of inflammatory mediators in young healthy men. Lipids 34, 317–324; Kew, S., Banerjee, T., Minihane, A.M., Finnegan, Y.E., Muggli, R., Albers, R., Williams, C.M., Calder, P.C., 2003b. Lack of effect of foods enriched with plant- or marine-derived n-3 fatty acids on human immune function. Am. J. Clin. Nutr. 77, 1287–1295; Lee, T.H., Hoover, R.L., Williams, J.D., Sperling, R.I., Ravalese, J., Spur, B.W., Robinson, D.R., Corey, E.J. Lewis, R.A., Austen, K.F., 1985. Effects of dietary enrichment with eicosapentaenoic acid and docosahexaenoic acid on in vitro neutrophil and monocyte leukotriene generation and neutrophil function. N. Engl. J. Med. 312, 1217–1224; Rees, D., Miles, E.A., Banerjee, T., Wells, S.J., Roynette, C.E., Wahle, K.W.J.W., Calder, P.C., 2006. Dose-related effects of eicosapentaenoic acid on innate immune function in healthy humans: a comparison of young and older men. Am. J. Clin. Nutr. 83, 331–342; Schmidt, E.B., Varming, K., Møller, J.M., Bü low Pedersen, I., Madsen, P., Dyerberg, J., 1996. No effect of a very low dose of n-3 fatty acids on monocyte function in healthy humans. Scand. J. Clin. Lab. Invest. 56, 87–92; Thies, F., Nebe-von-Caron, G., Powell, J.R.,Yaqoob, P., Newsholme, E.A., Calder, P.C., 2001. Dietary supplementation with γ-linolenic acid or fish oil decreases T lymphocyte proliferation in healthy older humans. J. Nutr. 131, 1918–1927; Yaqoob, P., Pala, H.S., Cortina-Borja, M., Newsholme, E.A., Calder, P.C., 2000. Encapsulated fish oil enriched in α-tocopherol alters plasma phospholipid and mononuclear cell fatty acid compositions but not mononuclear cell functions. Eur. J. Clin. Invest. 30, 260–274.*

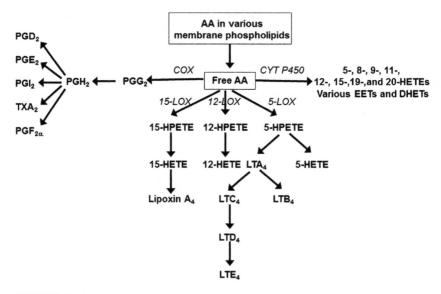

FIGURE 8.5 **Overview of the pathways of eicosanoid synthesis from arachidonic acid (AA).** *COX*, Cyclooxygenase; *CYT P450*, cytochrome P450 enzymes; *DHET*, dihydroxyeicosatrienoic acid; *EET*, epoxyeicosatrienoic acid; *HETE*, hydroxyeicosatetraenoic acid; *HPETE*, hydroperoxyeicosatetraenoic acid; *LOX*, lipoxygenase; *LT*, leukotriene; *PG*, prostaglandin; *TX*, thromboxane. Note that not all enzymes are named and that not all metabolites are shown.

PUFAs are substrates for such oxidative modifications and therefore give rise to bioactive metabolites, the processes involved and the activities of the metabolites produced are best described for AA.

Eicosanoids are oxidized derivatives of 20-carbon PUFAs. DGLA, AA, and EPA have each been shown to produce eicosanoids, as has mead acid (20:3n-9). Eicosanoids include PGs, thromboxanes (TXs), LTs, and lipoxins (LXs). The initial substrate for eicosanoid synthesis is a membrane phospholipid. Because of its prevalence in the phospholipids of membranes of cells involved in eicosanoid synthesis, AA is usually the major substrate (Fig. 8.5). Prior to synthesis of eicosanoids, AA is released from the sn-2 position of membrane phospholipids by the action of phospholipase $A_2$ enzymes, which are activated by various stimuli. AA may also be released from the sn-2 position of diacylglycerols by the sequential action of diacylglycerol lipase (to produce a 2-AA-glycerol) and monoacylglycerol lipase. The free AA then acts as a substrate for COX, LOX, or cytochrome P450 enzymes (Fig. 8.5). COX enzymes lead to PGs and TXs, LOX enzymes to LTs and LXs, and cytochrome P450 enzymes to hydroxyeicosatetraenoic and epoxyeicosatrienoic acids (Fig. 8.5). AA metabolism produces 2-series PGs and TXs and 4-series LTs and LXs. COX-1 is considered to be involved in cellular housekeeping functions, while COX-2 is induced by immune and inflammatory stimuli. Eicosanoids are produced in a cell and stimulus specific manner (e.g., neutrophils and mast cells produce significant amounts of $PGD_2$ while monocytes and macrophages produce significant amounts of $PGE_2$) and act through binding to specific receptors, usually G protein-coupled receptors (Calder et al., 2013). 2-series PGs and the 4-series LTs are well characterized mediators and regulators of immune responses and inflammation (Kalinski, 2012; Kroetz and Zeldin, 2002; Lewis et al., 1990; Tilley et al., 2001). For example, $PGE_2$ regulates T-cell, B-cell, and monocyte/macrophage responses. In general, AA-derived eicosanoids promote inflammatory processes (Calder, 2015). As a result of this, many agents have been developed to control their production and action, including aspirin, nonsteroidal antiinflammatory drugs, and COX-2 inhibitors.

Using the same series of enzymatic reactions described for AA, DGLA, and EPA are also converted to eicosanoids. DGLA generates 1-series PGs through the COX pathways and 3-series LT-like compounds through the LOX pathways, while EPA generates 3-series PGs and 5-series LTs through the COX and LOX pathways, respectively. These eicosanoids have a slightly different structure from those produced from AA (they have a different number of double bonds in their structure) and this influences their biological activity and potency. In general, they are weaker in action, probably due to poorer interaction with the eicosanoid receptors (Wada et al., 2007), although this is not always the case (Tull et al., 2009).

In the last 15 years or so a new series of lipid mediators produced from the n-3 PUFAs EPA and DHA by several cell types, including macrophages and neutrophils, has been described (Bannenberg and Serhan, 2010; Serhan, 2017; Serhan and Chiang, 2013; Serhan et al., 2008). Because these mediators are able to dampen ongoing inflammatory processes in experimental models (i.e., they can "resolve" inflammation) they are termed specialized proresolving mediators or SPMs. Much effort has gone into elucidating the structures, actions, and mechanisms of action of SPMs. SPMs include E-series resolvins produced from EPA, and D-series resolvins, protectins, and maresins produced from DHA. The synthesis of resolvins, protectins, and maresins involves the COX and LOX pathways (Fig. 8.6), with different epimers being produced in the presence and absence of aspirin. These pathways operate in a transcellular manner with the early steps occurring in one cell type and the latter in another. More recently it has been shown that analogous resolvin, protectin, and maresin-like compounds are produced from the n-3 PUFA DPA and from the n-6 PUFA osbond acid (22:5n-6) (Weylandt, 2016). SPMs produced from EPA are eicosanoids, while those produced from DHA, DPA, and osbond acid are docosanoids.

Eighteen-carbon PUFAs are also metabolized by COX and LOX. This is best described for LA, where the products are various epimers of 9- and 13-hydroxyoctadecadienoic acid (HODE). These are frequently proinflammatory in nature (Vangaveti et al., 2016).

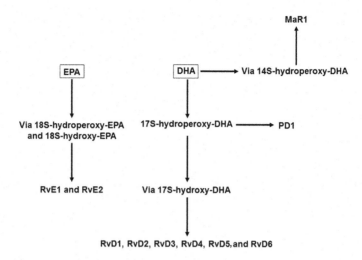

FIGURE 8.6    Overview of the conversion of eicosapentaenoic acid (EPA) and docosa-hexaenoic acid (DHA) to specialized proresolving mediators. *MaR*, Maresin; *PD*, protectin D; *RvD*, resolvin D; *RvE*, resolvin E.

## SUMMARY, DISCUSSION, AND CONCLUSIONS

PUFAs are major constituents of the phospholipids and the membranes of cells involved in the immune response. Typically, AA comprises 15%–20% of fatty acids present. The high PUFA content of immune cell phospholipids is linked to maintenance of membrane order and to the function of membrane lipids and proteins. PUFAs are constituents of signaling molecules generated from cell membrane phospholipids and are substrates for the synthesis of oxidized derivatives that serve as mediators and regulators of the immune and inflammatory responses. Because it is usually the most common PUFA in the membrane, AA is the dominant substrate for synthesis of such lipid mediators. The PUFA content of immune cell phospholipids and membranes can be modified by changing the supply of individual PUFAs. This has been well demonstrated in cell-culture experiments, in animal feeding studies and in human trials using supplements. Increasing the supply of GLA results in more DGLA in immune cell phospholipids. Increasing the supply of ALA, SDA, or EPA results in more EPA and DPA in immune cell phospholipids. Enrichment in DHA appears to require provision of preformed DHA. Cell-culture experiments demonstrate that lymphocytes, macrophages, and neutrophils can interconvert some PUFAs and that this process is enhanced when the cells are activated through extracellular stimulation. PUFA elongation and Δ5-desaturation are well described and some experiments suggest that some immune cells express Δ6-desaturase activity, especially when activated. The changes in

PUFA composition of immune cells seen with dietary change in experimental animals and humans probably reflect altered supply of PUFAs to the cells through the bloodstream rather than changes in PUFA metabolism in the cells themselves. However, it is possible that immune cells modify PUFA metabolism as part of their normal response to stimulation and that this is linked to regulating changes in the membrane and in lipid mediator production. The best evidence for this comes from a series of cell-culture studies conducted over 20 years ago (Anel et al., 1990a,b; Calder et al., 1994). Changes in the PUFA composition of membranes are linked to changes in immune function (Calder, 2008) and in inflammatory processes (Calder, 2015). In this regard, the n-3 PUFAs EPA and DHA appear to be very important, inducing changes in membrane function, cell signaling, gene expression, and lipid mediator production. EPA and DHA modulate the production of immune modulating and proinflammatory eicosanoids from AA and act as substrates for the generation of highly active SPMs (Calder, 2015). Many other PUFAs, including LA, DGLA, DPA, and osbond acid, are substrates for synthesis of bioactive lipid mediators. Incorporation of PUFAs into immune cells is likely to be cell-type, membrane-type, and phospholipid-type specific but this is not well studied. Fatty acids are used by cells of the immune system as energy sources but the extent to which PUFAs can fulfil this role is not well investigated. What is known is that the phenotype of immune cells (both macrophages and lymphocytes) that are dominated by glycolytic metabolism is different from the phenotype of immune cells that rely more on fatty acid oxidation. The extent to which alterations in the supply of PUFAs can modulate this phenotypic switch is poorly described and deserves more exploration.

# References

Anel, A., Naval, J., González, B., Torres, J.M., Mishal, Z., Uriel, J., Piñeiro, A., 1990a. Fatty acid metabolism in human lymphocytes. I. Time-course changes in fatty acid composition and membrane fluidity during blastic transformation of peripheral blood lymphocytes. Biochim. Biophys. Acta 1044, 323–331.

Anel, A., Naval, J., González, B., Uriel, J., Piñeiro, A., 1990b. Fatty acid metabolism in human lymphocytes. II. Activation of fatty acid desaturase-elongase systems during blastic transformation. Biochim. Biophys. Acta 1044, 332–339.

Ardawi, M.S.M., Newsholme, E.A., 1984. Metabolism of ketone bodies, oleate and glucose in lymphocytes of the rat. Biochem. J. 221, 255–260.

Bannenberg, G., Serhan, C.N., 2010. Specialized pro-resolving lipid mediators in the inflammatory response: an update. Biochim. Biophys. Acta 1801, 1260–1273.

Brouard, C., Pascaud, M., 1990. Effects of moderate dietary supplementations with n-3 fatty acids on macrophage and lymphocyte phospholipids and macrophage eicosanoid synthesis in the rat. Biochim. Biophys. Acta 1047, 19–28.

Browning, L.M., Walker, C.G., Mander, A.P., West, A.L., Madden, J., Gambell, J.M., Young, S., Wang, L., Jebb, S.A., Calder, P.C., 2012. Incorporation of eicosapentaenoic and docosahexaenoic acids into lipid pools when given as supplements providing doses equivalent to typical intakes of oily fish. Am. J. Clin. Nutr. 96, 748–758.

Calder, P.C., 1995. Fuel utilisation by cells of the immune system. Proc. Nutr. Soc. 54, 65–82.

Calder, P.C., 2008. The relationship between the fatty acid composition of immune cells and their function. Prostagl. Leuk. Essent. Fatty Acids 79, 101–108.

Calder, P.C., 2015. Marine omega-3 fatty acids and inflammatory processes: effects, mechanisms and clinical relevance. Biochim. Biophys. Acta 1851, 469–484.

Calder, P.C., Ahluwalia, N., Albers, R., Bosco, N., Bourdet-Sicard, R., Haller, D., Holgate, S.T., Jonsson, L.S., Latulippe, M.E., Marcos, A., Moreines, J., M'Rini, C., Muller, M., Pawelec, G., van Neerven, R.J.J., Watzl, B., Zhao, J., 2013. A consideration of biomarkers to be used for evaluation of inflammation in human nutritional studies. Brit. J. Nutr. 109 (Suppl. 1), S1–S34.

Calder, P.C., Bond, J.A., Harvey, D.J., Gordon, S., Newsholme, E.A., 1990. Uptake and incorporation of saturated and unsaturated fatty acids into macrophage lipids and their effect upon macrophage adhesion and phagocytosis. Biochem. J. 269, 807–814.

Calder, P.C., Yaqoob, P., Harvey, D.J., Watts, A., Newsholme, E.A., 1994. The incorporation of fatty acids by lymphocytes and the effect on fatty acid composition and membrane fluidity. Biochem. J. 300, 509–518.

Caughey, G.E., Mantzioris, E., Gibson, R.A., Cleland, L.G., James, M.J., 1996. The effect on human tumor necrosis factor a and interleukin 1b production of diets enriched in n-3 fatty acids from vegetable oil or fish oil. Am. J. Clin. Nutr. 63, 116–122.

Chapkin, R.S., Akoh, C.C., Lewis, R.E., 1992. Dietary fish oil modulation of in vivo peritoneal macrophage leukotriene production and phagocytosis. J. Nutr. Biochem. 3, 599–604.

Chapkin, R.S., Coble, K.J., 1991. Utilization of gammalinolenic acid by mouse peritoneal macrophages. Biochim. Biophys. Acta. 1085, 365–370.

Chapkin, R.S., Miller, C.C., 1990. Chain elongation of eicosapentaenoic acid in the macrophage. Biochim. Biophys. Acta 1042, 265–267.

Chapkin, R.S., Miller, C.C., Somers, S.D., Erickson, K.L., 1988b. Utilization of dihomo-gamma-linolenic acid (8,11,14-eicosatrienoic acid) by murine peritoneal macrophages. Biochim. Biophys. Acta 959, 322–331.

Chapkin, R.S., Somers, S.D., Erickson, K.L., 1988a. Inability of murine peritoneal macrophages to convert linoleic acid into arachidonic acid: evidence of chain elongation. J. Immunol. 140, 2350–2355.

Chapkin, R.S., Somers, S.D., Schumacher, L., Erickson, K.L., 1988c. Fatty acid composition of macrophage phospholipids in mice fed fish or borage oil. Lipids 23, 380–383.

Chiaranunt, P., Ferrara, J.L., Byersdorfer, C.A., 2015. Rethinking the paradigm: how comparative studies on fatty acid oxidation inform our understanding of T cell metabolism. Mol. Immunol. 68, 564–574.

Chilton-Lopez, Surette, M.E., Swan, D.D., Fonteh, A.N., Johnson, M.M., Chilton, F.H., 1996. Metabolism of gammalinolenic acid in human neutrophils. J. Immunol. 156, 2941–2947.

Endres, S., Ghorbani, R., Kelley, V.E., Georgilis, K., Lonnemann, G., van der Meer, J.M.W., Cannon, J.G., Rogers, T.S., Klempner, M.S., Weber, P.C., Schaeffer, E.J., Wolff, S.M., Dinarello, C.A., 1989. The effect of dietary supplementation with n-3 polyunsaturated fatty acids on the synthesis of interleukin-1 and tumor necrosis factor by mononuclear cells. N. Engl. J. Med. 320, 265–271.

Faber, J., Berkhout, M., Vos, A.P., Sijben, J.W., Calder, P.C., Garssen, J., van Helvoort, A., 2011. Supplementation with a fish oil-enriched, high-protein medical food leads to rapid incorporation of EPA into white blood cells and modulates immune responses within one week in healthy men and women. J. Nutr. 141, 964–970.

Fisher, M., Levine, P.H., Weiner, B.H., Johnson, M.H., Doyle, E.M., Ellis, P.A., Hoogasian, J.J., 1990. Dietary n-3 fatty acid supplementation reduces superoxide production and chemiluminescence in a monocyte-enriched preparation of leukocytes. Am. J. Clin. Nutr. 51, 804–808.

Fritsche, K., 2007. Important differences exist in the dose-response relationship between diet and immune cell fatty acids in humans and rodents. Lipids 42, 961–979.

Gibney, M.J., Hunter, B., 1993. The effects of short- and long-term supplementation with fish oil on the incorporation of n-3 polyunsaturated fatty acids into cells of the immune system in healthy volunteers. Eur. J. Clin. Nutr. 47, 255–259.

Hardardóttir, I., Kinsella, J.E., 1992. Increasing the dietary (n-3) to (n-6) polyunsaturated fatty acid ratio increases tumor necrosis factor production by murine resident peritoneal macrophages without an effect on elicited peritoneal macrophages. J. Nutr. 122, 1942–1951.

Healy, D.A., Wallace, F.A., Miles, E.A., Calder, P.C., Newsholme, P., 2000. The effect of low to moderate amounts of dietary fish oil on neutrophil lipid composition and function. Lipids 35, 763–768.

Hwang, D.H., Kim, J.A., Lee, J.Y., 2016. Mechanisms for the activation of toll-like receptor 2/4 by saturated fatty acids and inhibition by docosahexaenoic acid. Eur. J. Pharmacol. 785, 24–35.

Johnson, M.M., Swan, D.D., Surette, M.E., Stegner, J., Chilton, T., Fonteh, A.N., Chilton, F.H., 1997. Dietary supplementation with gamma-linolenic acid alters fatty acid content and eicosanoid production in healthy humans. J. Nutr. 127, 1435–1444.

Kalinski, P., 2012. Regulation of immune responses by prostaglandin E2. J. Immunol. 188, 21–28.

Kelley, D.S., Nelson, G.J., Love, J.E., Branch, L.B., Taylor, P.C., Schmidt, P.C., Mackey, B.E., Iacono, J.M., 1993. Dietary alpha-linolenic acid alters tissue fatty acid composition, but not blood lipids, lipoproteins or coagulation status in humans. Lipids 28, 533–537.

Kelley, D.S., Taylor, P.C., Nelson, G.J., Schmidt, P.C., Ferretti, A., Erickson, K.L., Yu, R., Chandra, R.K., 1999. Docosahexaenoic acid ingestion inhibits natural killer cell activity and production of inflammatory mediators in young healthy men. Lipids 34, 317–324.

Kew, S., Banerjee, T., Minihane, A.M., Finnegan, Y.E., Muggli, R., Albers, R., Williams, C.M., Calder, P.C., 2003b. Lack of effect of foods enriched with plant- or marine-derived n-3 fatty acids on human immune function. Am. J. Clin. Nutr. 77, 1287–1295.

Kew, S., Mesa, M.D., Tricon, S., Buckley, R., Minihane, A.M., Yaqoob, P., 2004. Effects of oils rich in eicosapentaenoic and docosahexaenoic acids on immune cell composition and function in healthy humans. Am. J. Clin. Nutr. 79, 674–681.

Kew, S., Wells, S., Thies, F., McNeill, G.P., Quinlan, P.T., Clark, G.T., Dombrowsky, H., Postle, A.D., Calder, P.C., 2003a. The effect of eicosapentaenoic acid on rat lymphocyte proliferation depends upon its position in dietary triacylglycerols. J. Nutr. 133, 4230–4238.

Kroetz, D.L., Zeldin, D.C., 2002. Cytochrome P450 pathways of arachidonic acid metabolism. Curr. Opin. Lipidol. 13, 273–283.

Lee, J.Y., Sohn, K.H., Rhee, S.H., Hwang, D., 2001. Saturated fatty acids, but not unsaturated fatty acids, induce the expression of cyclooxygenase-2 mediated through toll-like receptor 4. J. Biol. Chem. 276, 16683–16689.

Lee, T.H., Hoover, R.L., Williams, J.D., Sperling, R.I., Ravalese, J., Spur, B.W., Robinson, D.R., Corey, E.J., Lewis, R.A., Austen, K.F., 1985. Effects of dietary enrichment with eicosapentaenoic acid and docosahexaenoic acid on in vitro neutrophil and monocyte leukotriene generation and neutrophil function. N. Engl. J. Med. 312, 1217–1224.

Lewis, R.A., Austen, K.F., Soberman, R.J., 1990. Leukotrienes and other products of the 5-lipoxygenase pathway: biochemistry and relation to pathobiology in human diseases. N. Engl. J. Med. 323, 645–655.

Lochner, M., Berod, L., Sparwasser, T., 2015. Fatty acid metabolism in the regulation of T cell function. Trends Immunol. 36, 81–91.

Lokesh, B.R., Hsieh, H.L., Kinsella, J.E., 1986. Alterations in the lipids and prostaglandins in mouse spleen following the ingestion of menhaden oil. Ann. Nutr. Metab. 30, 357–364.

Luostarinen, R., Saldeen, T., 1996. Dietary fish oil decreases superoxide generation by human neutrophils: relation to cyclooxygenase pathway and lysosomal enzyme release. Prostal. Leuk. Essent. Fatty Acids 55, 167–172.

Maloney, E., Sweet, I.R., Hockenbery, D.M., Pham, M., Rizzo, N.O., Tateya, S., Handa, P., Schwartz, M.W., Kim, F., 2009. Activation of NF-kappaB by palmitate in endothelial cells:

a key role for NADPH oxidase-derived superoxide in response to TLR4 activation. Arterioscler. Thromb. Vasc. Biol. 29, 1370–1375.

Mantzioris, E., James, M.J., Gibson, R.A., Cleland, L.G., 1994. Dietary substitution with an alpha-linolenic acid-rich vegetable oil increases eicosapentaenoic acid concentrations in tissues. Am. J. Clin. Nutr. 59, 1304–1309.

Miles, E.A., Banerjee, T., Dooper, M.W.B.W., M'Rabet, L., Graus, Y.M.F., Calder, P.C., 2004. The influence of different combinations of γ-linolenic acid, stearidonic acid and EPA on immune function in healthy young male subjects. Br. J. Nutr. 91, 893–903.

Mills, E.L., O'Neill, L.A., 2016. Reprogramming mitochondrial metabolism in macrophages as an anti-inflammatory signal. Eur. J. Immunol. 46, 13–21.

Mølvig, J., Pociot, F., Worsaae, H., Wogensen, L.D., Baek, L., Christensen, P., Mandrup-Poulsen, T., Andersen, K., Madsen, P., Dyerberg, J., 1991. Dietary supplementation with omega 3 polyunsaturated fatty acids decreases mononuclear cell proliferation and interleukin 1 beta content but not monokine secretion in healthy and insulin dependent diabetic individuals. Scand. J. Immunol. 34, 399–410.

Namgaladze, D., Brüne, B., 2016. Macrophage fatty acid oxidation and its roles in macrophage polarization and fatty acid-induced inflammation. Biochim. Biophys. Acta 1861, 1796–1807.

Newsholme, P., Newsholme, E.A., 1989. Rates of utilisation of glucose, glutamine and oleate and formation of end products by mouse peritoneal macrophages in culture. Biochem. J. 261, 211–218.

Peterson, L.D., Jeffery, N.M., Thies, F., Sanderson, P., Newsholme, E.A., Calder, P.C., 1998. Eicosapentaenoic and docosahexaenoic acids alter rat spleen leukocyte fatty acid composition and prostaglandin E2 production but have different effects on lymphocyte functions and cell-mediated immunity. Lipids 33, 171–180.

Rees, D., Miles, E.A., Banerjee, T., Wells, S.J., Roynette, C.E., Wahle, K.W.J.W., Calder, P.C., 2006. Dose-related effects of eicosapentaenoic acid on innate immune function in healthy humans: a comparison of young and older men. Am. J. Clin. Nutr. 83, 331–342.

Schmidt, E.B., Varming, K., Møller, J.M., Bülow Pedersen, I., Madsen, P., Dyerberg, J., 1996. No effect of a very low dose of n-3 fatty acids on monocyte function in healthy humans. Scand. J. Clin. Lab Invest. 56, 87–92.

Serhan, C.N., 2017. Discovery of specialized pro-resolving mediators marks the dawn of resolution physiology and pharmacology. Mol. Aspects Med. 58, 1–11.

Serhan, C.N., Chiang, N., 2013. Resolution phase lipid mediators of inflammation: agonists of resolution. Curr. Opin. Pharmacol. 13, 632–640.

Serhan, C.N., Chiang, N., van Dyke, T.E., 2008. Resolving inflammation: dual anti-inflammatory and pro-resolution lipid mediators. Nat. Rev. Immunol. 8, 349–361.

Sperling, R.I., Benincaso, A.I., Knoell, C.T., Larkin, J.K., Austen, K.F., Robinson, D.R., 1993. Dietary ω-3 polyunsaturated fatty acids inhibit phosphoinositide formation and chemotaxis in neutrophils. J. Clin. Invest. 91, 651–660.

Spolarics, Z., Lang, C.H., Bagby, G.J., Spitzer, J.J., 1991. Glutamine and fatty acid oxidation are the main sources of energy in Kupffer and endothelial cells. Am. J. Physiol. 261, G185–G190.

Surette, M.E., Edens, M., Chilton, F.H., Tramposch, K.M., 2004. Dietary echium oil increases plasma and neutrophil long-chain (n-3) fatty acids and lowers serum triacylglycerols in hypertriglyceridemic humans. J. Nutr. 134, 1406–1411.

Surette, M.E., Whelan, J., Lu, G., Hardard'ottir, I., Kinsella, J.E., 1995. Dietary n-3 polyunsaturated fatty acids modify Syrian hamster platelet and macrophage phospholipid fatty acyl composition and eicosanoid synthesis: a controlled study. Biochim. Biophys. Acta 1255, 185–191.

Thies, F., Nebe-von-Caron, G., Powell, J.R., Yaqoob, P., Newsholme, E.A., Calder, P.C., 2001. Dietary supplementation with γ-linolenic acid or fish oil decreases T lymphocyte proliferation in healthy older humans. J. Nutr. 131, 1918–1927.

Tilley, S.L., Coffman, T.M., Koller, B.H., 2001. Mixed messages: modulation of inflammation and immune responses by prostaglandins and thromboxanes. J. Clin. Invest. 108, 15–23.

Tull, S.P., Yates, C.M., Maskrey, B.H., O'Donnell, V.B., Madden, J., Grimble, R.F., Calder, P.C., Nash, G.B., Rainger, G.E., 2009. Omega-3 fatty acids and inflammation: novel interactions reveal a new step in neutrophil recruitment. PLoS Biol. 7, e1000177.

Van den Bossche, J., O'Neill, L.A., Menon, D., 2017. Macrophage immunometabolism: where are we (going)? Trends Immunol. 38, 395–406.

Vangaveti, V.N., Jansen, H., Kennedy, R.L., Malabu, U.H., 2016. Hydroxyoctadecadienoic acids: oxidised derivatives of linoleic acid and their role in inflammation associated with metabolic syndrome and cancer. Eur. J. Pharmacol. 785, 70–76.

Vats, D., Mukundan, L., Odegaard, J.I., Zhang, J., Smith, K.L., Morel, C.R., Wagner, R.A., Greaves, D.R., Murray, P.J., Chawla, A., 2006. Oxidative metabolism and PGC-1beta attenuate macrophage-mediated inflammation. Cell. Metab. 4, 13–24.

Wada, M., DeLong, C.J., Hong, Y.H., Rieke, C.J., Song, I., Sidhu, R.S., Yuan, C., Warnock, M., Schmaier, A.H., Yokoyama, C., Smyth, E.M., Wilson, S.J., FitzGerald, G.A., Garavito, R.M., Sui, X., Regan, J.W., Smith, W.L., 2007. Enzymes and receptors of prostaglandin pathways with arachidonic acid-derived versus eicosapentaenoic acid-derived substrates and products. J. Biol. Chem. 282, 22254–22266.

Wallace, F.A., Miles, E.A., Evans, C., Stock, T.E., Yaqoob, P., Calder, P.C., 2001. Dietary fatty acids influence the production of Th1- but not Th2-type cytokines. J. Leuk. Biol. 69, 449–457.

Wallace, F.A., Neely, S.J., Miles, E.A., Calder, P.C., 2000. Dietary fats affect macrophage-mediated cytotoxicity towards tumour cells. Immunol. Cell Biol. 78, 40–48.

Wang, R., Green, D.R., 2012. Metabolic checkpoints in activated T cells. Nat. Immunol. 13, 907–915.

Weylandt, K.H., 2016. Docosapentaenoic acid derived metabolites and mediators: the new world of lipid mediator medicine in a nutshell. Eur. J. Pharmacol. 785, 108–115.

Whelan, J., Surette, M.E., Hardardóttir, I., Lu, G., Golemboski, K.A., Larsen, E., Kinsella, J.E., 1993. Dietary arachidonate enhances tissue arachidonate levels and eicosanoid production in Syrian hamsters. J. Nutr. 123, 2174–2185.

Yaqoob, P., Newsholme, E.A., Calder, P.C., 1994. Fatty acid oxidation by lymphocytes. Biochem. Soc. Trans. 22, 116s.

Yaqoob, P., Newsholme, E.A., Calder, P.C., 1995. Influence of cell culture conditions on diet-induced changes in lymphocyte fatty acid composition. Biochim. Biophys. Acta 1255, 333–340.

Yaqoob, P., Pala, H.S., Cortina-Borja, M., Newsholme, E.A., Calder, P.C., 2000. Encapsulated fish oil enriched in α-tocopherol alters plasma phospholipid and mononuclear cell fatty acid compositions but not mononuclear cell functions. Eur. J. Clin. Invest. 30, 260–274.

Zhao, G., Etherton, T.D., Martin, K.R., West, S.G., Gillies, P.J., Kris-Etherton, P.M., 2004. Dietary alpha-linolenic acid reduces inflammatory and lipid cardiovascular risk factors in hypercholesterolemic men and women. J. Nutr. 134, 2991–2997.

# Polyunsaturated Fatty Biosynthesis and Metabolism in Reproductive Tissues

*D. Claire Wathes, Zhangrui Cheng*

Royal Veterinary College, North Mymms, Hatfield, Herts, United Kingdom

## GENERAL INTRODUCTION

Lipids have many roles within the body, providing energy, as structural components of membranes and by acting as signalling molecules. All of these aspects are important for normal reproductive function. Details of polyunsaturated fatty acid (PUFA) biosynthesis and metabolism in mammals are provided in Chapter 2. In brief, members of the n-6 family of

Polyunsaturated Fatty Acid Metabolism. http://dx.doi.org/10.1016/B978-0-12-811230-4.00009-0

PUFAs are derived from linoleic acid (18:2 n-6, LA) by a process of desaturation and elongation, while n-3 family members are similarly derived from α-linolenic acid (18:3n-3, ALA). The main enzymes involved in these processes are Δ6- and Δ5-desaturases (FADS2 and FADS1) and elongases (ELOVL2 and EVOLV5). The main sources of LA are vegetable oils, while ALA is present in green leafy vegetables, marine algae, seeds (e.g., linseed) and nuts and the longer chain n-3 PUFAs eicosapentaenoic acid (EPA) and docosahexaenoic acid (DHA) are found at high concentrations in fish oils. Although a significant proportion of the dietary PUFAs become saturated during digestion, the proportions of different PUFAs found in cell membranes throughout the body do generally reflect the amounts consumed in the diet (Fischer, 1989; Sprecher, 2000; Wathes et al., 2007). This chapter begins with brief summaries of mechanisms of action of PUFAs, which may influence reproduction, considers why results are often inconclusive, and then uses three example systems to investigate aspects of metabolism in reproductive tissues. These are placental transfer, male fertility, and uterine prostaglandin (PG) synthesis.

## MECHANISMS OF ACTION

An individual cell may contain more than 1000 different lipid species (Wenk, 2010). Triacylglycerol is the primary form used for energy storage, but PUFAs can be also be metabolized to supply energy. PUFAs are also incorporated into cell membrane phospholipids where they influence membrane properties, including fluidity, thickness, and formation of microdomains. One way that n-3 PUFAs in particular can affect cell signaling is by modifying the structure and composition of lipid rafts. These are localized regions in plasma membranes that are rich in cholesterol, sphingolipids, and phospholipids: specific proteins localize to these regions, so promoting efficient signaling by clustering of relevant proteins (Pike, 2006). When n-3 PUFAs, in particular DHA, become incorporated into membrane phospholipids they cause lipid raft regions to merge, with an associated depletion of cholesterol and sphingolipids and partitioning of some proteins away from the raft (Turk and Chapkin, 2013; Williams et al., 2012).

PUFAs can regulate many cellular processes through mechanisms independent of the lipid mediators. They act directly as ligands for a number of transcription factors, including PPARs and RXR (Nakamura et al., 2014). PPARs are a subfamily of nuclear receptors with three known subtypes α, β, and γ. These are expressed in a tissue-specific manner and play a key role in energy metabolism influencing beta-oxidation, ketogenesis, and glucose metabolism (Nakamura et al., 2014; Wahli and Michalik, 2012). PPARs also influence steroidogenesis (by regulating genes involved in the biosynthesis and metabolism of cholesterol and fatty acids) and insulin sensitivity (by regulating adipocyte size and spacing) (Michalik and Wahli, 2008; Nakamura et al., 2014). Both n-3 and n-6 PUFAs and their metabolites can also activate PPARs to inhibit NFkB signaling to decrease cytokine production and these pathways may play important roles in regulating inflammation and homeostasis (Calder, 2015). These processes can all affect fertility, which is strongly influenced by both metabolic and inflammatory processes. RXR has a broad spectrum of activity by forming a heterodimeric partner with other nuclear receptors including PPARs, liver-X-receptor, retinoic acid receptor, thyroid hormone receptor, and vitamin D receptor. Many of these receptors play critical roles during development, metabolic disease, and cancer (Pérez et al., 2012).

A tiny proportion of the PUFAs consumed in the diet is metabolized into signaling molecules, in particular eicosanoids. These have important regulatory functions, which control key reproductive processes, such as ovulation and parturition. Eicosanoids also influence many aspects of immunology and inflammation, also crucial in reproduction (Wathes et al., 2007).

## PROBLEMS OF INTERPRETATION

Before considering PUFA metabolism with respect to reproduction it is helpful to consider briefly possible reasons for the frequent inconsistencies in the results reported. PUFA levels in tissues of the body generally reflect those in the current diet, but PUFAs are also stored in fat depots and membranes, from which they may later be mobilized. Release of prestored PUFAs can therefore influence the availability at much later time periods. Food sources contain mixtures of different PUFAs, making it hard to formulate a diet in which only one PUFA is increased or decreased, particularly as other aspects, such as protein and energy also need to be balanced between treatments. The ratio of n-3 to n-6 is important and there are also dose responses for individual PUFAs, which can alter their effects from stimulation to inhibition. Experiments performed in vitro can use pure oils or simple mixtures but these cannot accurately reflect the complex biology of the whole body. In most human studies a source of PUFA, for example fish oil capsules, is often supplemented but individuals on the trial will continue to eat a range of other foodstuffs. For ruminant species (e.g., cattle, sheep) the diet requires a fodder component, such as grass or silage, in which PUFA levels may differ considerably. In ruminants the ingested food is also subjected to biohydrogenation by rumen microbes, which can have a major impact on the absorption level (Santos et al., 2008). Dietary PUFAs are also taken up in a tissue-specific manner. The reaction of a particular cell type will depend not only on the relative proportions of different PUFAs and other lipids stored within it, but is also regulated by signals from the periphery.

Another consideration is the age and physiological state of the animal, which affects lipid metabolism in general and thus the balance of storage and release. This is particularly pertinent during pregnancy and lactation. Recent advances in genomics have shown further complications but also some solutions. On the one hand, studies on human populations are revealing genetic diversity in the enzymes which metabolize PUFAs, which can affect the outcomes measured (Moltó-Puigmartí et al., 2014). On the other hand, the development of transgenic animals offers the possibility of modifying the relative concentrations of different PUFAs in the body without some of the problems mentioned earlier, as both treated and control animals can be fed the same diet. Unless the gene deletion is targeted this will, however, still affect all body systems making it hard to judge if effects on reproduction are direct or indirect. A further particular problem relating to work on reproduction is that the main outcome of interest is often birth of a healthy offspring. The effects of dietary PUFAs on fertility are often quite subtle and many of the published experiments have not included enough animals to provide sufficient statistical power for reliability.

## PLACENTAL TRANSFER

### Placental Transport Mechanisms

PUFAs are essential to support fetal growth and LCPUFAs are of particular importance for the visual system and in cognitive development (Calder et al., 2010). The fetus must acquire sufficient of the essential fatty acids (EFA) LA and ALA indirectly from the maternal diet via placental exchange. The placenta lacks FADS1 and FADS2 used in the conversion of EFA into LCPUFAs and the fetus itself only has limited desaturase activity (Chambaz et al., 1985), so most LCPUFAs are also transferred to the fetus via the placenta. Kuipers et al. (2012) estimated that a typical human infant at normal term had accreted about 21 g LA, 7.5 g arachidonic acid (AA) and 3 g DHA. The mother adapts her own metabolism to meet this fetal demand. Fat depots are accumulated in the initial two thirds of gestation and mobilized through lipolysis toward the end of pregnancy, when most fetal growth occurs. The majority of studies have either investigated women directly or used rodent models to determine how this lipid transfer is regulated.

Most PUFAs present in maternal plasma during pregnancy are associated with lipoprotein triglycerides, with a small proportion present in the form of free fatty acids (FFA) (Herrera, 2002). Maternal triglycerides do not cross the placenta intact. They initially bind to lipoprotein receptors on placental trophoblast cells, which in women are bathed in maternal blood (Fig. 9.1). Human placental tissue expresses very low-density lipoprotein receptor in greater abundance than low-density lipoprotein receptor and an LDL receptor-related protein (Herrera, 2002). The bound triglycerides are then hydrolyzed by lipoprotein lipase and endothelial lipase, which are present on the syncytiotrophoblast to release nonesterified fatty acids (NEFA) into the trophoblast cells (Jones et al., 2014). Endothelial lipase also has a high phospholipase activity, suggesting that phospholipids may also supply fatty acids to the fetus (Larqué et al., 2011). Placental lipase activity increases during the third trimester to increase availability of NEFA at this time (Dutta-Roy, 2000; Rodríguez-Cruz et al., 2016). Although of lesser importance, uptake of maternal FFA by the placenta can also take place through facilitated membrane translocation involving

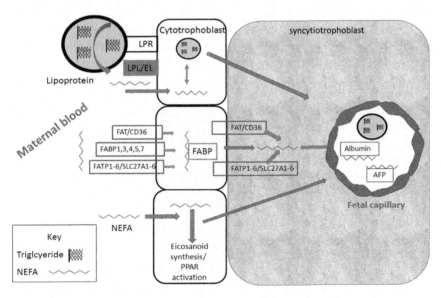

FIGURE 9.1 **Uptake of PUFAs by the human placenta.** Most PUFAs in maternal plasma are associated with lipoprotein triglycerides. These bind to lipoprotein receptors (LPR) on placental trophoblast cells and are then hydrolyzed by lipoprotein lipase (LPL) and endothelial lipase (EL) to release nonesterified fatty acids (NEFA) into the cytotrophoblast. Uptake of NEFA can also be by passive diffusion or through facilitated translocation involving membrane-associated fatty acid binding proteins (FABP), fatty acid translocase (FAT/CD36), and fatty acid transport proteins (FATP1-6/SLC27A1-6). PUFAs within trophoblast cells may be converted into eicosanoids or act as transcription regulators. PUFAs may be released from the syncytiotrophoblast into the fetal circulation as triglycerides associated with lipoproteins or as NEFAs, involving FAT, FATP, or passive diffusion. These are then transported to the fetal liver bound to carrier proteins, such as alpha-fetoprotein (AFP) and serum albumin.

membrane-associated fatty acid binding/transport proteins. These include plasma membrane fatty-acid-binding protein (FABPpm), placental plasma membrane FABP (p-FABPpm), fatty acid translocase (FAT/CD36), and fatty acid transport proteins (FATP1-6/SLC27A1-6) (Dutta-Roy, 2000; Hanebutt et al., 2008). Multiple isoforms of FABP have now been identified of which FABP1, 3, 4, 5, and 7 are present in human and rodent placentae (Islam et al., 2014; Rodríguez-Cruz et al., 2016). Placentae from *Fabp3* knockout mice showed reduced ability to transport both LA and ALA (Islam et al., 2014). Once within the placental cells, the fatty acids can bind to cytosolic FABPs. Using tracer studies in the pregnant rat, López-Luna et al. (2016) showed that rapid initial uptake of PUFAs by the placenta was due to triacylglycerol (TAG) but from 2 h onward an increasing proportion of label was associated with phospholipids. Some PUFAs may also cross into the cytotrophoblast by passive diffusion.

Less is known about how PUFAs are subsequently moved out of the syncytiotrophoblast into the fetal circulation. Jones et al. (2014) suggested that this process involved either FAT, FATP, or passive diffusion. Following oral administration of $^{14}$C labeled fatty acids to pregnant rats, López-Luna et al. (2016) found that some PUFAs had already reached the fetal circulation within 30 min, increasing steadily up to 8 h. The evidence indicated that a substantial proportion of the FA were released from the placenta in an esterified form in the TAG fraction associated with specific lipoproteins. Some were also released as NEFA, which were then transported to the fetal liver bound to carrier proteins, such as alpha-fetoprotein and serum albumin (Parmelee et al., 1978).

The end result is that LCPUFAs cross the placenta in a directional manner and show a linear correlation between concentrations in maternal and fetal cord blood at term (Berghaus et al., 2000; Elias and Innis, 2001; Haggarty et al., 1997). There is, however, some selectivity with preferential transport of AA and DHA (Dutta-Roy, 2000). Expression of placental mRNA for *FATP1* and *FATP4* was correlated with the DHA percentage in placental phospholipids and cord blood and it was suggested that these enzymes facilitate placental transfer of DHA and EPA by incorporating them into placental phospholipids via esterification coupled influx (Haggarty et al., 1997; Larqué et al., 2011). Furthermore, López-Luna et al. (2016) found differential transport of radiolabeled DHA in the rat.

Placental LCPUFAs are also able to act in an autocrine manner to regulate their own uptake, transport, and metabolism (Jones et al., 2014). There is evidence that this may be mediated, at least in part, through their ability to act as ligands for PPARs, with all three PPAR isoforms ($\alpha$, $\beta/\delta$, and $\gamma$) present in the placenta (Hewitt et al., 2006). Treatment of mice with rosiglitazone, a PPARγ agonist, increased placental mRNA expression of *Fabppm*, *Fat*, *Fatp1*, and *Fatp4* but decreased expression of *Fatp2*, *Fatp3*, and *Fatp6* (Schaiff et al., 2007).

## Lipid Transfer in the Preimplantation Ruminant Conceptus

The discussion so far has focused on the human and rodent hemo-chorial placentae. Lipid transfer differs in species with epitheliochorial placentae, such as ruminants, in which the fetal chorion does not have direct contact with maternal blood. The preimplantation ruminant con-ceptus is reliant for nutrition on histotroph secreted by the endometrial glands, which contain a rich variety of nutrients including lipids (Bazer et al., 2011). These lipids are probably derived from droplets which accu-mulate in endometrial epithelial cells under the influence of progesterone (Brinsfield and Hawk, 1973) and reach the lumen through export of exo-somes, microvesicles, carrier proteins, and lipoproteins (Burns et al., 2014). Albumin, apolipoprotein-A1 (the main protein component of HDL), and FABPs are all present in luminal fluid and could act as lipid carriers (Forde et al., 2014; Koch et al., 2010). During conceptus elongation there was increased expression of an extensive number of genes involved in the uptake, metabolism, and biosynthesis of fatty acids and PGs (Ribeiro et al., 2016). The list included *SLC27A2*, *SLC27A6*, *FADS1*, *FADS2*, and the phospholipases *PLA2G7* and *PLA2G12A*. Bioinformatic analysis sug-gested that activation of PPARγ in the trophectoderm was playing a key coordinating role in this process by binding lipid ligands.

Although the main focus on the nutritional role of histotroph has been during the preimplantation period, there is increasing evidence that it con-tinues to play an important role throughout gestation in a variety of species including some with hemochorial placentae (Enders and Carter, 2006). In ruminants there is, however, also lipid transfer by the more conventional hemotrophic route in the caruncles. Unlike the human or rat placenta, the sheep placenta is able to convert 18:2n-6 and 18:3n-3 to LCPUFAs (Noble et al., 1985). PUFA concentrations in fetal ovine blood were consistently lower than in maternal blood, but feeding an 18:2n-6-supplemented diet (Soypreme) in late gestation increased the concentrations of AA in fetal allantochorion, liver, and plasma (Elmes et al., 2004).

## MALE FERTILITY

### PUFAs in the Testis and Spermatozoa

The second example for the role of PUFAs in reproduction comes from a consideration of their importance in the development of normal gametes. The testes and spermatozoa of all mammals and birds investigated con-tain large amounts of LCPUFAs. PUFAs constitute 30%–50% of the total FAs present in mammalian spermatozoa, mainly in membrane phospho-lipids (Poulos et al., 1973). Spermatozoa, along with neurons and retinal photoreceptors, contain particularly high concentrations of DHA (Saether et al., 2007). AA, docosapentaenoic acid (C22:4n-5) and docosatetranoic acid (22:4 n-6) are also abundant in sperm membranes, with some differences in relative proportions between species (Esmaeli et al., 2015; Fair et al., 2014; Jones et al., 1979; Surai et al., 1998). Altering the PUFA content of the diet generally results in concomitant changes in the n-6 and n-3 composition of spermatozoa as demonstrated in a variety of species, for example boar (Maldjian et al., 2005), ram (Fair et al., 2014), and cockerel (Cerolini et al., 2005).

Fatty acids accumulate in seminiferous tubules by a combination of passive diffusion and facilitated transport by FAT/CD36, which is highly expressed in Sertoli cells (Regueira et al., 2015). They are then passed to the developing spermatogonia, which progress through the stages of spermatogenesis to become spermatozoa. The release of sperm from the testis to the epididymis places a continual drain on the supply of PUFA, particularly in stud males used frequently for breeding. Once they have been released spermatozoa can also take up PUFAs bound to albumin from the surrounding fluid (Alvarez and Storey, 1995) (Fig. 9.2).

The proportions of different PUFAs incorporated into the sperm plasma membrane influences the membrane physiology (Stubbs and Smith, 1984; Van Tran et al., 2016). In order to achieve fertilization, the individual

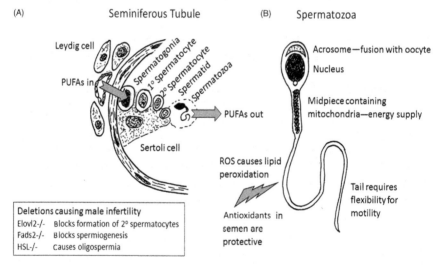

FIGURE 9.2  **Role of PUFAs in male fertility.** (A) Spermatogenesis requires a constant supply of PUFAs, particularly LCPUFAs, such as DHA and AA. These PUFAs become incorporated into the sperm membranes so as mature spermatozoa are released from the seminiferous tubules this drains the testis. Deletions of several enzymes used in the biosynthesis of LCPUFAs causes sterility. (B) The PUFAs in spermatozoa are essential for cell signaling, membrane fluidity, and as an energy supply. Their presence at high concentrations makes the spermatozoa particularly vulnerable to lipid peroxidation, so it is essential that the semen or extender used for sperm storage in vitro contain a sufficient supply of antioxidants. *ROS,* reactive oxygen species.

sperm must exhibit appropriate motility, undergo the acrosome reaction and then fuse with the oocyte. The DHA content influences the flexibility of the tail (Connor et al., 1998), sperm motility (Aksoy et al., 2006) and the acrosome reaction (Roqueta-Rivera et al., 2011). A comparison of semen from different boars found differences in sperm total n-3 PUFA, DHA and the ratio of n-6:n-3 PUFA between boars with normal and low motility sperm (Am-in et al., 2011).

A number of studies, reviewed by Esmaeli et al. (2015), have compared the PUFA profiles of spermatozoa between normal men and those with different causes of infertility, providing further evidence that low levels of DHA and lower n-3:n-6 ratios are deleterious. The benefits of dietary supplementation with various PUFAs on sperm quality in both humans and livestock species remains, however, equivocal (Van Tran et al., 2017). In part this is likely due to the problems highlighted above, such as the precise content of the diet, the length of time it is fed, and the base level of PUFAs in the body at the start of the trial. The time factor is particularly pertinent for spermatozoa as spermatogenesis takes around 30–75 days in mammals, depending on species, with around a further 10 days required for sperm to transit through the epididymis (França et al., 2005). A study in rams suggested that supplementary n-3 PUFA needed to be fed during the early stages of spermatogenesis to be effective (Samadian et al., 2010). Yan et al. (2013) fed male rats diets with six different n-3:n-6 ratios ranging from 0.13 to 2.85, achieved through altered proportions of soybean and flaxseed oils. These diets were fed to young, postpubertal animals for 60 days. The animals receiving the 1.52 ratio performed best in terms of testis development and sperm morphology. Their litter size following mating to females on normal commercial diets was also highest, averaging 12.3 pups born in comparison to only 8.9 for the 0.13 ratio diet.

The requirements of specific PUFAs for normal spermatogenesis and sperm function was confirmed using various knockout mice. Ablation of *FADS2* abolishes the initial enzymatic step in PUFA synthesis from LA or ALA. Both male and female $Fads2^{-/-}$ mice had normal viability and lifespan but were sterile with impairment of Sertoli cell polarity, a disrupted blood-testis barrier and spermatogenesis arrested at late spermiogenesis (Roqueta-Rivera et al., 2010; Stoffel et al., 2008). Normal fertility was restored by feeding a diet supplemented with DHA, whereas AA was less effective (Roqueta-Rivera et al., 2010). Hormone sensitive lipase (HSL) is a key enzyme involved in mobilization of fatty acids from intracellular stores. Mice deficient in HSL were oligospermic and had reduced testicular concentrations of LA and ALA, increased intermediaries 22:4n-6 and 22:5n-3 and normal concentrations of 22:5n-6 (docosapentaenoic acid) and 22:6n-3 (DHA) (Casado et al., 2013; Osuga et al., 2000). Expression of *Scd-1*, *Fads1*, and *Fads2* was increased in the testes of these mice while *Elovl2* was decreased. While these data indicate that PUFA biosynthesis was altered, these HSL deficient mice also have markedly altered cholesterol metabolism, which may contribute to their infertility. Zadravec et al. (2011) confirmed that ELOVL2 was a key enzyme controlling the amounts of n-6 LCPUFAs with 24–30 carbon atoms in testis. Spermatogenesis in testes of *Elovl2* null mice $(-/-)$ was arrested at the stage of primary spermatocytes, although Leydig and Sertoli cell morphology appeared normal. Heterozygotes $(+/-)$ had impaired formation of spermatids producing spermatozoa with rounded, condensed heads and only 10% were fertile. In contrast to the *Fads2* deficient mice, fertility of the *Elovl2*$^{+/-}$ males could not be restored by dietary supplementation with DHA, supporting a key role for LC n-6 as well as n-3 PUFAs in sperm maturation. Casado et al. (2016) also examined HSL deficiency in mouse testis and suggested that the changes in the phospholipid and sterol composition of lipid rafts in HSL$^{-/-}$ mice might contribute to their sterility.

Membrane fluidity also affects the sensitivity of spermatozoa to chilling and freezing, both important in assisted reproduction (Arav et al., 2000). A number of studies, reviewed by Van Tran et al. (2016), examined the effects of PUFAs on the characteristics of frozen-thawed sperm either by feeding animals different PUFA diets before semen collection or by altering the composition of the extender used during processing. Some, but by no means all, of the studies, which added more n-3 PUFA to extenders reported some benefits, particularly if combined with α-tocopherol, a form of vitamin E, to provide enhanced antioxidant capacity.

## PUFAs and Lipid Peroxidation

A further key issue for male fertility is that LCPUFAs are particularly vulnerable to attack by reactive oxygen species, which can initiate a damaging lipid peroxidation cascade (Wathes et al., 2007). This can seriously compromise the functional integrity of spermatozoa with their high PUFA content (Alvarez and Storey, 1995; Jones et al., 1979). The cytoplasmic extrusion associated with sperm morphogenesis depletes these cells of their internal store of antioxidant enzymes. The fluids bathing spermatozoa during their passage through the male reproductive tract are therefore rich in highly specialized secreted forms of antioxidants (Agarwal et al., 2004). There is a negative correlation between products of lipid peroxidation in sperm suspensions with semen quality in both man and domestic animals (Gomez et al., 1998; Kasimanickam et al., 2006). Exposure of human spermatozoa to the PUFAs LA, AA, or DHA all triggered free radical generation, lipid peroxidation and DNA damage (Aitken et al., 2006). Feeding more dietary PUFA reduced both the antioxidant status and semen quality (concentration and volume) in chickens but adding vitamin E as an antioxidant was able to reverse the negative impact of PUFA supplementation (Cerolini et al., 2005; Zanini et al., 2003). This means that the balance between PUFAs and the availability of antioxidants in the testis and male reproductive tract will together influence the outcome in terms of fertility.

## UTERINE ACTIVITY

The final aspect of PUFAs and reproduction included in this review relates to the uterus, with a focus on their key role in eicosanoid synthesis. This is crucial for many aspects of fertility as uterine-derived PGs regulate the length of the oestrous cycle, the receptiveness of the endometrium to the embryo, the timing and progression of labor, and the response of the uterus to infection (Wathes et al., 2007). Dietary changes, which alter circulating PUFA concentrations, are reflected in the endometrium and therefore have the opportunity to alter the balance of eicosanoids produced (Bilby et al., 2006; Childs et al., 2008; Elmes et al., 2004).

### Control of Eicosanoid Synthesis

The family of eicosanoids includes PGs, leukotrienes, thromboxanes, lipoxins, neuroprotectins, and resolvins which are all derived from 20 carbon PUFAs (Serhan et al., 2008; Sprecher, 2000; Wathes et al., 2007) (Fig. 9.3). These physiologically active compounds are produced in a tissue-specific manner according to the availability of precursor lipids and enzymes present. The actions of different PUFAs on PG synthesis have been studied in detail, revealing a system of extreme complexity. The n-6 and n-3 PUFA families each produce their own specific metabolites (Sprecher, 2000) but they compete with each other for both cellular membrane lipid incorporation and metabolic enzymes (Wathes et al., 2007). This competition can influence the amounts of different LCPUFAs produced from ALA and LA via the sequential actions of desaturases and elongases (FADS2, ELOVL5, and FADS1). Dietary supplementation with longer chain n-6 (GLA, AA) or n-3 (SDA, EPA, DHA) PUFAs can bypass these rate-limiting steps.

The initial step in PG synthesis is through release of stored PUFAs from membrane phospholipids via the action of phospholipase A2 or the coordinate actions of phospholipase C and diglyceride lipase (Irvine, 1982). The enzymes PTGS1 and PTGS2 (also known as COX-1 and COX-2, for cyclooxygenase or "COX") catalyze the metabolism of DGLA, AA, and EPA into 1-, 2-, and 3 series PGs, respectively. The 5-lipoxygenase (LOX) pathway generates 4-series leukotrienes (LTs) from AA and 5-series LTs from EPA. The 15-LOX and 5-LOX pathways catalyze AA sequentially to produce 4-series lipoxins (LXs), EPA to 5-series LXs, or DHA into neuroprotectins, another family of lipid mediators. Both aspirin-dependent and -independent pathways generate E-series resolvins (RvEs) from EPA and D series resolvins (RvDs) from DHA (Serhan et al., 2008).

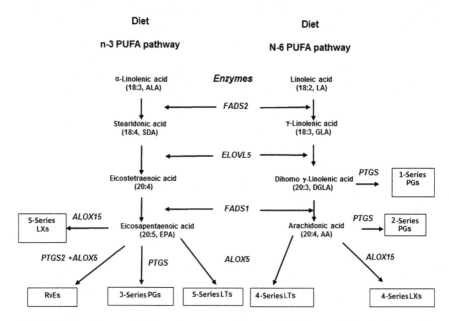

FIGURE 9.3   **The metabolic pathways for n-3 and n-6 PUFA metabolism in the synthesis of eicosanoids.** These lead to the production of lipid mediators with proinflammatory, antiinflammatory, and proresolution effects: *LX*, lipoxin; *LT*, leukotriene; *PG*, prostaglandin; *RvE*, resolvin; *FADS2*, Δ5-Desaturase; *FADS1*, Δ5-Desaturase; *ALOX5*, arachidonate 5 lipoxygenase; *ALOX9*, arachidonate 9 lipoxygenase, lipoxygenase; *PTGS*, cyclooxygenase.

The pattern of PUFA-derived mediators produced in any situation is tightly regulated and can be altered by a variety of mechanisms. PTGS1 and PTGS2 have similar actions but are encoded by different genes that are regulated differentially in a cell-specific manner (Smith and Lagenbach, 2001). For both enzymes AA is the preferred substrate, so EPA metabolism to 3-series PGs is poor, and EPA also inhibits PTGS1 activity. Endometrial PG synthesis is therefore influenced by both PTGS expression and the expression of different PG synthases, which control the relative amounts of PGE and PGF (Cheng et al., 2013; Fortier et al., 2008). This is critical in terms of endometrial function as E and F series PGs have opposing effects with regard to luteolysis and inflammation (Herath et al., 2009). The n-3 PUFAs EPA and DHA are also known to have antiinflammatory properties (Miles and Calder, 2012). The incorporation of n-3 LCPUFAs into cell membranes decreases the availability of AA, so reducing the production of proinflammatory eicosanoids. In addition, both EPA and DHA are precursors for resolvins that are actively antiinflammatory and inflammation-resolving.

Both n-3 and n-6 PUFAs utilize the same enzyme systems for their metabolism and PG production and they compete for both the enzymes and membrane lipid incorporation. The metabolic pathways are, therefore, influenced by presence of the precursors and their metabolites. Our previous studies have shown that PUFAs influence endometrial PG production by: (1) regulating PTGS expression (Cheng et al., 2005a; Sheldrick et al., 2007), (2) altering the proportions of 1-, 2-, and 3-series PGs (Cheng et al., 2005a), (3) changing the PGE:PGF ratios (Cheng et al., 2004), and (4) changing the responsiveness of uterine cells to a challenge with oxytocin (OT) or bacterial lipopolysaccharide (Cheng et al., 2004). Furthermore, AA supplementation in cultured ovine uterine epithelial cells increased not only 2-series PG production but also 1- and 3-series PG production. As AA is not the precursor for 1- and 3-series PG production, this indicated an induction of the enzymes for PG synthesis (Cheng et al., 2005a). Further study confirmed that AA supplementation indeed stimulated mRNA expression of both *PTGS1* and *PTGS2* whereas SDA and EPA decreased *PTGS1* mRNA expression but did not change *PTGS2* expression (Cheng et al., 2013). The effect of EPA on uterine PG production was moderate, but when it was combined with AA, the stimulatory effect of AA on PG production was attenuated by up to 80%. The ratio of $PGE_2$ to $PGF_{2\alpha}$ was not affected by AA or EPA alone, but was raised when these PUFAs were combined (Cheng et al., 2013). The ruminant uterine endometrium lacks FADS2 activity so, when LA is supplemented in vitro, it is not metabolized further and LA accumulation inhibits PG production (Cheng et al., 2001, 2004, 2005a). It can be seen from this brief summary that the actual amounts of different PGs produced by endometrial cells at any one time can be influenced by the PUFA composition of the tissue at many different levels.

## Luteolysis and Uterine Involution Postpartum

In most mammalian species pulsatile release of $PGF_{2\alpha}$ from the endometrium is the key factor driving the onset of luteolysis: this process needs to be suppressed in early pregnancy to maintain the corpus luteum (Wathes and Lamming, 1995). It has been suggested that n-3 PUFA supplementation may be beneficial for embryo survival, particularly in cattle, by suppressing luteolytic $PGF_{2\alpha}$ production (Santos et al., 2008). This idea was supported by Zachut et al. (2011) who found that dairy cows fed a high n-3 PUFA diet containing flaxseed exhibited longer time intervals between receiving a single $PGF_{2\alpha}$ injection to initiate luteolysis and the subsequent manifestation of oestrus behavior. Similarly, when ALA was fed to nonpregnant sheep the length of the luteal phase was prolonged slightly, by about one day (Naddafy et al., 2005). Interpreting the physiological benefit is not, however, straightforward as uterine PG production ($PGE_2$, $PGF_{2\alpha}$, and $PGI_2$,) in ruminants is upregulated by conceptus-derived IFNτ (Lewis, 1989; Spencer et al., 2013; Ullbrich et al., 2012). IFNτ is the key signaling molecule in ruminants for the maternal recognition of pregnancy. These PGs are now thought to comprise an essential component of the bidirectional signaling between the endometrium and embryo in early pregnancy. The upregulation of $PGF_{2\alpha}$ does not, however, cause luteolysis under these circumstances as the IFNτ also prevents upregulation of endometrial OT receptors, so the $PGF_{2\alpha}$ is released steadily rather than in a pulsatile fashion (Wathes and Lamming, 1995).

Gulliver et al. (2012) reviewed 10 studies in which cows received various PUFA supplements followed by an OT challenge during the luteal phase of the oestrous cycle to assess the release of the PGF metabolite 15-keto-dihydro-PGF2 alpha (PGFM) into the peripheral circulation. Of these, three studies reported no treatment effect. In a further three the results differed according to either the progesterone profile or the day of the cycle. Petit et al. (2002) found that n-3 PUFAs increased the baseline for PGFM but reduced the response to OT. In contrast, Childs et al. (2008) found a dose-responsive rise in OT-stimulated PGFM release in beef heifers supplemented with fish oil. There were two reports that PGFM release was lower with n-3 supplementation (fishmeal or linseed) and two that it was higher after sunflower seeds or Megalac (both high in n-6). These studies are supported by a more recent trial by Dirandeh et al. (2015) and provide reasonable agreement that an increased intake of n-6 PUFA increases the amount of PGFM released, presumably by providing more AA as precursor. The response to n-3 PUFA was, however, much more variable. This may in part be due to the different supplements used, which included both linseed and fish oils. The timing of the challenge was also critical as the response was strongly influenced by how close the animal was to natural luteolysis (Childs et al., 2008; Robinson et al., 2002). Several studies have collected bovine endometrium following dietary intervention with n-3 or n-6 PUFA and found that animals had altered gene expression of a number of transcription factors, PG and steroid synthetic enzymes, and receptors and immune regulators (Coyne et al., 2008; Dirandeh et al., 2015; Waters et al., 2012, 2014). Despite the variability in PGFM released this work supports the notion that the n-3 PUFA intake can undoubtedly influence the bovine endometrium, although the significance for fertility remains uncertain.

Feeding dietary PUFA supplements to cows has also been considered as a possible means of reducing the impact of uterine infection after calving. In almost all dairy cows after calving the uterus becomes contaminated with bacteria, which can lead to metritis or endometritis if not cleared efficiently (Sheldon et al., 2009). Uterine involution is normally associated with an upregulation of endometrial PGF production for about 3–4 weeks after calving but this is maintained for longer when the uterus becomes infected (Bekana et al., 1996). Feeding fat sources rich in n-6 PUFA during late gestation and early lactation has generally increased uterine PG secretion after calving (Filley et al., 2000; Grant et al., 2005; Santos et al., 2008), reduced the interval to uterine involution and decreased the incidence of puerperal metritis and clinical endometritis (Dirandeh et al., 2013; Juchem et al., 2010). Mattos et al. (2004) fed a fish oil supplement from 3 weeks before until 3 weeks after calving and found that this reduced concentrations of plasma PGFM during the 60 h immediately after parturition. This might not, however, be beneficial as contractions at this time are important to clear the cell debris from the postpartum uterus.

## PUFAs and Preterm Labor

Spontaneous preterm labor is a worldwide problem that affects about 10% of all human pregnancies and is associated with 70% of neonatal deaths (Challis et al., 2002). An increased production of $PGF_{2\alpha}$ and $PGE_2$ by the feto-placental unit (including amnion, chorion, and decidual endometrium) is a key component of labor as these PGs are major contributors to cervical dilation, membrane rupture, and myometrial contraction (Challis et al., 2000). The levels of PTGS2 increase either just before or during labor and inhibitors of PG synthesis prolong gestation (Olson and Fazio, 2003). PUFAs may thus be able to influence the timing of parturition through alterations to PG synthesis as discussed earlier. The risk of preterm delivery can indeed be increased by high consumption of n-6 PUFAs. When sheep were fed a diet rich in n-6 PUFAs during late gestation this increased both their production of $PGE_2$ and $PGF_{2\alpha}$ and the risk of preterm labor (Cheng et al., 2005b; Elmes et al., 2004, 2005). Plasma concentrations of AA in women increase throughout pregnancy with the highest levels of both LA and AA being observed during labor (Ashby et al., 1997; Ogburn et al., 1980) and women who delivered prematurely were shown to have higher n-6 PUFA concentrations (Reece et al., 1997).

More emphasis has, however, focused on the role of n-3 PUFAs consumed during pregnancy. Several studies in rats showed that high n-3 PUFA diets increased both gestational length and birth weight (Leat and Northrop, 1981; Olsen et al., 1990; Waltman et al., 1977). This was, however, associated with a prolonged and difficult labor (Mathias et al., 1987). Various human populations with a high fish consumption were also reported to have longer pregnancies (Olsen et al., 1991) and two of seven trials on sows found increased gestation lengths when animals were fed linseed or fish oil (Tanghe and De Smet, 2013). Possible mechanisms suggested to explain these effects included a shift from synthesis of 2-series to the less biologically active 3-series PGs (Leaver et al., 1991), reduced expression of myometrial *PTGS2* mRNA (Ma et al., 2000), decreased FADS1 activity resulting in lower circulating concentrations of AA (Stark et al., 2002), and direct effects on $Ca^{2+}$ channels and cell signaling in the myometrium (Allen and Harris, 2001). The latter suggestion was supported by an in vitro study showing that both AA and DHA reduced myometrial membrane OT receptor concentrations (Kim et al., 2012).

Dietary supplementation of pregnant women with n-3 PUFAs derived from fish oil was therefore suggested as a potential strategy to reduce the risk of premature delivery. A large number of trials have now been published but with mixed results. Olsen et al. (2000) gave a fish oil supplement to women with a previous preterm delivery and found a reduced risk from 33% to 21% in comparison with a placebo group. Borod et al. (1999) provided women with DHA enriched eggs in the third trimester and found fewer low birth weight and preterm babies. Some studies reported that high fish oil intakes caused small increases in gestation length of between 1.6 and 6 days accompanied by a small increase in birth weight of about 47 g (Olsen et al., 1992; Horvath et al., 2007; Makrides et al., 2006; Smuts et al., 2003; Szajewska et al., 2006) but another showed no effect on either birth weight or gestation length (Helland et al., 2001). The weight change is probably partly, but not entirely, accounted for by the increased gestation length (Moltó-Puigmartí et al., 2014). In contrast, supplementation of seafood to an American population actually decreased birth weight and did not reduce the risk of preterm birth (Oken et al., 2004). These discrepancies are, perhaps, not surprising given the variability in both the starting populations and the content, dose, and duration of the dietary interventions used. A more recent study reported opposite direction correlations between maternal DHA and AA intake, gestation length and birth weight but these effects were dependent on maternal genotype classified according to a SNP in the *FADS1/FADS2* gene cluster (Moltó-Puigmartí et al., 2014). This SNP explained 28% of the variance of serum phospholipid AA and up to 12% of its precursor acids, resulting in mothers with differing n-3 to n-6 ratios (Koletzko et al., 2008a). Allen and Harris (2001) suggested that there were also possible differences in the actions of DHA and EPA, which were present in varying proportions in the different supplements used. Another comprehensive review concluded that results of n-3 PUFA supplementation were inconclusive, with more work required to establish both the mechanism of action and the efficacy in preventing prematurity (Lewin et al., 2005). A further important consideration is that the perinatal DHA status also appears to affect some key aspects of postnatal development including visual function, risk of childhood asthma, adipose tissue deposition, and insulin resistance (Demmelmair and Koletzko, 2015; Mennitti et al., 2015). Taking the requirements for visual and cognitive development into account, Koletzko et al. (2008b) recommended an average daily intake of 200 mg DHA for pregnant women, an amount that should also reduce the risk of preterm birth.

## CONCLUSIONS

PUFAs have multiple actions within the body that are essential for fertility and fetal development. Evidence for specific functions is generally based on in vitro studies that often fail to translate into consistent in vivo effects. There are many potential reasons for this but a pervading difficulty is to achieve predictable changes in concentrations of individual PUFAs and the ratios between them in particular tissues. Tissue contents do reflect dietary intake but are also influenced by metabolic status, genotype, and specific uptake mechanisms. Despite these reservations there is some evidence that n-3 PUFAs can benefit aspects of fertility, in particular by improving semen quality and reducing the risk of preterm labor. Further validation is, however, required to ensure that the supposed benefits are of sufficient size and consistency to be cost effective in practice.

## Acknowledgments

We thank our colleagues and students at the Royal Veterinary College who have contributed to our own research in this area. RVC manuscript number: PPH_01470.

## References

Agarwal, A., Nallella, K.P., Allamaneni, S.S., Said, T.M., 2004. Role of antioxidants in treatment of male infertility: an overview of the literature. Reprod. Biomed. Online 8, 616–627.

Aitken, R.J., Wingate, J.K., De Iuliis, G.N., Koppers, A.J., McLaughlin, E.A., 2006. Cis-unsaturated fatty acids stimulate reactive oxygen species generation and lipid peroxidation in human spermatozoa. J. Clin. Endocrinol. Metab. 91, 4154–4163.

Aksoy, Y., Aksoy, H., Altinkaynak, K., Aydin, H.R., Ozkan, A., 2006. Sperm fatty acid composition in subfertile men. Prostaglandins Leuk. Essent. Fatty Acids 75, 75–79.

Allen, K.G.D., Harris, M.A., 2001. The role of n-3 fatty acids in gestation and parturition. Exp. Biol. Med. 226, 498–506.

Alvarez, J.G., Storey, B.T., 1995. Differential incorporation of fatty acids into and peroxidative loss of fatty acids from phospholipids of human spermatozoa. Mol. Reprod. Dev. 42, 334–346.

Am-in, N., Kirkwood, R.N., Techakumphu, M., Tantasuparuk, W., 2011. Lipid profiles of sperm and seminal plasma from boars having normal or low sperm motility. Theriogenology 75, 897–903.

Arav, A., Pearl, M., Zeron, Y., 2000. Does membrane lipid profile explain chilling sensitivity and membrane lipid phase transition of spermatozoa and oocytes? Cryo Lett. 21, 179–186.

Ashby, A.M., Robinette, B., Kay, H.H., 1997. Plasma and erythrocyte profiles of nonesterified polyunsaturated fatty acids during normal pregnancy and labor. Am. J. Perinatol. 14, 623–629.

Bazer, F.W., Wu, G., Johnson, G.A., Kim, J., Song, G., 2011. Uterine histotroph and conceptus development: select nutrients and secreted phosphoprotein 1 affect mechanistic target of rapamycin cell signaling in ewes. Biol. Reprod. 85, 1094–1107.

Bekana, M., Jonsson, P., Kindahl, H., 1996. Intrauterine bacterial findings and hormonal profiles in post-partum cows with normal puerperium. Acta Vet. Scand. 37, 251–263.

Berghaus, T.M., Demmelmair, H., Koletzko, B., 2000. Essential fatty acids and their long-chain polyunsaturated metabolites in maternal and cord plasma triglycerides during late gestation. Biol. Neonate 77, 96–100.

Bilby, T.R., Guzeloglu, A., MacLaren, L.A., Staples, C.R., Thatcher, W.W., 2006. Pregnancy, bovine somatotropin, and dietary n-3 fatty acids in lactating dairy cows: II. endometrial gene expression related to maintenance of pregnancy. J. Dairy Sci. 89, 3375–3385.

Borod, E., Atkinson, R., Barclay, W.R., Carlson, S.E., 1999. Effects of third trimester consumption of eggs high in docosahexaenoic acid on docosahexaenoic acid status and pregnancy. Lipids Suppl. 34, S231.

Brinsfield, T.H., Hawk, H.W., 1973. Control by progesterone of the concentration of lipid droplets in epithelial cells of the sheep endometrium. J. Anim. Sci. 36, 919–922.

Burns, G., Brooks, K., Wildung, M., Navakanitworakul, R., Christenson, L.K., Spencer, T.E., 2014. Extracellular vesicles in luminal fluid of the ovine uterus. PLoS One 9, e90913.

Calder, P.C., 2015. Marine omega-3 fatty acids and inflammatory processes: effects, mechanisms and clinical relevance. Biochim. Biophys. Acta 1851, 469–484.

Calder, P.C., Kremmyda, L.S., Vlachava, M., Noakes, P.S., Miles, E.A., 2010. Is there a role for fatty acids in early life programming of the immune system? Proc. Nutr. Soc. 69, 373–830.

Casado, M.E., Pastor, O., García-Seisdedos, D., Huerta, L., Kraemer, F.B., Lasunción, M.A., Martín-Hidalgo, A., Busto, R., 2016. Hormone-sensitive lipase deficiency disturbs lipid composition of plasma membrane microdomains from mouse testis. Biochim. Biophys. Acta 1861 (9 Pt A), 1142–1150.

Casado, M.E., Pastor, O., Mariscal, P., Canfrán-Duque, A., Martínez-Botas, J., Kraemer, F.B., Lasunción, M.A., Martín-Hidalgo, A., Busto, R., 2013. Hormone-sensitive lipase deficiency disturbs the fatty acid composition of mouse testis. Prostaglandins Leuk. Essent. Fatty Acids 88, 227–233.

Cerolini, S., Surai, P.F., Speake, B.K., Sparks, N.H., 2005. Dietary fish and evening primrose oil with vitamin E effects on semen variables in cockerels. Br. Poult. Sci. 46, 214–222.

Challis, J.R.G., Matthews, S.G., Gibb, W., Lye, S.J., 2000. Endocrine and paracrine regulation of birth at term and preterm. Endocr. Rev. 21, 514–550.

Challis, J.R., Sloboda, D.M., Alfaidy, N., Lye, S.J., Gibb, W., Patel, F.A., Whittle, W.L., Newnham, J.P., 2002. Prostaglandins and mechanisms of preterm birth. Reproduction 124, 1–17.

Chambaz, J., Ravel, D., Manier, M.C., Pepin, D., Mulliez, N., Bereziat, G., 1985. Essential fatty acids interconversion in the human fetal liver. Biol. Neonate 47, 136–140.

Cheng, Z., Abayasekara, D.R., Ward, F., Preece, D.M., Raheem, K.A., Wathes, D.C., 2013. Altering n-3 to n-6 polyunsaturated fatty acid ratios affects prostaglandin production by ovine uterine endometrium. Anim. Reprod. Sci. 143, 38–47.

Cheng, Z., Abayasekara, D.R., Wathes, D.C., 2005a. The effect of supplementation with n-6 polyunsaturated fatty acids on 1-, 2- and 3-series prostaglandin F production by ovine uterine epithelial cells. Biochim. Biophys. Acta 1736, 128–135.

Cheng, Z., Elmes, M., Kirkup, S.E., Abayasekara, D.R., Wathes, D.C., 2004. Alteration of prostaglandin production and agonist responsiveness by n-6 polyunsaturated fatty acids in endometrial cells from late-gestation ewes. J. Endocrinol. 182, 249–256.

Cheng, Z., Elmes, M., Kirkup, S.E., Chin, E.C., Abayasekara, D.R.E., Wathes, D.C., 2005b. The effect of a diet supplemented with n-6 polyunsaturated fatty acid (PUFA) linoleic acid on prostaglandin production in early and late pregnant ewes. J. Endocrinol. 184, 167–178.

Cheng, Z., Robinson, R.S., Pushpakumara, P.G., Mansbridge, R.J., Wathes, D.C., 2001. Effect of dietary polyunsaturated fatty acids on uterine prostaglandin synthesis in the cow. J. Endocrinol. 171, 463–473.

Childs, S., Hennessy, A.A., Sreenan, J.M., Wathes, D.C., Cheng, Z., Stanton, C., Diskin, M.G., Kenny, D.A., 2008. Effect of level of dietary n-3 polyunsaturated fatty acid supplementation on systemic and tissue fatty acid concentrations and on selected reproductive variables in cattle. Theriogenology 70, 595–611.

Connor, W.E., Lin, D.S., Wolf, D.P., Alexander, M., 1998. Uneven distribution of desmosterol and docosahexaenoic acid in the heads and tails of monkey sperm. J. Lipid Res. 39, 1404–1411.

Coyne, G.S., Kenny, D.A., Childs, S., Sreenan, J.M., Waters, S.M., 2008. Dietary n-3 polyunsaturated fatty acids alter the expression of genes involved in prostaglandin biosynthesis in the bovine uterus. Theriogenology 70, 772–782.

Demmelmair, H., Koletzko, B., 2015. Importance of fatty acids in the perinatal period. World Rev. Nutr. Diet. 112, 31–47.

Dirandeh, E., Towhidi, A., Pirsaraei, Z.A., Saberifar, T., Akhlaghi, A., Roodbari, A.R., 2015. The endometrial expression of prostaglandin cascade components in lactating dairy cows fed different polyunsaturated fatty acids. Theriogenology 83, 206–212.

Dirandeh, E., Towhidi, A., Zeinoaldini, S., Ganjkhanlou, M., Ansari Pirsaraei, Z., Fouladi-Nashta, A., 2013. Effects of different polyunsaturated fatty acid supplementations during the postpartum periods of early lactating dairy cows on milk yield, metabolic responses, and reproductive performances. J. Anim. Sci. 91, 713–721.

Dutta-Roy, A.K., 2000. Cellular uptake of long-chain fatty acids: role of membrane-associated fatty-acid-binding/transport proteins. Cell. Mol. Life Sci. 57, 1360–1372.

Elias, S.L., Innis, S.M., 2001. Infant plasma trans, n-6, and n-3 fatty acids and conjugated linoleic acids are related to maternal plasma fatty acids, length of gestation, and birth weight and length. Am. J. Clin. Nutr. 73, 807–814.

Elmes, M., Green, L.R., Poore, K., Newman, J., Burrage, D., Abayasekara, D.R.E., Cheng, Z., Hanson, M.A., Wathes, D.C., 2005. Raised dietary n-6 polyunsaturated fatty intake increases 2-series prostaglandin production during labour in the ewe. J. Physiol. 562, 583–592.

Elmes, M., Tew, P., Cheng, Z., Kirkup, S.E., Abayasekara, D.R.E., Calder, P.C., Hanson, M.A., Wathes, D.C., Burdge, G.C., 2004. The effect of dietary supplementation with linoleic acid to late gestation ewes on the fatty acid composition of maternal and fetal plasma and tissues and the synthetic capacity of the placenta for 2-series prostaglandins. Biochim. Biophys. Acta 1686, 139–147.

Enders, A.C., Carter, A.M., 2006. Comparative placentation: some interesting modifications for histotrophic nutrition: a review. Placenta 27 (Suppl. A), S11–S16.

Esmaeli, V., Shahverdi, H., Moghadasian, M.H., Alizadeh, A.R., 2015. Dietary fatty acids affect semen quality: a review. Andrology 3, 450–457.

Fair, S., Doyle, D.N., Diskin, M.G., Hennessy, A.A., Kenny, D.A., 2014. The effect of dietary n-3 polyunsaturated fatty acids supplementation of rams on semen quality and subsequent quality of liquid stored semen. Theriogenology 8, 210–219.

Filley, S.J., Turner, H.A., Stormshak, F., 2000. Plasma fatty acids, prostaglandin F2alpha metabolite, and reproductive response in postpartum heifers fed rumen bypass fat. J. Anim. Sci. 78, 139–144.

Fischer, S., 1989. Dietary polyunsaturated fatty acids and eicosanoid formation in humans. Adv. Lipid Res. 23, 169–198.

Fortier, M.A., Krishnaswamy, K., Danyod, G., Boucher-Kovalik, S., Chapdalaine, P., 2008. A postgenomic integrated view of prostaglandins in reproduction: implications for other body systems. J. Physiol. Pharmacol. 59 (Suppl. 1), 65–89.

Forde, N., McGettigan, P.A., Mehta, J.P., O'Hara, L., Mamo, S., Bazer, F.W., Spencer, T.E., Lonergan, P., 2014. Proteomic analysis of uterine fluid during the pre-implantation period of pregnancy in cattle. Reproduction 147, 575–587.

França, L.R., Avelar, G.F., Almeida, F.F., 2005. Spermatogenesis and sperm transit through the epididymis in mammals with emphasis on pigs. Theriogenology 63, 300–318.

Gomez, E., Irvine, D.S., Aitken, R.J., 1998. Evaluation of a spectrophotometric assay for the measurement of malondialdehyde and 4-hydroxyalkenals in human spermatozoa: relationships with semen quality and sperm function. Int. J. Androl. 21, 81–94.

Grant, M.H., Alexander, B.M., Hess, B.W., Bottger, J.D., Hixon, D.L., Van Kirk, E.A., Nett, T.M., Moss, G.E., 2005. Dietary supplementation with safflower seeds differing in fatty acid composition differentially influences serum concentrations of prostaglandin F metabolite in postpartum beef cows. Reprod. Nutr. Dev. 45, 721–727.

Gulliver, C.E., Friend, M.A., King, B.J., Clayton, E.H., 2012. The role of omega-3 polyunsaturated fatty acids in reproduction of sheep and cattle. Anim. Reprod. Sci. 131, 9–22.

Haggarty, P., Page, K., Abramovich, D.R., Ashton, J., Brown, D., 1997. Long-chain polyunsaturated fatty acid transport across the perfused human placenta. Placenta 18, 635–642.

Hanebutt, F.L., Demmelmair, H., Schiessl, B., Larqué, E., Koletzko, B., 2008. Long-chain polyunsaturated fatty acid (LC-PUFA) transfer across the placenta. Clin. Nutr. 27, 685–693.

Helland, I.B., Saugstad, O.D., Smith, L., Saarem, K., Solvoll, K., Ganes, T., Drevon, C.A., 2001. Similar effects on infants of n-3 and n-6 fatty acids supplementation to pregnant and lactating women. Pediatrics 108, E82.

Herath, S., Lilly, S.T., Fischer, D.P., Williams, E.J., Dobson, H., Bryant, C.E., Sheldon, I.M., 2009. Bacterial lipopolysaccharide induces an endocrine switch from prostaglandin F2alpha to prostaglandin E2 in bovine endometrium. Endocrinology 150, 1912–1920.

Herrera, E., 2002. Lipid metabolism in pregnancy and its consequences in the fetus and newborn. Endocrin 19, 43–55.

Hewitt, D.P., Mark, P.J., Waddell, B.J., 2006. Placental expression of peroxisome proliferator-activated receptors in rat pregnancy and the effect of increased glucocorticoid exposure. Biol. Reprod. 74, 23–28.

Horvath, A., Koletzko, B., Szajewska, H., 2007. Effect of supplementation of women in high-risk pregnancies with long-chain polyunsaturated fatty acids on pregnancy outcomes and growth measures at birth: a meta-analysis of randomized controlled trials. Br. J. Nutr. 98, 253–259.

Irvine, R.F., 1982. The enzymology of stimulated inositol lipid turnover. Cell Calcium 3, 295–309.

Islam, A., Kagawa, Y., Sharifi, K., Ebrahimi, M., Miyazaki, H., Yasumoto, Y., Kawamura, S., Yamamoto, Y., Sakaguti, S., Sawada, T., Tokuda, N., Sugino, N., Suzuki, R., Owada, Y., 2014. Fatty acid binding protein 3 is involved in n-3 and n-6 PUFA transport in mouse trophoblasts. J. Nutr. 144, 1509–1516.

Jones, R., Mann, T., Sherins, R.J., 1979. Peroxidative breakdown of phospholipids in human spermatozoa: spermicidal effects of fatty acid peroxides and protective action of seminal plasma. Fertil. Steril. 31, 531–537.

Jones, M.L., Mark, P.J., Waddell, B.J., 2014. Maternal dietary omega-3 fatty acids and placental function. Reproduction 147, R143–R152.

Juchem, S.O., Cerri, R.L., Villaseñor, M., Galvão, K.N., Bruno, R.G., Rutigliano, H.M., DePeters, E.J., Silvestre, F.T., Thatcher, W.W., Santos, J.E., 2010. Supplementation with calcium salts of linoleic and trans-octadecenoic acids improves fertility of lactating dairy cows. Reprod. Domest. Anim. 45, 55–62.

Kasimanickam, R., Pelzer, K.D., Kasimanickam, V., Swecker, W.S., Thatcher, C.D., 2006. Association of classical semen parameters, sperm DNA fragmentation index, lipid peroxidation and antioxidant enzymatic activity of semen in ram-lambs. Theriogenology 65, 1407–1421.

Kim, P.Y., Zhong, M., Kim, Y.S., Sanborn, B.M., Allen, K.G., 2012. Long chain polyunsaturated fatty acids alter oxytocin signaling and receptor density in cultured pregnant human myometrial smooth muscle cells. PLoS One 7, e41708.

Koch, J.M., Ramadoss, J., Magness, R.R., 2010. Proteomic profile of uterine luminal fluid from early pregnant ewes. J. Proteome Res. 9, 3878–3885.

Koletzko, B., Demmelmair, H., Schaeffer, L., Illig, T., Heinrich, J., 2008a. Genetically determined variation in polyunsaturated fatty acid metabolism may result in different dietary requirements. Nestle Nutr. Workshop Ser. Pediatr. Prog. 62, 35–44.

Koletzko, B., Lien, E., Agostoni, C., Böhles, H., Campoy, C., Cetin, I., Decsi, T., Dudenhausen, J.W., Dupont, C., Forsyth, S., Hoesli, I., Holzgreve, W., Lapillonne, A., Putet, G., Secher, N.J., Symonds, M., Szajewska, H., Willatts, P., Uauy, R., World Association of Perinatal Medicine Dietary Guidelines Working Group, 2008b. The roles of long-chain polyunsaturated fatty acids in pregnancy, lactation and infancy: review of current knowledge and consensus recommendations. J. Perinat. Med. 36, 5–14.

Kuipers, R.S., Luxwolda, M.F., Offringa, P.J., Boersma, E.R., Dijck-Brouwer, D.A., Muskiet, F.A., 2012. Gestational age dependent changes of the fetal brain, liver and adipose tissue fatty acid compositions in a population with high fish intakes. Prostaglandins Leuk. Essent. Fatty Acids 86, 189–199.

Larqué, E., Demmelmair, H., Gil-Sánchez, A., Prieto-Sánchez, M.T., Blanco, J.E., Pagán, A., Faber, F.L., Zamora, S., Parrilla, J.J., Koletzko, B., 2011. Placental transfer of fatty acids and fetal implications. Am. J. Clin. Nutr. 94 (Suppl.), 1908S–1913S.

Leat, W.M.F., Northrop, C.A., 1981. Effect of linolenic acid on gestation and parturition in the rat. Prog. Lipid Res. 20, 819–821.

Leaver, H.A., Howie, A., Wilson, N.H., 1991. The biosynthesis of the 3-series prostaglandins in rat uterus after alpha-linolenic acid feeding: mass spectroscopy of prostaglandins E and F produced by rat uteri in tissue culture. Prostaglandins Leuk. Essent. Fatty Acids 42, 217–224.

Lewin, G.A., Schachter, H.M., Yuen, D., Merchant, P., Mamaladze, V., Tsertsvadze, A., 2005. Effects of omega-3 fatty acids on child and maternal health. Evid. Rep. Technol. Assess. (Summ) 118, AHRQ Publication No. 05-E025-1. Agency for Healthcare Research and Quality, Rockville, MD, United States.

Lewis, G.S., 1989. Prostaglandin secretion by the blastocyst. J. Reprod. Fertil. Suppl. 37, 261–267.

López-Luna, P., Ortega-Senovilla, H., López-Soldado, I., Herrera, E., 2016. Fate of orally administered radioactive fatty acids in the late-pregnant rat. Am. J. Physiol. Endocrinol. Metab. 310, E367–E377.

Ma, X.H., Wu, W.X., Brenna, J.T., Nathanielsz, P.W., 2000. Maternal intravenous administration of long chain n-3 polyunsaturates to the pregnant ewe in late gestation results in specific inhibition of prostaglandin h synthase (PGHS) 2, but not PGHS1 and oxytocin receptor mRNA in myometrium during betamethasone-induced labor. J. Soc. Gynecol. Investig. 7, 233–237.

Makrides, M., Duley, L., Olsen, S.F., 2006. Marine oil, and other prostaglandin precursor, supplementation for pregnancy uncomplicated by pre-eclampsia or intrauterine growth restriction. Cochrane Database Syst. Rev. 19, CD003402.

Maldjian, A., Pizzi, F., Gliozzi, T., Cerolini, S., Penny, P., Noble, R., 2005. Changes in sperm quality and lipid composition during cryopreservation of boar semen. Theriogenology 63, 411–421.

Mathias, M.M., Tsai, A., Harris, M.A., Mcgregor, J.A., 1987. Oral administration of menhaden oil alters gestation in rats. In: Lands, W.E.M. (Ed.), Poyunsaturated Fatty Acids and Eicosanoids. American Oil Chemists' Society Press, Champaign, IL, pp. 508–512.

Mattos, R., Staples, C.R., Arteche, A., Wiltbank, M.C., Diaz, F.J., Jenkins, T.C., Thatcher, W.W., 2004. The effects of feeding fish oil on uterine secretion of PGF2alpha, milk composition, and metabolic status of periparturient Holstein cows. J. Dairy Sci. 87, 921–932.

Mennitti, L.V., Oliveira, J.L., Morais, C.A., Estadella, D., Oyama, L.M., Oller do Nascimento, C.M., Pisani, L.P., 2015. Type of fatty acids in maternal diets during pregnancy and/or lactation and metabolic consequences of the offspring. J. Nutr. Biochem. 26, 99–111.

Michalik, L., Wahli, W., 2008. PPARs mediate lipid signaling in inflammation and cancer. PPAR Res. 2008, 134059.

Miles, E.A., Calder, P.C., 2012. Influence of marine n-3 polyunsaturated fatty acids on immune function and a systematic review of their effects on clinical outcomes in rheumatoid arthritis. Br. J. Nutr. 107 (Suppl. 2), S171–S184.

Moltó-Puigmartí, C., van Dongen, M.C., Dagnelie, P.C., Plat, J., Mensink, R.P., Tan, F.E., Heinrich, J., Thijs, C., 2014. Maternal but not fetal FADS gene variants modify the association between maternal long-chain PUFA intake in pregnancy and birth weight. J. Nutr. 144, 1430–1437.

Naddafy, J.M., Chin, E.C., Cheng, Z., Brickell, J.S., Wathes, D.C., Abayasekara, D.R.E., 2005. Effects of dietary polyunsaturated fatty acids (PUFAs) on prostaglandin and progesterone secretion in the ewe. Reproduction Abstr Ser No 32 Abstract O05.

Nakamura, M.T., Yudell, B.E., Loor, J.J., 2014. Regulation of energy metabolism by long-chain fatty acids. Prog. Lipid Res. 53, 124–144.

Noble, R.C., Shand, J.H., Christie, W.W., 1985. Synthesis of C20 and C22 polyunsaturated fatty acids by the placenta of the sheep. Biol. Neonate 47, 333–338.

Ogburn, P.L., Johnson, S.B., Williams, P.P., Holman, R.T., 1980. Levels of free fatty acids and arachidonic acid in pregnancy and labor. J. Lab. Clin. Med. 95, 943–949.

Oken, E., Kleinman, K.P., Olsen, S.F., Rich-Edwards, J.W., Gillman, M.W., 2004. Associations of seafood and elongated n-3 fatty acid intake with fetal growth and length of gestation: results from a US pregnancy cohort. Am. J. Epidemiol. 160, 774–783.

Olsen, S.F., Hansen, H.S., Jensen, B., 1990. Fish oil versus arachis oil food supplementation in relation to pregnancy duration in rats. Prostaglandins Leuk. Essent. Fatty Acids 40, 255–260.

Olsen, S.F., Hansen, H.S., Sommer, S., Jensen, B., Sorensen, T.I., Secher, N.J., Zachariassen, P., 1991. Gestational age in relation to marine n-3 fatty acids in maternal erythrocytes: a study of women in the Faroe Islands and Denmark. Am. J. Obstet. Gynecol. 164, 1203–1209.

Olsen, S.F., Sørensen, J.D., Secher, N.J., Hedegaard, M., Henriksen, T.B., Hansen, H.S., Grant, A., 1992. Randomised controlled trial of effect of fish-oil supplementation on pregnancy duration. Lancet 339, 1003–1007.

Olsen, S.F., Secher, N.J., Tabor, A., Weber, T., Walker, J.J., Gluud, C., 2000. Randomised clinical trials of fish oil supplementation in high risk pregnancies, Fish Oil Trials In Pregnancy (FOTIP) Team. BJOG 107, 382–395.

Olson, M.A., Fazio, R.H., 2003. Relations between implicit measures of prejudice:what are we measuring? Psychol. Sci. 14, 636–639.

Osuga, J., Ishibashi, S., Oka, T., Yagyu, H., Tozawa, R., Fujimoto, A., Shionoiri, F., Yahagi, N., Kraemer, F.B., Tsutsumi, O., Yamada, N., 2000. Targeted disruption of hormone-sensitive lipase results in male sterility and adipocyte hypertrophy, but not in obesity. Proc. Natl. Acad. Sci. USA 97, 787–792.

Parmelee, D.C., Evenson, M.A., Deutsch, H.F., 1978. The presence of fatty acids in human alpha-fetoprotein. J. Biol. Chem. 253, 2114–2119.

Pérez, E., Bourguet, W., Gronemeyer, H., de Lera, A.R., 2012. Modulation of RXR function through ligand design. Biochim. Biophys. Acta 1821, 57–69.

Petit, H.V., Dewhurst, R.J., Scollan, N.D., Proulx, J.G., Khalid, M., Haresign, W., Twagiramungu, H., Mann, G.E., 2002. Milk production and composition, ovarian function and prostaglandin secretion of dairy cows fed omega-3 fatty acids. J. Dairy Sci. 85, 889–899.

Pike, L.J., 2006. Rafts defined: a report on the keystone symposium on lipid rafts and cell function. J. Lipid Res. 47, 1597–1598.

Poulos, A., Darin-Bennett, A., White, I.G., 1973. The phospholipid-bound fatty acids and aldehydes of mammalian spermatozoa. Comp. Biochem. Physiol. B 46, 541–549.

Reece, M.S., McGregor, J.A., Allen, K.G., Harris, M.A., 1997. Maternal and perinatal long-chain fatty acids: possible roles in preterm birth. Am. J. Obstet. Gynecol. 176, 907–914.

Regueira, M., Riera, M.F., Galardo, M.N., Camberos Mdel, C., Pellizzari, E.H., Cigorraga, S.B., Meroni, S.B., 2015. FSH and bFGF regulate the expression of genes involved in Sertoli cell energetic metabolism. Gen. Comp. Endocrinol. 222, 124–133.

Ribeiro, E.S., Santos, J.E., Thatcher, W.W., 2016. Role of lipids on elongation of the preimplantation conceptus in ruminants. Reproduction 152, R115–R126.

Robinson, R.S., Pushpakumara, P.G., Cheng, Z., Peters, A.R., Abayasekara, D.R., Wathes, D.C., 2002. Effects of dietary polyunsaturated fatty acids on ovarian and uterine function in lactating dairy cows. Reproduction 124, 119–131.

Rodríguez-Cruz, M., González, R.S., Maldonado, J., López-Alarcón, M., Bernabe-García, M., 2016. The effect of gestational age on expression of genes involved in uptake, trafficking and synthesis of fatty acids in the rat placenta. Gene 591, 403–410.

Roqueta-Rivera, M., Abbott, T.L., Sivaguru, M., Hess, R.A., Nakamura, M.T., 2011. Deficiency in the omega-3 fatty acid pathway results in failure of acrosome biogenesis in mice. Biol. Reprod. 85, 721–732.

Roqueta-Rivera, M., Stroud, C.K., Haschek, W.M., Akare, S.J., Segre, M., Brush, R.S., Agbaga, M.P., Anderson, R.E., Hess, R.A., Nakamura, M.T., 2010. Docosahexaenoic acid supplementation fully restores fertility and spermatogenesis in male delta-6 desaturase-null mice. J. Lipid Res. 51, 360–367.

Saether, T., Tran, T.N., Rootwelt, H., Grav, H.J., Christophersen, B.O., Haugen, T.B., 2007. Essential fatty acid deficiency induces fatty acid desaturase expression in rat epididymis, but not in testis. Reproduction 133, 467–477.

Samadian, F., Towhidi, A., Rezayazdi, K., Bahreini, M., 2010. Effects of dietary n-3 fatty acids on characteristics and lipid composition of ovine sperm. Animal 4, 2017–2022.

Santos, J.E., Bilby, T.R., Thatcher, W.W., Staples, C.R., Silvestre, F.T., 2008. Long chain fatty acids of diet as factors influencing reproduction in cattle. Reprod. Domest. Anim. 43 (Suppl. 2), 23–30.

Schaiff, W.T., Knapp, F.F., Barak, Y., Biron-Shental, T., Nelson, D.M., Sadovsky, Y., 2007. Ligand-activated peroxisome proliferator activated receptor gamma alters placental morphology and placental fatty acid uptake in mice. Endocrinology 148, 3625–3634.

Serhan, C.N., Chiang, N., Van Dyke, T.E., 2008. Resolving inflammation: dual anti-inflammatory and pro-resolution lipid mediators. Nat. Rev. Immunol. 8, 349–361.

Sheldon, I.M., Cronin, J., Goetze, L., Donofrio, G., Schuberth, H.J., 2009. Defining postpartum uterine disease and the mechanisms of infection and immunity in the female reproductive tract in cattle. Biol. Reprod. 81, 1025–1032.

Sheldrick, E.L., Derecka, K., Marshall, E., Chin, E.C., Hodges, L., Wathes, D.C., Abayasekara, D.R., Flint, A.P., 2007. Peroxisome-proliferator-activated receptors and the control of levels of prostaglandin-endoperoxide synthase 2 by arachidonic acid in the bovine uterus. Biochem. J. 406, 175–183.

Smith, W.L., Lagenbach, R., 2001. Why there are two cylcooxygenase enzymes. J. Clin. Invest. 107, 1491–1495.

Smuts, C.M., Huang, M., Mundy, D., Plasse, T., Major, S., Carlson, S.E., 2003. A randomized trial of docosahexaenoic acid supplementation during the third trimester of pregnancy. Obstet. Gynecol. 101, 469–479.

Spencer, T.E., Forde, N., Dorniak, P., Hansen, T.R., Romero, J.J., Lonergan, P., 2013. Conceptus-derived prostaglandins regulate gene expression in the endometrium prior to pregnancy recognition in ruminants. Reproduction 146, 377–387.

Sprecher, H., 2000. Metabolism of highly unsaturated n-3 and n-6 fatty acids. Biochim. Biophys. Acta 1486, 219–231.

Stark, K.D., Mulvad, G., Pedersen, H.S., Park, E.J., Dewailly, E., Holub, B.J., 2002. Fatty acid compositions of serum phospholipids of postmenopausal women: a comparison between Greenland Inuit and Canadians before and after supplementation with fish oil. Nutrition 18, 627–630.

Stoffel, W., Holz, B., Jenke, B., Binczek, E., Günter, R.H., Kiss, C., Karakesisoglou, I., Thevis, M., Weber, A.A., Arnhold, S., Addicks, K., 2008. Delta6-desaturase (FADS2) deficiency unveils the role of omega3- and omega6-polyunsaturated fatty acids. EMBO J. 27, 2281–2292.

Stubbs, C.D., Smith, A.D., 1984. The modification of mammalian membrane polyunsaturated fatty acid composition in relation to membrane fluidity and function. Biochim. Biophys. Acta 779, 89–137.

Surai, P.F., Blesbois, E., Grasseau, I., Chalah, T., Brillard, J.P., Wishart, G.J., Cerolini, S., Sparks, N.H., 1998. Fatty acid composition, glutathione peroxidase and superoxide dismutase activity and total antioxidant activity of avian semen. Comp. Biochem. Physiol. B Biochem. Mol. Biol. 120, 527–533.

Szajewska, H., Horvath, A., Koletzko, B., 2006. Effect of n-3 long-chain polyunsaturated fatty acid supplementation of women with low-risk pregnancies on pregnancy outcomes and growth measures at birth: a meta-analysis of randomized controlled trials. Am. J. Clin. Nutr. 83, 1337–1344.

Tanghe, S., De Smet, S., 2013. Does sow reproduction and piglet performance benefit from the addition of n-3 polyunsaturated fatty acids to the maternal diet? Vet. J. 197, 560–569.

Van Tran, L., Malla, B.A., Sharma, A.N., Kumar, S., Tyagi, N., Tyagi, A.K., 2016. Effect of omega-3 and omega-6 polyunsaturated fatty acid enriched diet on plasma IGF-1 and testosterone concentration, puberty and semen quality in male buffalo. Anim. Reprod. Sci. 173, 63–72.

Turk, H.F., Chapkin, R.S., 2013. Membrane lipid raft organization is uniquely modified by n-3 polyunsaturated fatty acids. Prostaglandins Leuk. Essent. Fatty Acids 88, 43–47.

Van Tran, L., Malla, B.A., Kumar, S., Tyagi, A.K., 2017. Polyunsaturated fatty acids in male ruminant reproduction: a review. Asian-Australas J. Anim. Sci. 30 (5), 622–637.

Ullbrich, S.E., Wolf, E., Bauersachs, S., 2012. Hosting the preimplantation embryo: potentials and limitations of different approaches for analysing embryo-endometrium interactions in cattle. Reprod. Fertil. Dev. 25, 62–70.

Wahli, W., Michalik, L., 2012. PPARs at the crossroads of lipid signaling and inflammation. Trends Endocrinol. Metab. 23, 351–363.

Waltman, R., Tricomi, V., Shabanah, E.H., Arenas, R., 1977. Prolongation of gestation time in rats by unsaturated fatty acids. Am. J. Obstet. Gynecol. 127, 626–627.

Waters, S.M., Coyne, G.S., Kenny, D.A., MacHugh, D.E., Morris, D.G., 2012. Dietary n-3 polyunsaturated fatty acid supplementation alters the expression of genes involved in the control of fertility in the bovine uterine endometrium. Physiol. Genomics 44, 878–888.

Waters, S.M., Coyne, G.S., Kenny, D.A., Morris, D.G., 2014. Effect of dietary n-3 polyunsaturated fatty acids on transcription factor regulation in the bovine endometrium. Mol. Biol. Rep. 41, 2745–2755.

Wathes, D.C., Abayasekara, D.R., Aitken, R.J., 2007. Polyunsaturated fatty acids in male and female reproduction. Biol. Reprod. 77, 190–201.

Wathes, D.C., Lamming, G.E., 1995. The oxytocin receptor, luteolysis and the maintenance of pregnancy. J. Reprod. Fertil. Suppl. 49, 53–67.

Wenk, M.R., 2010. Lipidomics: new tools and applications. Cell 143, 888–895.

Williams, J.A., Batten, S.E., Harris, M., Rockett, B.D., Shaikh, S.R., Stillwell, W., Wassall, S.R., 2012. Docosahexaenoic and eicosapentaenoic acids segregate differently between raft and nonraft domains. Biophys. J. 103, 228–237.

Yan, L., Bai, X.L., Fang, Z.F., Che, L.Q., Xu, S.Y., Wu, D., 2013. Effect of different dietary omega-3/omega-6 fatty acid ratios on reproduction in male rats. Lipids Health Dis. 12, 33.

Zachut, M., Arieli, A., Moallem, U., 2011. Incorporation of dietary n-3 fatty acids into ovarian compartments in dairy cows and the effects on hormonal and behavioral patterns around estrus. Reproduction 141, 833–840.

Zadravec, D., Tvrdik, P., Guillou, H., Haslam, R., Kobayashi, T., Napier, J.A., Capecchi, M.R., Jacobsson, A., 2011. ELOVL2 controls the level of n-6 28:5 and 30:5 fatty acids in testis, a prerequisite for male fertility and sperm maturation in mice. J. Lipid Res. 52, 245–255.

Zanini, S.F., Torres, C.A., Bragagnolo, N., Turatti, J.M., Silva, M.G., Zanini, M.S., 2003. Evaluation of the ratio of omega-6: omega-3 fatty acids and vitamin E levels in the diet on the reproductive performance of cockerels. Arch. Tierernahr. 57, 429–442.

# 10

# The Effect of Dietary Modification on Polyunsaturated Fatty Acid Biosynthesis and Metabolism

*Beverly S. Muhlhausler*\*,\*\*

\*FOODplus Research Centre, The University of Adelaide, Adelaide, Australia; \*\*Healthy Mothers, Babies and Children Theme, South Australian Health and Medical Research Institute (SAHMRI), Adelaide, SA, Australia

## INTRODUCTION

The polyunsaturated fatty acids (PUFA), comprising the omega-6 and omega-3 subgroups, have received considerable attention, largely owing to their important effects on human health. The omega-3 long chain

Polyunsaturated Fatty Acid Metabolism. http://dx.doi.org/10.1016/B978-0-12-811230-4.00010-7

polyunsaturated fatty acids (LCPUFA), eicosapentaenoic acid (EPA, 20:5n-3), and docosahexaenoic acid (DHA, 22:6n-3), in particular have a number of reported health benefits in humans, especially in relation to cardiovascular (Calder, 2004) and inflammatory (Makrides et al., 2010; Proudman et al., 2013) conditions. This has led to recommendations from health agencies worldwide to increase dietary intake of these fatty acids (Flock et al., 2013). EPA and DHA are mainly derived preformed through consumption of either fish or fish-oil supplements. However, they can also be synthesized *de novo* through conversion of the plant-derived shorter-chain omega-3 PUFA precursor, alpha-linolenic acid (ALA, 18:3n-3), although it is thought that the efficiency of this process in humans is generally very low (Burdge and Calder, 2005).

In addition to dietary intakes of omega-3 fats, omega-3 status is also heavily influenced by the level of intake of omega-6 PUFA. This is due to the competition between omega-6 and omega-3 PUFA for both enzymatic conversion of the short-chain precursors, linoleic acid (LA, 18:2n-6), and ALA, to their respective long-chain derivatives, and for incorporation into cell membranes (Sprecher et al., 1995). This is significant, since the incorporation of omega-3 LCPUFA into the cellular membrane is a necessary step in the synthesis of downstream lipid mediators that give rise to the physiological, and ultimately health, effects of these fatty acids.

The purpose of this chapter is to build on the overview of PUFA metabolism presented in detail in Chapter 2, by describing the impact of variations in LA and ALA intakes on the metabolic pathway and LCPUFA synthesis.

## PUFA METABOLISM

The PUFA synthesis pathway is described in detail in Chapter 2.

## REGULATION OF PUFA METABOLISM

### Regulation of Enzyme Expression

Animal studies have demonstrated that the rate of synthesis of long-chain PUFA from LA and ALA is primarily regulated by the levels of substrate (i.e., LA and ALA), rather than changes in the expression of the regulatory desaturase and elongase enzymes. This is demonstrated by the results of a study by Tu et al. (2010), in which adult rats were fed either a high PUFA (0.6% energy as ALA, 6% energy as LA) or low

PUFA (0.03% energy as ALA, 0.4% energy as LA) reference diet or diets containing between 0.2% and 3% energy as ALA, with LA intake kept constant at 1% energy. This study showed that the concentrations of EPA and DPA in rodent tissues increased with increasing dietary ALA intake. However, while the hepatic expression of *Fads2* and *Elovl2* were significantly higher in rats fed the low PUFA compared to high PUFA reference diet, there were no differences in the expression of these genes in the livers of rats with increasing intakes of ALA within the normal physiological range of PUFA intakes (Tu et al., 2010) (Fig. 10.1A and B).

These results are consistent with those reported by others (Cho et al., 1999a,b), and suggest that while gene expression of the desaturase and elongase enzymes can be upregulated in response to dietary PUFA deficiency, within the normal physiological range of PUFA intakes it is primarily the level of substrate (in this case ALA), rather than changes in the expression of the regulatory enzymes, that determines the efficiency of PUFA LA conversion to AA and ALA conversion to EPA, DPA, and DHA, respectively.

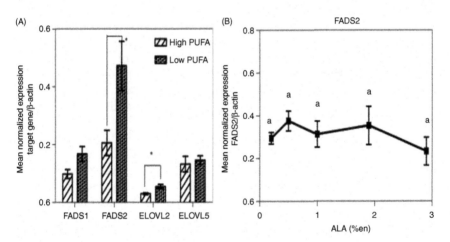

**FIGURE 10.1**    (A) The upregulation of hepatic gene expression of the delta-6-desaturase (*Fads2*) and elongase 2 (*Elovl2*) on hepatic gene expression in rats fed a low-PUFA compared to a high PUFA diet. (B) The lack of any significant difference in the expression of the delta-6-desaturase (*Fads2*) with increasing ALA intakes between 0.2% and 3% energy (%en) in rats where the LA content of the diet was maintained at 1% energy. *ALA*, Alpha-linolenic acid; *PUFA*, polyunsaturated fatty acids. *Source: Adapted from Tu, W.C., Cook-Johnson, R.J., James, M.J., Muhlhausler, B.S., Gibson, R.A., 2010. Omega-3 long chain fatty acid synthesis is regulated more by substrate levels than gene expression. Prostaglandins Leukot. Essent. Fatty Acids 83, 61–68.*

## Impact of ALA and LA Intakes

The fact that the same regulatory enzymes are utilized for both omega-3 and omega-6 PUFA metabolism results in competition between LA and ALA for the desaturase and elongase enzymes required for their conversion to their downstream LCPUFA (Lands et al., 1990; Sprecher et al., 1995). While the desaturase enzymes have a higher affinity for ALA compared to LA, high levels of LA in the background diet (such as those in typical Western diets), can reduce the efficiency of endogenous synthesis of omega-3 LCPUFA from ALA (Sprecher et al., 1995). William Lands was one of the first lipid researchers to evaluate the impact of altering dietary LA and ALA intakes in both rats and humans on tissue omega-3 LCPUFA accumulation, and these experiments provided clear evidence of the powerful effect of the competition between these fatty acids on tissue fatty acid levels (Lands et al., 1992, 1990). The role of ALA, LA, and total PUFA intakes on EPA, DPA, and, finally, DHA concentrations in tissues are discussed in later sections.

## Animal Studies

Animal studies have highlighted that levels of omega-3 LCPUFA in tissues can be regulated by simply altering the balance of LA and ALA in the diet and that a dose-response relationship generally exists between dietary ALA intake and the omega-3 LCPUFA content of blood and tissues (Blank et al., 2002; Gibson et al., 2013; Tu et al., 2010). Total PUFA intake (i.e., the sum of omega-3 and omega-6 PUFA intakes) also has a significant influence on the relationship between PUFA intake and tissue LCPUFA accumulation.

The most comprehensive study to date to investigate the regulation of PUFA synthesis was conducted by Gibson et al. (2013), and involved assigning rats to one of 54 different diets containing differing levels of LA, ALA, and total PUFA. The results of this study provide an excellent illustration of several key aspects of the PUFA metabolic pathway described in Chapter 2 including the impact of LA and ALA intakes, the ratio of LA to ALA in the diet, and the total dietary PUFA content. We will first concentrate the effects observed for EPA and DPA, because DHA exhibits a somewhat different pattern in comparison to the other omega-3 LCPUFA. The authors found that, not surprisingly, the ALA content of the plasma phospholipids of the rats was directly related to ALA content of the diet. However, the extent of the incorporation was generally low. In addition, ALA incorporation decreased as the dietary LA:ALA ratio increased, and when this ratio exceeded 7:1, there was almost complete inhibition of ALA incorporation (Gibson et al., 2013). This finding demonstrates that ALA incorporation, as well as conversion/metabolism,

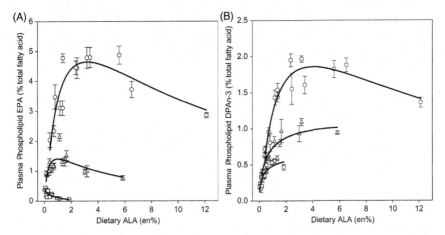

FIGURE 10.2 Concentrations of (A) EPA and (B) DPA in plasma phospholipids of rats fed diets containing increasing amounts of ALA (as a percent of total dietary energy). The LA:ALA ratio was either a low (0.5–0.8, *circular symbols*), moderate (1.9–2.6, *triangular symbols*), or high 7.4–11.3, *square symbols*). *ALA*, Alpha-linolenic acid; *EPA*, eicosapentaenoic acid. Data are mean ± SD ($n$ = 5 rats per group). *Source: Adapted from Gibson, R.A., Neumann, M.A., Lien, E.L., Boyd, K.A., Tu, W.C., 2013. Docosahexaenoic acid synthesis from alpha-linolenic acid is inhibited by diets high in polyunsaturated fatty acids. Prostaglandins Leukot Essent. Fatty Acids 88, 139–146.*

is inhibited in the presence of excess dietary LA, due to competition between the fatty acids for uptake into the membrane.

In the case of EPA and DPA, the impact of the same increase in dietary ALA content differed significantly according to the LA:ALA ratio of the diet (Gibson et al., 2013). Thus, at low ratios (<0.8:1), the EPA and DPA content of plasma phospholipids increased in direct proportion to the increase in dietary ALA intake, although the slope of the increase was much lower when ALA intake was >3% energy (Fig. 10.2). Increasing the dietary LA:ALA ratio even modestly (to ~2:1) resulted a reduction of 50%–60% in the accumulation of both EPA and DPA, and at LA:ALA ratios (>7:1) there was virtually no accumulation of EPA in plasma phospholipids (Fig. 10.2). These results provide a clear demonstration that high LA intakes significantly inhibit the capacity of ALA both to be incorporated into tissue phospholipids and, perhaps more importantly, be converted to its downstream omega-3 long-chain derivatives. The results of these investigations provide critical insights into the regulation of PUFA synthesis in vivo, and support for the PUFA metabolic pathway illustrated in Chapter 2.

## Regulation of DHA Synthesis

As depicted in the PUFA metabolic pathway, the conversion of ALA to DHA requires additional steps in comparison to EPA and DPA, and this has implications for how DHA concentrations respond to increases in dietary ALA intakes. In rodent studies, while the EPA content of blood and tissues increases linearly with increasing ALA intake, DHA content plateaus at relatively low dietary ALA intakes (~1% energy) (Tu et al., 2010). The rodent study by Gibson et al. (2013) discussed earlier also demonstrated that DHA accumulation into plasma phospholipids was dependent on the level of both ALA and LA in the diet. Thus, at low ALA intakes DHA accumulation was linearly related to ALA intake; however, the level of ALA intake at which peak DHA accumulation decreased as the LA:ALA ratio increased (1% energy ALA at low LA:ALA ratios, 0.75% energy ALA at moderate LA:ALA ratios, and 0.3% energy ALA at high LA:ALA ratios). In all cases, once this peak level of DHA accumulation was obtained, DHA accumulation into plasma phospholipids actually decreased with further increases in ALA intake (Fig. 10.3; Gibson et al., 2013).

Collectively, these findings clearly demonstrate that synthesis of DHA from ALA is highly sensitive to the fatty acid composition of the background diet, and rapidly blocked by even relatively modest increases in LA and/or total PUFA intakes. This is not unexpected if we review the PUFA metabolic pathway, and recall that the delta-6-desaturase is used once in the synthesis of EPA and DPA from ALA, but twice in the synthesis of DHA, such that increasing the amount of ALA in the system effectively limits the availability of the desaturase enzyme for converting EPA through to DHA (Sprecher et al., 1995). It is also possible that the additional β-oxidation step required for DHA synthesis could be inhibited at high LA/total PUFA intakes. This is supported by the results of a study by James et al. (2003) that bypassed the first delta-6-desaturase, by supplementing subjects with steradoic acid (SDA). The study showed that SDA supplementation was also not associated with an increase in plasma DHA, suggesting that other enzymes in the pathway besides the delta-6-desaturase are also rate-limiting to conversion (James et al., 2003).

## Human Studies

There is evidence that the balance of ALA and LA in the diet also has a significant impact on omega-3 LCPUFA status in humans. In addition to

FIGURE 10.3   **Effect of increasing dietary ALA concentrations on DHA levels in plasma** ▶
**phospholipids of rats.** Diets contained low (11.8% energy, *circular symbols*), medium (22.2 en%, *triangular symbols*), and high (39.4 en%, *square symbols*) levels of total dietary fat. The data are shown for three different LA:ALA ratios: (A) low 0.5:1–0.8:1, (B) moderate 1.9:1–2.6:1, and (C) high 7.4:1–11.3:1. Data are mean ± SD (*n* = 5 rats per group). *Source: Adapted from Gibson, R.A., Neumann, M.A., Lien, E.L., Boyd, K.A., Tu, W.C., 2013. Docosahexaenoic acid synthesis from alpha-linolenic acid is inhibited by diets high in polyunsaturated fatty acids. Prostaglandins Leukot. Essent. Fatty Acids 88, 139–146.*

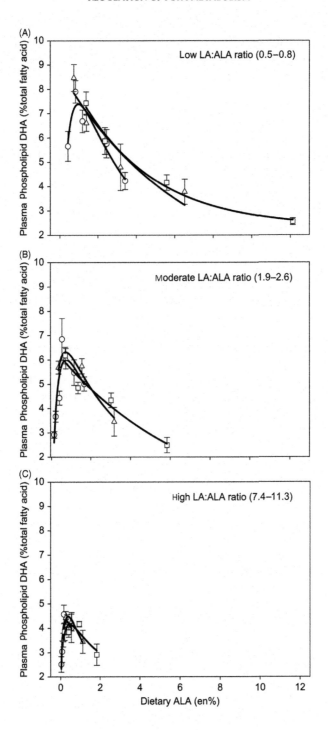

work in animal models, Lands et al. (1990) conducted comparable studies in humans, and these demonstrated that increases in dietary LA intake were associated with reductions in omega-3 LCPUFA status, implying that excessive intakes of dietary omega-6 PUFA limit the ability of ALA and omega-3 LCPUFA consumed in the diet to increase omega-3 LCPUFA status. The findings from both human and animal studies that higher omega-6 PUFA intakes were associated with a poorer conversion of ALA to EPA, DPA, and DHA, led to the suggestion that lowering dietary LA intake could potentially improve omega-3 LCPUFA status without a need to increase omega-3 LCPUFA intake (Lands et al., 1990).

## The Impact of Altering ALA and LA Intakes on Omega-3 LCPUFA Status in Humans

While there are relatively few studies that have specifically examined the impact of altering dietary LA and/or ALA intakes, there is evidence of a role for both increased ALA and decreased LA intake in improving omega-3 LCPUFA status. In infants, feeding a formula containing a lower LA content increases the efficiency of cellular DHA incorporation, resulting in a higher DHA status (Clark et al., 1992). A systematic review of studies in human adults involving manipulations of the ALA and LA intakes in the diet and their effect on omega-3 status concluded that increasing ALA intake was quite effective at increasing EPA concentrations (most studies didn't measure DPA), but that this needed to be done in conjunction with a reduction in the intake of omega-6 LA to less than 2.5% energy in order to also increase DHA concentrations (Wood et al., 2015). In this systematic review, nine studies increased dietary ALA while maintaining LA intake, and six of these reported a significant increase in EPA content in plasma/erythrocyte phospholipids, with a median increase of 60% (range 30.3%–366%) (Finnegan et al., 2003; Goyens and Mensink, 2005; Goyens et al., 2006; Kontogianni et al., 2013; Li et al., 1999; Wallace et al., 2003). However, none reported increases in DHA status. In addition, six of the seven studies, which decreased LA while maintaining ALA intake, reported a significant increase in EPA status, with a median increase of 33% (range 3.5%–51.3%) (Goyens and Mensink, 2005; Goyens et al., 2006; Macintosh et al., 2013; Renaud et al., 1995; Taha et al., 2014; Wood et al., 2014). Three of these studies also reported increases in DHA (median 13.9%, range 7.9%–24.7%) (Macintosh et al., 2013; Renaud et al., 1995; Taha et al., 2014). Six of the seven studies that both increased ALA and decreased LA intake reported a significant increase in EPA (Finnegan et al., 2003; Hagfors et al., 2005; Hussein et al., 2005; Kontogianni et al., 2013; Macintosh et al., 2013; Mantzioris et al., 1994; Taha et al., 2014) and three reported increases in both EPA and DHA status (Hagfors et al., 2005; Macintosh et al., 2013;

Taha et al., 2014). These findings support the concept that reducing levels of LA (and total PUFA), as well as increasing ALA are necessary in order to optimise the metabolic conversion of ALA in humans.

## Oleic Acid Metabolism

In addition to ALA and LA, the oleic acid (18:1n-9) can also act as a substrate for the desaturase enzymes and can be converted to eicosatrienoic acid (also known as mead acid). This occurs primarily when dietary intakes of ALA and LA are low and consequently elevated levels of mead acid have been historically be used as an index of essential fatty deficiency (Holman, 1960). However, more recent studies in rats have suggested that mead acid is also present at high levels in the absence of essential fatty acid deficiency (Choque et al., 2015), and can inhibit the synthesis of inflammatory mediators in animal models (Cleland et al., 1996; James et al., 1993). Thus, the traditional view of mead acid may require reevaluation.

## Public Health Perspective

The impact of high intakes of LA on PUFA metabolism and, ultimately, the relationship between omega-3 intakes and omega-3 status has particular significance given the substantial increases in omega-6 PUFA intakes in Western countries worldwide over the past century (Blasbalg et al., 2011; Lands, 2008). As seen earlier, the high levels of LA intake have the potential to limit both the synthesis of omega-3 LCPUFA from ALA and the uptake of these long-chain fatty acids into tissue membranes. This is important, because these fatty acids need to be incorporated into tissue membranes to be able to be released into the free fatty acid pool. It is these omega-3 and omega-6 free fatty acids that ultimately determine their bioactivity, because these are converted to downstream lipid mediators, collectively known as oxylipins, which have a broad range of actions on various physiological systems, most notably immune and inflammatory pathways (Samuelsson et al., 1987; Serhan, 2006). Traditionally, it was considered that all omega-6 oxylipins were proinflammatory and/or prothrombotic, while all omega-3 oxylipins had the opposite physiological effects, and that this accounted for the generally is somewhat too simplistic. Nevertheless, it still appears that the omega-3-derived oxylipins are generally less potent inflammatory mediators than those oxylipins derived from omega-6 fatty acids, and indeed many omega-3-derived oxylipins have immune-resolving, antiinflammatory and/or neuroprotective effects. As a result, an excess of omega-6-derived oxylipins has been associated with increased inflammation, pain, and increased risk of allergic and cardiometabolic diseases (Spiteller, 1998).

## SUMMARY AND CONCLUSION

This chapter has described our current understanding of PUFA metabolism, and highlighted the key aspects of this metabolic pathway. First, the omega-3 and omega-6 fats utilize the same set of desaturase and elongase enzymes in their synthetic pathway, and this leads to competition between LA and ALA for conversion to their longer-chain fatty acid derivatives. Second, the synthesis of longer-chain fats appears to be driven more by substrate levels than altered expression of the desaturase and elongase enzymes. Third, the synthesis of DHA requires additional steps in comparison to that of the other omega-3 LCPUFA, specifically a second utilization of the delta-6-desaturase and a β-oxidation step. An understanding of this pathway can be used as a basis for explaining the previously somewhat confusing results of studies involving manipulations of dietary fatty acid intake. Specifically, why a higher total PUFA intake limits both the incorporation of ALA into the membrane and the capacity for metabolic conversion of ALA to EPA, DPA, and DHA, a higher LA:ALA content in the diet is associated with a reduced efficiency of omega-3 LCPUFA synthesis, and why DHA accumulation into tissues rapidly plateaus as ALA intakes increase, whereas accumulation of EPA and DPA increases linearly to much higher levels of ALA intake. While not as comprehensive in terms of the range of diets examined, and less tightly controlled than animal studies, the results of human studies involving manipulations of dietary ALA and LA are broadly in line with the findings from studies in rodents, pigs, and nonhuman primates, and suggest that lowering LA intakes has the capacity to substantially increase the efficiency of ALA conversion in both infants and adults.

### Acknowledgments

BSM is supported by a Career Development Fellowship from the National Health and Medical Research Council of Australia (NHMRC, APP1083009).

### References

Blank, C., Neumann, M.A., Makrides, M., Gibson, R.A., 2002. Optimizing DHA levels in piglets by lowering the linoleic acid to α-linolenic acid ratio. J. Lipid Res. 43, 1537–1543.

Blasbalg, T.L., Hibbeln, J.R., Ramsden, C.E., Majchrzak, S.F., Rawlings, R.R., 2011. Changes in consumption of omega-3 and omega-6 fatty acids in the United States during the 20th century. Am. J. Clin. Nutr. 93, 950–962.

Burdge, G.C., Calder, P.C., 2005. Conversion of alpha-linolenic acid to longer-chain polyunsaturated fatty acids in human adults. Reprod. Nutr. Dev. 45, 581–597.

Calder, P.C., 2004. n-3 Fatty acids and cardiovascular disease: evidence explained and mechanisms explored. Clin. Sci. 107, 1–11.

Cho, H.P., Nakamura, M., Clarke, S.D., 1999a. Cloning, expression, and fatty acid regulation of the human Delta-5 desaturase. J. Biol. Chem. 274, 37335–37339.

Cho, H.P., Nakamura, M.T., Clarke, S.D., 1999b. Cloning, expression, and nutritionalregulation of the mammalian Delta-6 desaturase. J. Biol. Chem. 274, 471–477.

Choque, B., Catheline, D., Delplanque, B., Guesnet, P., Legrand, P., 2015. Dietary linoleic acid requirements in the presence of α-linolenic acid are lower than the historical 2% of energy intake value, study in rats. Br. J. Nutr. 113, 1056–1068.

Clark, K.J., Makrides, M., Neumann, M.A., Gibson, R.A., 1992. Determination of the optimal ratio of linoleic acid to alpha-linolenic acid in infant formulas. J. Pediatr. 120, S151–S158.

Cleland, L.G., Gibson, R.A., Neumann, M.A., Hamazaki, T., Akimoto, K., James, M.J., 1996. Dietary (n-9) eicosatrienoic acid from a cultured fungus inhibits leukotriene B4 synthesis in rats and the effect is modified by dietary linoleic acid. J. Nutr. 126, 1534–1540.

Finnegan, Y.E., Minihane, A.M., Leigh-Firbank, E.C., Kew, S., Meijer, G.W., Muggli, R., Calder, P.C., Williams, C.M., 2003. Plant- and marine-derived n-3 polyunsaturated fatty acids have differential effects on fasting and postprandial blood lipid concentrations and on the susceptibility of LDL to oxidative modification in moderately hyperlipidemic subjects 77, 783–795.

Flock, M.R., Harris, W.S., Kris-Etherton, P.M., 2013. Long-chain omega-3 fatty acids: time to establish a dietary reference intake. Nutr. Rev. 71, 692–707.

Gibson, R.A., Neumann, M.A., Lien, E.L., Boyd, K.A., Tu, W.C., 2013. Docosahexaenoic acid synthesis from alpha-linolenic acid is inhibited by diets high in polyunsaturated fatty acids. Prostaglandins Leukot. Essent. Fatty Acids 88, 139–146.

Goyens, P.L., Mensink, R.P., 2005. The dietary alpha-linolenic acid to linoleic acid ratio does not affect the serum lipoprotein profile in humans. J. Nutr. 135, 2799–2804.

Goyens, P.L., Spilker, M.E., Zock, P.L., Katan, M.B., Mensink, R.P., 2006. Conversion of alpha-linolenic acid in humans is influenced by the absolute amounts of alpha-linolenic acid and linoleic acid in the diet and not by their ratio. Am. J. Clin. Nutr. 84, 44–53.

Hagfors, L., Nilsson, I., Skoldstam, L., Johansson, G., 2005. Fat intake and composition of fatty acids in serum phospholipids in a randomized, controlled, Mediterranean dietary intervention study on patients with rheumatoid arthritis. Nutr. Metab. 2, 26.

Holman, R.T., 1960. The ratio of trienoic: tetraenoic acids in tissue lipids as a measure of essential fatty acid requirement. J. Nutr. 70, 405–410.

Hussein, N., Ah-Sing, E., Wilkinson, P., Leach, C., Griffin, B.A., Millward, D.J., 2005. Long-chain conversion of [13C]linoleic acid and α-linolenic acid in response to marked changes in their dietary intake in men. J. Lipid Res. 46, 269–280.

James, M.J., Gibson, R.A., Neumann, M.A., Cleland, L.G., 1993. Effect of dietary supplementation with n-9 eicosatrienoic acid on leukotriene B4 synthesis in rats: a novel approach to inhibition of eicosanoid synthesis. J. Exp. Med. 178, 2261–2265.

James, M.J., Ursin, V.M., Cleland, L.G., 2003. Metabolism of stearidonic acid in human subjects: comparison with the metabolism of other n-3 fatty acids. Am. J. Clin. Nutr. 77, 1140–1145.

Kontogianni, M.D., Vlassopoulos, A., Gatzieva, A., Farmaki, A.E., Katsiougiannis, S., Panagiotakos, D.B., Kalogeropoulos, N., Skopouli, F.N., 2013. Flaxseed oil does not affect inflammatory markers and lipid profile compared to olive oil, in young, healthy, normal weight adults. Metabol. Clin. Exp. 62, 686–693.

Lands, B., 2008. A critique of paradoxes in current advice on dietary lipids. Prog. Lipid Res. 47, 77–106.

Lands, W.E., Libelt, B., Morris, A., Kramer, N.C., Prewitt, T.E., Bowen, P., Schmeisser, D., Davidson, M.H., Burns, J.H., 1992. Maintenance of lower proportions of (n-6) eicosanoid precursors in phospholipids of human plasma in response to added dietary (n-3) fatty acids. Biochim. Biophys. Acta 1180, 147–162.

Lands, W.E., Morris, A., Libelt, B., 1990. Quantitative effects of dietary polyunsaturated fats on the composition of fatty acids in rat tissues. Lipids 25, 506–516.

Li, D., Sinclair, A., Wilson, A., Nakkote, S., Kelly, F., Abedin, L., Mann, N., Turner, A., 1999. Effect of dietary α-linolenic acid on thrombotic risk factors in vegetarian men. Am. J. Clin. Nutr. 69, 872–882.

Macintosh, B.A., Ramsden, C.E., Faurot, K.R., Zamora, D., Mangan, M., Hibbeln, J.R., Douglas Mann, J., 2013. Low-n-6 and low-n-6 plus high-n-3 diets for use in clinical research. Br. J. Nutr. 110, 559–568.

Makrides, M., Gibson, R.A., McPhee, A.J., Yelland, L., Quinlivan, J., Ryan, P., 2010. Effect of DHA supplementation during pregnancy on maternal depression and neurodevelopment of young children: a randomized controlled trial. JAMA 304, 1675–1683.

Mantzioris, E., James, M., Gibson, R., Cleland, L., 1994. Dietary substitution with an a-linolenic acid-rick vegetable oil increases eicosapentaenoic acid concentrations in tissues. Am. J. Clin. Nutr. 59, 1304–1309.

Proudman, S.M., James, M.J., Spargo, L.D., Metcalf, R.G., Sullivan, T.R., Rischmueller, M., Flabouris, K., Wechalekar, M.D., Lee, A.T., Cleland, L.G., 2013. Fish oil in recent onset rheumatoid arthritis: a randomised, double-blind controlled trial within algorithm-based drug use. Ann. Rheum. Dis. 74 (1), 89–95.

Renaud, S., De Lorgeril, M., Delaye, J., Guidollet, J., Jacquard, F., Mamelle, N., Martin, J.L., Monjaud, I., Salen, P., Toubol, P., 1995. Cretan Mediterranean diet for prevention of coronary heart disease. Am. J. Clin. Nutr. 61, 1360S–1367S.

Samuelsson, B., Dahlen, S.E., Lindgren, J.A., Rouzer, C.A., Serhan, C.N., 1987. Leukotrienes and lipoxins: structures, biosynthesis, and biological effects. Science 237, 1171–1176.

Serhan, C.N., 2006. Novel chemical mediators in the resolution of inflammation: resolvins and protectins. Anesthesiol. Clin. 24, 341–364.

Spiteller, G., 1998. Linoleic acid peroxidation: the dominant lipid peroxidation process in low density lipoprotein: and its relationship to chronic diseases. Chem. Phys. Lipids 95, 105–162.

Sprecher, H., Luthria, D., Mohammed, B., Baykousheva, S., 1995. Reevaluation of the pathways for the biosynthesis of polyunsaturated fatty acids. J. Lipid Res. 36, 2471–2477.

Taha, A.Y., Cheon, Y., Faurot, K.F., Macintosh, B., Majchrzak-Hong, S.F., Mann, J.D., Hibbeln, J.R., Ringel, A., Ramsden, C.E., 2014. Dietary omega-6 fatty acid lowering increases bioavailability of omega-3 polyunsaturated fatty acids in human plasma lipid pools. Prostaglandins Leukot. Essent. Fatty Acids 90, 151–157.

Tu, W.C., Cook-Johnson, R.J., James, M.J., Muhlhausler, B.S., Gibson, R.A., 2010. Omega-3 long chain fatty acid synthesis is regulated more by substrate levels than gene expression. Prostaglandins Leukot. Essent. Fatty Acids 83, 61–68.

Wallace, F.A., Miles, E.A., Calder, P.C., 2003. Comparison of the effects of linseed oil and different doses of fish oil on mononuclear cell function in healthy human subjects. Br. J. Nutr. 89, 679–689.

Wood, K.E., Lau, A., Mantzioris, E., Gibson, R.A., Ramsden, C.E., Muhlhausler, B.S., 2014. A low omega-6 polyunsaturated fatty acid (n-6 PUFA) diet increases omega-3 (n-3) long chain PUFA status in plasma phospholipids in humans. Prostaglandins Leukot. Essent. Fatty Acids 90 (4), 133–138.

Wood, K.E., Mantzioris, E., Gibson, R.A., Ramsden, C.E., Muhlhausler, B.S., 2015. The effect of modifying dietary LA and ALA intakes on omega-3 long chain polyunsaturated fatty acid (n-3 LCPUFA) status in human adults: a systematic review and commentary. Prostaglandins Leukot. Essent. Fatty Acids 95, 47–55.

# 11

# Omega-3 Polyunsaturated Fatty Acid Metabolism in Vegetarians

*Graham C. Burdge\*, Christiani J. Henry\*\**

\*University of Southampton, Southampton, United Kingdom;
\*\*Clinical Nutrition Research Centre, Centre for Translational Medicine,
National University of Singapore, Singapore

## INTRODUCTION

Vegetarianism encompasses several dietary patterns including the complete exclusion of meat, fish, and dairy produce (veganism), exclusion of meat and fish with inclusion of dairy products and eggs (ovo-lacto-vegetarianism), the exclusion of all meat except fish (pesco-vegetarianism), and only excluding meat from the diet. Vegetarian diets have been associated with specific health benefits (Chiu et al., 2014; Dinu et al., 2016; Glick-Bauer and Yeh, 2014; Huang et al., 2016; Orlich et al., 2015; Sabate and Wien, 2015; Tonstad et al., 2013; Wang et al., 2015). However, exclusion of major food groups from the diet incurs a potential risk of marginal status

Polyunsaturated Fatty Acid Metabolism. http://dx.doi.org/10.1016/B978-0-12-811230-4.00011-9

or deficiency of nutrients that are found predominately or exclusively in the excluded foods. For example, exclusion of meat and fish from the diet may lead to suboptimal intakes of vitamin $B_{12}$ and of the longer-chain n-3 polyunsaturated fatty acids (LCPUFA), eicosapentaenoic acid (EPA, 20:5n-3) and docosahexaenoic acid (DHA, 22:6n-3). Lower EPA and DHA intakes and status have been associated with increased risk of cardiovascular and inflammatory disease (Calder, 2015; Calder and Yaqoob, 2009; Harris and Von Schacky, 2004). Thus there is an apparent paradox between the health benefits associated with vegetarian diets and increased disease risk associated with low EPA and DHA intakes. One potential explanation for the apparent health benefits of vegetarian diets despite low EPA and DHA intakes is that conversion of α-linolenic acid (ALA, 18:3n-3) to n-3 LCPUFA increases when intakes of preformed EPA and DHA are low. Moreover, the developing central nervous system requires adequate accumulation of DHA for optimal function (Lauritzen et al., 2016). Thus there is a potential risk of low maternal DHA status leading to developmental deficits in children of vegetarian mothers. The purpose of this review is to examine the evidence of the effect of vegetarian diets on EPA and DHA status in adults and to assess whether EPA and DHA synthesis may be increased in individuals who consume a vegetarian diet.

# EPA AND DHA STATUS IN VEGETARIANS AND VEGANS

Relatively few studies have assessed the effect of vegetarian diets on EPA and DHA status. These include studies in geographically diverse cohorts, in men, women, or both sexes combined, that differ in dietary classification, and have assessed the levels of these fatty acids in different lipid compartments (plasma or serum phospholipids, total serum fatty acids, or erythrocyte membrane phospholipids). Nevertheless, despite such differences in study design, significant differences between vegetarians, vegans, and omnivores have been reported consistently.

## EPA and DHA Status in Vegetarian or Vegan Men

The majority of studies on n-3 LCPUFA status in vegetarians have been carried out on men. The findings of individual studies are detailed in Table 11.1. Huang et al. (2013) compared vegetarian Chinese men with omnivores, mean age 40 years. Vegetarians were defined as excluding meat and fish, but consuming dairy products. The proportions of ALA (35%), EPA (63%), and DHA (63%) were lower in vegetarians than in omnivores. Yu et al. (2014) used the same definition to assess n-3 LCPUFA status in male Chinese vegetarians. They found that compared to omnivores,

**TABLE 11.1**  Summary of Studies That Have Compared Omega-3 PUFA Status in Omnivores and Vegetarians

| Study | Subjects and nationality (n) | ALA | EPA | DHA |
|---|---|---|---|---|
| **MEN** | | | | |
| Huang et al. (2013) | Chinese | Plasma phospholipids (% total fatty acids; mean (SD)) | | |
| | Omnivores (128) | 0.65 (3.21) | 2.98 (2.5) | 5.61 (1.36) |
| | Vegetarians (103) | 0.42 (0.28) | 1.11 (1.77) | 2.1 (1.04) |
| | Difference between groups | $P = 0.017$ | $P < 0.001$ | $P < 0.001$ |
| Yu et al. (2014) | Chinese | Plasma phospholipids (% total fatty acids; mean (SD)) | | |
| | Omnivores (106) | 0.19 (0.12) | 1.34 (1.13) | 4.66 (1.45) |
| | Vegetarians (89) | 0.39 (0.29) | 0.83 (1.2) | 2.12 (0.85) |
| | Difference between groups | $P < 0.0001$ | $P = 0.003$ | $P < 0.0001$ |
| Mann et al. (2006) | Australian | Plasma phospholipids (% total fatty acids; mean (SD)) | | |
| | High meat eating (16) | 0.2 (0.1) | 1.1 (0.5) | 3.4 (1.0) |
| | Moderate meat eating (53) | 0.2 (0.1) | 1.0 (0.3) | 3.3 (0.8) |
| | Lacto-ovo-vegetarian (40) | 0.3 (0.1) | 0.7 (0.3) | 2.2 (0.7) |
| | Vegan (17) | 0.3 (0.1) | 0.6 (0.3) | 2.0 (0.4) |
| | Difference: meat eaters versus vegetarians | $P < 0.01$ | $P < 0.01$ | $P < 0.01$ |

*(Continued)*

**TABLE 11.1** Summary of Studies That Have Compared Omega-3 PUFA Status in Omnivores and Vegetarians (*Cont.*)

| Study | Subjects and nationality (n) | ALA | EPA | DHA |
|---|---|---|---|---|
| Welch et al. (2010) | UK | Plasma phospholipids ($\mu$mol/L; mean (SD)) | | |
| | Fish eaters (2257) | 10.9 (5.7) | 57.5 (43.2) | 239.7 (106.2) |
| | Meat eaters (359) | 11.8 (7.0) | 47.4 (30.3) | 215.6 (96.4) |
| | Vegetarians (25) | 13.6 (10.1) | 55.9 (45.3) | 222.2 (138.4) |
| | Vegans (5) | 15.8 (9.7) | 65.1 (45.5) | 195.0 (58.8) |
| | Difference between groups (ANOVA) | $P < 0.001$ | $P = 0.001$ | $P < 0.001$ |
| Melchert et al. (1987) | German | Total serum fatty acids (% total fatty acids; mean (SD)) | | |
| | Male nonvegetarians (37) | 1.39 (0.29) | ND | 2.15 (0.53) |
| | Male vegetarians (38) | 1.62 (1.89) | ND | 1.2 (0.46) |
| | Difference between groups | $P < 0.01$ | | $P < 0.001$ |
| *WOMEN* | | | | |
| Welch et al. (2010) | UK | Plasma phospholipids ($\mu$mol/L; mean (SD)) | | |
| | Fish eaters (1891) | 12.4 (6.1) | 64.7 (43.4) | 271.2 (113.1) |
| | Meat eaters (309) | 13.1 (7.3) | 57.1 (38.4) | 241.3 (109.6) |
| | Vegetarians (51) | 12.3 (4.8) | 55.1 (52.5) | 223.5 (137.8) |
| | Vegans (5) | 13.71 (8.1) | 50.0 (29.4) | 286.4 (211.7) |
| | Difference between groups (ANOVA) | NS | $P < 0.001$ | $P < 0.001$ |

| | | | | |
|---|---|---|---|---|
| Li et al. (1999) | Australian | Serum phospholipids (% total fatty acids; mean (SD)) | | |
| | Omnivores (24) | 0.23 (0.1) | 0.95 (0.46) | 3.51 (0.64) |
| | Vegetarian (50) | 0.24 (0.1) | 0.6 (0.29) | 2.8 (0.86) |
| | Difference between groups | NS | $P = 0.003$ | $P < 0.0001$ |
| Reddy et al. (1994) | UK | Plasma phospholipids (% total fatty acids; mean (SEM)) | | |
| | Caucasian omnivore women (24) | ND | 0.97 | 2.26 (0.19) |
| | South Asian women (24) | ND | 0.36 | 1.2 (0.09) |
| | Difference between groups (P) | | $P < 0.001$ | $P < 0.001$ |
| Melchert et al. (1987) | Female nonvegetarians (37) | 1.1 (0.35) | ND | 2.36 (0.59) |
| | Female vegetarians (38) | 1.35 (0.81) | ND | 1.42 (0.47) |
| | Difference between groups | $P < 0.01$ | | $P < 0.001$ |

### MEN PLUS WOMEN

| | | | | |
|---|---|---|---|---|
| Phinney et al. (1990) | North American | Serum phospholipids (% total fatty acids; mean (SEM)) | | |
| | Omnivore (100) | 0.21 (0.01) | 0.59 (0.03) | 3.59 (0.11) |
| | Semi-vegetarian (16) | 0.28 (0.03) | 0.67 (0.08) | 3.27 (0.37) |
| | Vegetarian (25) | 0.29 (0.02) | 0.64 (0.05) | 3.19 (0.29) |
| | | 1 versus 2 or 3 $P = .0004$ | NS | NS |
| Melchert et al. (1987) | Female nonvegetarians (37) | 1.1 (0.35) | ND | 2.36 (0.59) |
| | Female vegetarians (38) | 1.35 (0.81) | ND | 1.42 (0.47) |
| | Difference between groups | $P < 0.01$ | | $P < 0.001$ |

*(Continued)*

**TABLE 11.1** Summary of Studies That Have Compared Omega-3 PUFA Status in Omnivores and Vegetarians (*Cont.*)

| Study | Subjects and nationality (n) | ALA | EPA | DHA |
|---|---|---|---|---|
| Sanders et al. (1978) | UK | Plasma phosphatidylcholine (mg/g; mean (SEM)) | | |
| | Omnivores (12) | ND | 13.0 (0.9) | 40 (2.5) |
| | Vegans (12) | ND | 2.0 (0.2) | 14 (1.7) |
| Lee et al. (2000) | Chinese | Total serum fatty acids (% total fatty acids; mean (SD)) | | |
| | Omnivores (133) | 0.8 (0.6) | 1.3 (1.3) | 3.4 (2.2) |
| | Vegetarian (60) | 1.7 (2.0) | 0.2 (0.5) | 1.7 (2.5) |
| | Difference between groups | $P < 0.001$ | $P < 0.001$ | $P < 0.05$ |
| Kornsteiner et al. (2008) | Australian | Erythrocyte phospholipids (% total fatty acids; mean (SD)) | | |
| | Omnivores (23) | 0.37 (0.25) | 0.35 (0.14) | 1.81 (0.63) |
| | Lacto-ovo-vegetarians (25) | 0.24 (0.16) | 0.27 (0.1) | 1.28 (0.37) |
| | Vegans (37) | 0.28 (0.21) | 0.16 (0.06) | 0.87 (0.31) |
| | Occasional meat eaters (13) | 0.34 (0.17) | 0.34 (0.14) | 1.84 (0.68) |
| | Difference between groups | NS | 1 versus 3 $P < 0.001$; 2 versus 3 $P < 0.001$; 3 versus 4 $P < 0.01$ | 1 versus 2 $P < 0.05$; 1 versus 3 $P < 0.01$; 2 versus 3 $P < 0.001$; 3 versus4 $P < 0.01$ |
| Rosell et al. (2005) | UK | Total serum fatty acids (% total fatty acids; arithmetic mean (95% CI)) | | |
| | Meat eaters (196) | 1.3 (1.2–1.41) | 0.72 (0.65–0.8) | 1.69 (1.59–1.79) |
| | Vegetarians (231) | 1.39 (1.3–1.48) | 0.52 (0.48–0.57) | 1.16 (1.07–1.24) |
| | Vegans (232) | 1.41 (1.32–1.5) | 0.34 (0.31–0.37) | 0.7 (0.61–0.79) |
| | Difference between groups | NS | $P < 0.001$ all comparisons | $P < 0.001$ all comparisons |

vegetarians had higher proportions of ALA (105%), but lower proportions of EPA (38%) and DHA (55%). Mann et al. (2006) compared Australian men who were high meat eaters (greater than 285 g per day), moderate meat eaters (less than 260 g meat per day), ovo-lacto-vegetarian and vegan. The nonmeat eating groups had higher proportions of ALA (33%) and lower proportions of EPA (45%), 22:5n-3 (15%), and DHA (41%) in plasma phospholipids. Melchert et al. (1987) showed that the proportion of ALA in total serum fatty acids was 17% higher and the proportion of DHA was 44% lower in German male vegetarians than in omnivores.

## EPA and DHA Status in Vegetarian or Vegan Women

Li et al. (1999) compared n-3 LCPUFA status in Australian women (aged 20–45 years) who had not consumed red meat for at least 6 months and ate fish or chicken less than once per week (vegetarians) with women who ate red meat at least five times per week (omnivores). The proportions of ALA or DPAn-3 in plasma phospholipids did not differ significantly between groups. However, the proportions of EPA and DHA in plasma phospholipids were lower (37% and 20%, respectively) in vegetarians compared to omnivores. Melchert et al. (1987) found that in German women the proportion of ALA in total serum fatty acids was 23% higher in women and the proportion of DHA was 40% lower in German vegetarian women than in omnivores. In addition, Reddy et al. (1994) have shown that EPA and DHA were not detected in the diets of South Asian vegetarian women living in the UK. The proportion of EPA in plasma phospholipids was 63% and DHA 47% lower in these women than in Caucasian omnivorous women.

## EPA and DHA Status in Combined Groups of Male and Female Vegetarians or Vegans

Lee et al. (2000) reported a higher proportion of ALA (53%) and lower proportions of EPA (84%) and DHA (50%) in total serum lipids in a combined cohort of Chinese vegetarian men and women compared to omnivores. Rosell et al. (2005) found that the proportion of EPA in plasma phospholipids from UK men and women was lower in vegans compared to vegetarians (34%) and to omnivores (53%). Similarly, the proportion of DHA was lower in vegans compared to vegetarians (40%) and omnivores (59%). Others have also reported substantially lower concentrations of EPA (84%), DPAn-3 (27%) and DHA (65%) in plasma phospholipids from a group of UK vegan men and women compared to omnivores (Sanders et al., 1978). In a combined group of Australian men and women, the rank order of the proportion of ALA in erythrocytes was omnivores greater than occasional meat eaters greater than vegans greater than ovo-lacto-vegetarians (Kornsteiner et al., 2008). The rank order of the proportion of EPA in erythrocytes was omnivores greater than occasional meat eaters greater than ovo-lacto-vegetarians greater than vegans, while for DHA the rank order was omnivores similar to occasional meat eaters greater than ovo-lacto-vegetarians greater than vegans. However, others did not find significant differences in the proportions of EPA or DHA in serum phospholipids in a combined cohort of men and women who were either vegetarians (consumed meat or eggs less than once per month) or semivegetarians (consuming meat or eggs once per week or less), compared to omnivores (consumed meat or eggs at least twice per week), although the proportion of ALA was higher in the vegetarian groups (Phinney et al., 1990).

## EPA and DHA Intakes in Vegetarians and Vegans

Few of the studies that investigated the effect of vegetarian diets on EPA and DHA status also assessed the dietary intakes of these fatty acids. In a combined cohort of Australian men and women, EPA plus DHA intake was 74 mg/day in vegetarians compared to 318 mg/day in omnivores (Kornsteiner et al., 2008). In Australian vegetarian men, EPA plus DHA intake was 10 mg/day compared to omnivores 100–190 mg/day (Mann et al., 2006). EPA plus DHA intake was 27 mg/day in UK men and 14 mg/day in UK women compared to fish-eating omnivore men (320 mg/day) and women (260 mg/day) (Welch et al., 2010). Thus intakes of preformed EPA and DHA were approximately 70%–96% lower than the lowest recommended intake [250 mg/day (European Food Standards Authority, 2010)]. Estimated intakes of EPA plus DHA in vegans have been reported to be 40 mg/day in Australian men and women (Kornsteiner et al., 2008), undetectable in Australian men (Mann et al., 2006), and 10 mg/day in UK men and 20 mg/day in UK women (Welch et al., 2010). Such estimated intakes are from 84% to >99% lower than the lowest recommended intake. Moreover, the estimated intake of EPA plus DHA was 17 mg/day in Scottish vegan mothers, approximately 93% lower than the lowest dietary recommended intake, compared to 316 mg/day in omnivore mothers (Lakin et al., 1998). The findings of dietary analyses are consistent with lower EPA and DHA status in vegetarians and vegans and show that vegetarians and vegans do not meet recommended EPA and DHA intakes. One possible implication is that health benefits that have been associated with vegetarian diets (Key et al., 2006) may be offset by lower EPA plus DHA intakes.

20:3n-9 and 22:5n-6 are markers of essential fatty acid and DHA deficiency, respectively. The concentration of 20:3n-9 did not to differ significantly between vegetarians and omnivores (Sanders and Reddy, 1992). The concentration of 20:3n-9 has not been reported for vegans. The concentration of 22:5n-6 has been shown to be higher in ovo-lacto-vegetarians and vegans in some studies (Sanders et al., 1978), but not others (Huang et al., 2013; Lakin et al., 1998). This suggests that some groups of vegetarians may be at risk of DHA deficiency.

# CONVERSION OF ALA TO EPA AND DHA IN VEGETARIANS

All studies to date have reported EPA and DHA levels that were above detection limits, despite very low intakes. One possible explanation is that plasma EPA and DHA concentrations may, at least in part, reflect synthesis from ALA (Chapter 2). High ALA intake has been shown to be associated with increased plasma phospholipid EPA, but not DHA, concentration in men (reviewed in Burdge and Calder, 2006), and women are able to synthesise both EPA and DHA from ALA (Burdge and Wootton, 2002) and maintain higher DHA status than men (Lohner et al., 2013). Thus it is possible that higher ALA intake may increase conversion and so compensate for low EPA and DHA intakes.

The proportions of ALA, EPA, DPAn-3, and DHA in studies that reported measurements of plasma phospholipids are summarized in Fig. 11.1. Comparisons of the mean level for all studies within each category for each fatty acid showed that the proportion of DHA was significantly lower (50%), but there was no difference in the proportions of ALA, EPA, or DPAn-3, in vegetarians compared with omnivores in men (Fig. 11.1A). These findings are consistent with the observation that men are able to convert ALA to EPA and DPAn-3, but not DHA (Burdge et al., 2002). If so, then it is possible that ALA intake in men was sufficient to maintain plasma EPA and DPAn-3, but not DHA, concentrations. There were too few studies in women to allow statistical analysis (Fig. 11.1B). However, there was a nonsignificant trend ($P < 0.1$) toward approximately 30% lower proportion of DHA in plasma from vegetarian women compared to omnivores. This suggests that greater conversion of ALA to DHA in women compared to men (Burdge and Wootton, 2002) is unable to compensate completely for low intakes of preformed EPA and DHA. There were no significant differences between vegetarians and omnivores in combined groups of men and women (Fig. 11.1C). This may reflect, at least in part, differences in capacity for conversion of ALA to longer-chain n-3 PUFA between sexes leading to masking of any differences in EPA and DHA status.

ALA intakes of vegetarians and vegans were not significantly different to omnivores (Kornsteiner et al., 2008; Mann et al., 2006; Welch et al., 2010). Thus it is possible that vegetarians may have increased their capacity for EPA and DHA synthesis in order to compensate for low intakes of

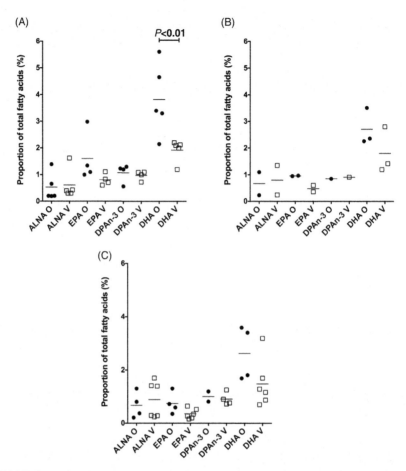

**FIGURE 11.1** **The proportions of n-3 LCPUFA in plasma phospholipids.** (A) Proportions in men, (B) in women, and (C) in men plus women. *Symbols* represent the mean reported by each study giving the proportion of each fatty acid; the *bar* represents the mean of all studies (Table 11.1). Statistical analysis was by students' unpaired *t* test. *O*, Omnivores; *V*, vegetarians.

preformed n-3 LCPUFA. The ratios of individual n-3 LCPUFA have been used to estimate the activity of one or more reactions in the essential fatty acid conversion pathway. This approach has important limitations because the calculation does not take into account differential synthesis and turnover of individual phospholipid molecular species and to date this approach has not been validated directly. Nevertheless, it may have utility for comparing groups where data have been collected under comparable conditions. The ratio of EPA to ALA may reflect the combined activities of Δ6 and Δ5 desaturases, and elongase 5, while the ratio of DHA to EPA may reflect the combined activities of elongases 2 and 5, and Δ6

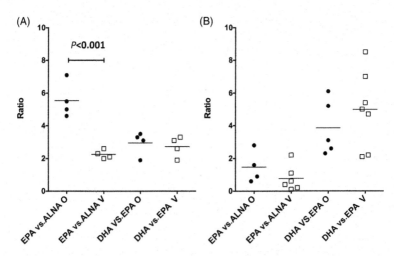

FIGURE 11.2 The ratio of EPA to ALA and of DHA to EPA for the data reported in Fig. 11.1. (A) Men; (B) men + women. *Symbols* represent the mean reported by each study; the *bar* represents the mean of all studies. Statistical analysis was by students' unpaired *t* test. *O*, Omnivores; *V*, vegetarians.

desaturase (Chapter 2). In men, the ratio of EPA to ALA was lower in vegetarians than omnivores (Fig. 11.2A), which would be consistent with lower conversion of ALA to EPA, while there was no difference in the ratio of DHA to EPA between groups (Fig. 11.2A). There were too few studies of women to allow statistical analysis of these ratios. There were no differences in the ratio of EPA to ALA or DHA to EPA between vegetarians and omnivores in combined groups of men and women (Fig. 11.2B). Together these findings are in agreement of those of stable isotope tracer studies (Hussein et al., 2005) which have shown that increased ALA intake does not increase capacity for conversion to longer-chain n-3 PUFA and supports the argument of Salem and Kuratko (2011) that there is no evidence for up-regulation of ALA conversion in vegetarians.

## CONCLUSIONS

Studies published to date show consistently that EPA and DHA status in vegetarians and vegans are lower than in omnivores. However, vegetarians and vegans do not appear to exhibit increased concentrations of fatty acids that are associated with essential fatty acid deficiency. The effects of low intakes of preformed EPA and DHA do not appear to be offset by increased conversion of ALA to n-3 LCPUFA. The current literature is relatively small, particularly with respect to studies of women. None of the reports published to date have provided a statement to indicate whether the number of subjects that have been studied was sufficient to provide an appropriate level of statistical power. Importantly, no study to date has investigated whether low EPA and DHA status in vegetarians or vegans is associated with adverse health outcomes.

## References

European Food Standards Authority, 2010. EFSA: scientific opinion on dietary reference values for fats, including saturated fatty acids, polyunsaturated fatty acids, monounsaturated fatty acids, trans fatty acids, and cholesterol. EFSA J. 8, 1461.

Burdge, G.C., Calder, P.C., 2006. Dietary alpha-linolenic acid and health-related outcomes: a metabolic perspective. Nutr. Res. Rev. 19, 26–52.

Burdge, G.C., Jones, A.E., Wootton, S.A., 2002. Eicosapentaenoic and docosapentaenoic acids are the principal products of alpha-linolenic acid metabolism in young men*. Br. J. Nutr. 88, 355–363.

Burdge, G.C., Wootton, S.A., 2002. Conversion of alpha-linolenic acid to eicosapentaenoic, docosapentaenoic and docosahexaenoic acids in young women. Br. J. Nutr. 88, 411–420.

Calder, P.C., 2015. Marine omega-3 fatty acids and inflammatory processes: effects, mechanisms and clinical relevance. Biochim. Biophys. Acta 1851, 469–484.

Calder, P.C., Yaqoob, P., 2009. Omega-3 polyunsaturated fatty acids and human health outcomes. Biofactors 35, 266–272.

Chiu, T.H., Huang, H.Y., Chiu, Y.F., Pan, W.H., Kao, H.Y., Chiu, J.P., Lin, M.N., Lin, C.L., 2014. Taiwanese vegetarians and omnivores: dietary composition, prevalence of diabetes and IFG. PLoS One 9, e88547.

Dinu, M., Abbate, R., Gensini, G.F., Casini, A., Sofi, F., 2016. Vegetarian, vegan diets and multiple health outcomes: a systematic review with meta-analysis of observational studies. Crit. Rev. Food Sci. Nutr., 0.

Glick-Bauer, M., Yeh, M.C., 2014. The health advantage of a vegan diet: exploring the gut microbiota connection. Nutrients 6, 4822–4838.

Harris, W.S., Von Schacky, C., 2004. The Omega-3 Index: a new risk factor for death from coronary heart disease? Prev. Med. 39, 212–220.

Huang, R.Y., Huang, C.C., Hu, F.B., Chavarro, J.E., 2016. Vegetarian diets and weight reduction: a meta-analysis of randomized controlled trials. J. Gen. Intern. Med. 31, 109–116.

Huang, T., Yu, X., Shou, T., Wahlqvist, M.L., Li, D., 2013. Associations of plasma phospholipid fatty acids with plasma homocysteine in Chinese vegetarians. Br. J. Nutr. 109, 1688–1694.

Hussein, N., Ah-Sing, E., Wilkinson, P., Leach, C., Griffin, B.A., Millward, D.J., 2005. Long-chain conversion of [13C]linoleic acid and alpha-linolenic acid in response to marked changes in their dietary intake in men. J. Lipid Res. 46, 269–280.

Key, T.J., Appleby, P.N., Rosell, M.S., 2006. Health effects of vegetarian and vegan diets. Proc. Nutr. Soc. 65, 35 41.

Kornsteiner, M., Singer, I., Elmadfa, I., 2008. Very low n-3 long-chain polyunsaturated fatty acid status in Austrian vegetarians and vegans. Ann. Nutr. Metab. 52, 37–47.

Lakin, V., Haggarty, P., Abramovich, D.R., Ashton, J., Moffat, C.F., McNeill, G., Danielian, P.J., Grubb, D., 1998. Dietary intake and tissue concentration of fatty acids in omnivore, vegetarian and diabetic pregnancy. Prostaglandins Leukot. Essent. Fatty Acids 59, 209–220.

Lauritzen, L., Brambilla, P., Mazzocchi, A., Harslof, L.B., Ciappolino, V., Agostoni, C., 2016. DHA effects in brain development and function. Nutrients, 8.

Lee, H.Y., Woo, J., Chen, Z.Y., Leung, S.F., Peng, X.H., 2000. Serum fatty acid, lipid profile and dietary intake of Hong Kong Chinese omnivores and vegetarians. Eur. J. Clin. Nutr. 54, 768–773.

Li, D., Ball, M., Bartlett, M., Sinclair, A., 1999. Lipoprotein(a), essential fatty acid status and lipoprotein lipids in female Australian vegetarians. Clin. Sci. (Lond.) 97, 175–181.

Lohner, S., Fekete, K., Marosvolgyi, T., Decsi, T., 2013. Gender differences in the long-chain polyunsaturated fatty acid status: systematic review of 51 publications. Ann. Nutr. Metab. 62, 98–112.

Mann, N., Pirotta, Y., O'Connell, S., Li, D., Kelly, F., Sinclair, A., 2006. Fatty acid composition of habitual omnivore and vegetarian diets. Lipids 41, 637–646.

Melchert, H.U., Limsathayourat, N., Mihajlovic, H., Eichberg, J., Thefeld, W., Rottka, H., 1987. Fatty acid patterns in triglycerides, diglycerides, free fatty acids, cholesteryl esters and phosphatidylcholine in serum from vegetarians and non-vegetarians. Atherosclerosis 65, 159–166.

Orlich, M.J., Singh, P.N., Sabate, J., Fan, J., Sveen, L., Bennett, H., Knutsen, S.F., Beeson, W.L., Jaceldo-Siegl, K., Butler, T.L., Herring, R.P., Fraser, G.E., 2015. Vegetarian dietary patterns and the risk of colorectal cancers. JAMA Intern. Med. 175, 767–776.

Phinney, S.D., Odin, R.S., Johnson, S.B., Holman, R.T., 1990. Reduced arachidonate in serum phospholipids and cholesteryl esters associated with vegetarian diets in humans. Am. J. Clin. Nutr. 51, 385–392.

Reddy, S., Sanders, T.A., Obeid, O., 1994. The influence of maternal vegetarian diet on essential fatty acid status of the newborn. Eur. J. Clin. Nutr. 48, 358–368.

Rosell, M.S., Lloyd-Wright, Z., Appleby, P.N., Sanders, T.A., Allen, N.E., Key, T.J., 2005. Long-chain n-3 polyunsaturated fatty acids in plasma in British meat-eating, vegetarian, and vegan men. Am. J. Clin. Nutr. 82, 327–334.

Sabate, J., Wien, M., 2015. A perspective on vegetarian dietary patterns and risk of metabolic syndrome. Br. J. Nutr. 113 (Suppl. 2), S136–143.

Salem, Jr., N., Kuratko, C.N., 2011. Lack of evidence for increased alpha-linolenic acid metabolism in vegetarians. Am. J. Clin. Nutr. 93, 1154–1155, author reply 1155–1156.

Sanders, T.A., Ellis, F.R., Dickerson, J.W., 1978. Studies of vegans: the fatty acid composition of plasma choline phosphoglycerides, erythrocytes, adipose tissue, and breast milk, and some indicators of susceptibility to ischemic heart disease in vegans and omnivore controls. Am. J. Clin. Nutr. 31, 805–813.

Sanders, T.A., Reddy, S., 1992. The influence of a vegetarian diet on the fatty acid composition of human milk and the essential fatty acid status of the infant. J. Pediatr. 120, S71–77.

Tonstad, S., Stewart, K., Oda, K., Batech, M., Herring, R.P., Fraser, G.E., 2013. Vegetarian diets and incidence of diabetes in the Adventist Health Study-2. Nutr. Metab. Cardiovasc. Dis. 23, 292–299.

Wang, F., Zheng, J., Yang, B., Jiang, J., Fu, Y., Li, D., 2015. Effects of vegetarian diets on blood lipids: a systematic review and meta-analysis of randomized controlled trials. J. Am. Heart Assoc. 4, e002408.

Welch, A.A., Shakya-Shrestha, S., Lentjes, M.A., Wareham, N.J., Khaw, K.T., 2010. Dietary intake and status of n-3 polyunsaturated fatty acids in a population of fish-eating and non-fish-eating meat-eaters, vegetarians, and vegans and the product-precursor ratio [corrected] of alpha-linolenic acid to long-chain n-3 polyunsaturated fatty acids: results from the EPIC-Norfolk cohort. Am. J. Clin. Nutr. 92, 1040–1051.

Yu, X., Huang, T., Weng, X., Shou, T., Wang, Q., Zhou, X., Hu, Q., Li, D., 2014. Plasma $n$-3 and $n$-6 fatty acids and inflammatory markers in Chinese vegetarians. Lipids Health Dis. 13, 151.

# Genetic Influences on Polyunsaturated Fatty Acid Biosynthesis and Metabolism

*Colette O' Neill\*, Anne M. Minihane\*\**

*Cork Centre for Vitamin D and Nutrition Research, University College Cork, Cork, Ireland; \*\*University of East Anglia, Norwich, United Kingdom*

## INTRODUCTION

There is a large body of evidence from prospective epidemiology and randomized controlled trials indicating that polyunsaturated fatty acid (PUFA) intake and status is an important determinant of cardio-metabolic

Polyunsaturated Fatty Acid Metabolism. http://dx.doi.org/10.1016/B978-0-12-811230-4.00012-0

and cognitive health (Janssen and Kiliaan, 2014; Mozaffarian and Wu, 2011; Riserus et al., 2009). As a result, national and global authorities typically provide recommendations for total PUFA as a percent of dietary energy, linoleic acid (LA), and alpha linolenic acid ($\alpha$LNA), which are the essential PUFAs, and for the marine n-3 PUFAs, eicosapentaenoic acid (EPA), and docosahexaenoic acid (DHA) (http://www.issfal.org/statements/pufa-recommendations). These dietary recommendations are based on average population responses. It is well recognized that intake-tissue status associations are highly heterogonous. The concentration of PUFAs in blood and tissue pools is not only dependent on dietary exposure but also on endogenous biosynthesis, and on the partitioning of fatty acid between the circulation and different tissues. In addition to concentrations, the impact of PUFAs on cell and tissue function, and therefore on the risk of disease, is dependent on their enzymatic and nonenzymatic oxidation into an array of bioactive metabolites. Much of the variation in these metabolic processes is due to genetic variability in rate-limiting enzymes, which may influence both the structure, function, and tissue location of the resultant protein and also the levels of the protein produced. Family studies indicate that 40%–70% of red blood cell fatty acid status is inherited (Lemaitre et al., 2008).

In the future, such knowledge of the influence of genotype on PUFA status and metabolism may be used to take a "personalized" preventive and therapeutic medicine approach with bespoke PUFA recommended intakes for individuals who are likely to be deficient and/or responsive, based on their genetic diagnostics.

# DNA AND GENETIC VARIABILITY

The origin of the science of genetics dates back to the 1850s with Charles Darwin, who described his theories of natural selection in *On the Origin of Species* and the notion that individuals most genetically suited to their environment are likely to survive and reproduce. In the 1860s Gregor Johann Mendel published his work on pea plants and described the concepts of Mendelian inheritance. However, it was not until 1953 that Watson, Crick, and Franklin identified DNA as the molecule that carries genetic information and the unit of inheritance. In 2004, the full sequence of human DNA was described as containing 3 billion nucleotides (which consist of a base, sugar, and phosphate group; Lander, 2011). These nucleotides are arranged into 23 chromosome pairs, with one from each pair inherited from each biological parent. About 2% of the DNA encodes for proteins, with these regions termed genes, with a total of ~22,000 genes in the human genome. The function of the remainder of the genome is poorly understood, and often referred to as "junk DNA" although future research is likely to identify its essentiality in modulating DNA replication and repair, protein synthesis, and overall cell function.

Only identical twins have identical DNA. In the general population 99.5% of the DNA is common between individuals with the remaining 0.5% defining an individual's phenotype (observable characteristics) and response to their environment. In 2015, output from the 1000 Genome Consortium, indicated that there are typically 88 million variants in a human genome (Auton et al., 2015). These variants influence the progression from health to disease and response to changes in lifestyle including altered food intake.

Genetic variation comes in many forms ranging from gross structural alterations affecting more than 1000 bases/nucleotides in the DNA sequence such as copy number variation, deletions, insertion, and inversion, to single nucleotide polymorphisms (SNPs), where a single base in a nucleotide is changed. Structural variants are rare relative to SNPs and underlie monogenic disorders, whereby a mutation in a single gene causes the diseases. In contrast SNPs, which constitute >90% of all genetic variability, are common and therefore most relevant for public health and will be the form of genetic variation considered in this chapter.

In addition to targeted genotyping, whereby SNPs in key genes involved in PUFA synthesis and metabolism are considered, genome-wide association studies (GWAS) have been used to identify functional gene variants (for a description of the principles of GWAS and GWAS conducted to date, please refer to the National Institute of Health, US on-line resources, https://www.genome.gov/20019523/, http://www.ebi.ac.uk/gwas/). Rather than focus in on individual genes, in GWAS a tagging-SNPs approach is used to collect genetic information from across the genome and relate it to a particular trait, for example, PUFA concentrations. Publicly available SNP databases such as the site curated by the National Center for Biotechnology Information, United States (www.ncbi.nlm.nih.gov/snp) are used to select SNPs of interest. Each SNP in the database has a unique reference SNP cluster ID (rs number) accession number which is typically cited in published papers.

## PUFA BIOCONVERSION AND THE FATTY ACID DESATURASE (FADS) GENOTYPE

In addition to habitual dietary intake, the tissue status of the long chain (LC) PUFAs, AA, EPA, docosapentaenoic acid (DPA), and DHA is influenced by the rate of bioconversion from LA and $\alpha$LNA, which involves several desaturation and elongation steps (see Chapter 2). The key rate-limiting enzymes in this pathway are the delta-5 (D5D) and delta-6 desaturases (D6D). The human desaturase complementary DNAs were first cloned in 1999 by Cho et al. (1999a,b) and were later identified as *FADS*-1 and *FADS*-2 in the human genome (Marquardt et al., 2000), located in a cluster on chromosome 11 (*11q12*-13.1). D5D and D6D are found in many human tissues but the liver is the site at which they are most highly expressed (Cho et al., 1999a,b). n-6 LA and n-3 αLNA are metabolized by the same series of enzymes. EPA and DHA are produced at limited conversion rates of 0.2%–6% for EPA and <0.1% for DHA in human males and postmenopausal females, with higher rates evident in premenopausal females (thought to be an evolutionary adaptation to meet pregnancy demands; Burdge, 2006). As will be described, variation across the *FADS* gene region appears to be important in modulating LC PUFA status and has emerged as an important common genetic determinant of inflammatory conditions, including cardiovascular diseases (CVD) and dementia. The basis of this association is the fact that AA, EPA, and DHA are the precursors for an array of pro- and antiinflammatory eicosanoids and pro-resolving lipids (Fig. 12.1).

Dietary composition has also been shown to influence the relationship between the *FADS* genotype and PUFAs and cardiovascular health status, as previously reviewed (O'Neill and Minihane, 2016). Although many SNPs in *FADS1* and *FADS2* have been described as influencing PUFA status and associated health outcomes, the majority of these are in linkage disequilibrium and co-inherited. The actual functional SNP(s) have not yet been identified.

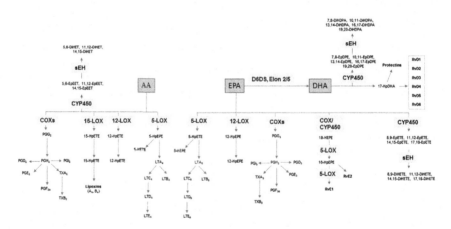

FIGURE 12.1  **Long chain PUFA metabolites (Buczynski et al., 2009; Funk, 2001; Serhan et al., 2008; Stables and Gilroy, 2011).** *AA*, Arachidonic acid; *COX*, cyclooxygenase; *CYP450*, cytochrome P450; *DHA*, docosahexaenoic acid; *EPA*, eicosapentaenoic acid; *HETE*, hydroxyeicosatetraenoic acid; *HPETE*, hydroperoxyeicosatetraenoic acids; *LOX*, lipoxygenase; *LT*, leukotriene; *PG*, prostaglandin; *Rv*, resolvin; *sEH*, soluble epoxide hydrolase; *TX*, thromboxane. DiHET, DiHETE, DiHDPA, EpDPE, EpEET, and EpETE are CYP450/sEH-derived oxylipins.

# IMPACT OF *FADS* GENOTYPE ON PUFA STATUS

Using both a candidate gene and GWAS approach, numerous studies have reported associations between variations in *FADS* and desaturase (product/precursor ratio of fatty acids) activity and fatty acid status in humans (examples outlined in Table 12.1). In 2006, using a candidate genotyping approach, Schaeffer et al. (2006) examined 18 SNPs and reconstructed haplotypes (groups of SNPs) in the *FADS1-2* cluster in 727 adults. A 5-locus *FADS* haplotype accounted for 27.7%, 5.2%, and 1.4% of the variation in AA, EPA, and DHA, respectively, in serum phospholipids. The minor alleles were associated with higher LA and αLNA and lower γ-linolenic acid, AA, EPA, and n-3 DPA concentrations, with no significant impact on DHA (Schaeffer et al., 2006).

Gieger et al. (2008) carried out the first GWAS ($n = 284$, 34–79-year-old males) to report a significant association between the rs174548 SNP and D5D product substrate ratio (Gieger et al., 2008). Since then a number of larger GWAS have reported significant associations between various *FADS* SNPs and PUFA status (Table 12.1). Using an elegant research approach, Ameur et al. (2012) performed genome-wide analysis in five European populations and data from the 1000 Genomes Project, and compared genomic data to that of early modern humans (archaic hominins) and distant primates, to examine the role of *FADS* genotype in human evolution (Ameur et al., 2012). The proportionally large human brain relative to body size is unique among primates. The human brain and central nervous system (CNS) is highly enriched in AA and DHA relative to other tissues. Therefore, in human evolution, a more efficient AA and DHA synthesis would be advantageous, in particular if dietary intake of LC PUFAs is limited. Two common *FADS* haplotype blocks (D and A, spanning *FADS1* and the first part of *FADS2*) were identified, which are defined by 28 closely linked SNPs (Ameur et al., 2012). The *FADS* D haplotype, unique to contemporary humans, was associated with lower levels of LA and αLNA and higher levels of EPA, gamma-linolenic acid (GLA), DHA, and AA. Individuals homozygous for haplotype D had 24% higher levels of DHA and 43% higher levels of AA, and higher liver expression of *FADS1* than those homozygous for haplotype A. This data clearly demonstrates the large size effect of *FADS* genotype on LC PUFA synthesis, and in vegan and vegetarians where intake is negligible, it is likely to be an important determinant of tissue EPA and DHA status.

However, as with other common genotypes, allele frequency varies between populations (Merino et al., 2011) and the penetrance of *FADS*, that is, its impact on PUFA status, is not homogenous, with factors such as habitual diet, health status, sex, and ethnicity likely to be important. Also the functional SNP(s) in the *FADS1–2* region and the molecular basis for genotype-PUFA associations have not been fully identified. Wang et al.

**TABLE 12.1** Candidate Gene Studies and Genome-Wide Association Studies (Gwas): Associations Between *Fads* Snps, Fatty Acid Status, and Select Health Outcomes

| Study | Subjects | Age (Mean ± SD or range) | Sex | SNPs | Outcomes | Results (impact on fatty acid status and other significant findings) |
|---|---|---|---|---|---|---|
| Schaeffer et al. (2006) | 727 | 41.6 ± 12.3 years, 20–64 years | Both | rs99780, rs174544, rs174545, rs174546, rs174553, rs174556, rs174561, rs174568, rs174570, rs174583, rs174589, rs174602, rs174620, rs2072114, rs3834458, rs482548, rs526126, and rs968567 | FA in serum phospholipids | SNPs showed strongest associations with AA ($P < 1.0 \times 10^{-13}$), also with LA, $\alpha$LNA, EPA ($P < 0.001$) |
| Baylin et al. (2007) | 1694 MI, 1694 controls | 58 ± 11 years | Both | rs3834458 | PUFA in plasma and adipose tissue. Risk of MI | EPA, LA, and AA were significantly lower in adipose tissue and plasma with increasing copy number of variant alleles ($P < 0.05$ for all). No association with MI |

| Reference | Sample size | Age | Population | SNPs | Trait | Findings |
|---|---|---|---|---|---|---|
| Malerba et al. (2008) | 658 | 59.7 ± 11.1 years | Both | rs174545, rs174556, rs174561, rs174570, rs174583, rs174589, rs174611, rs174627, rs498793, rs1000778, rs2524299, rs3834458, and rs17831757 | FA in serum phospholipids and erythrocytes in CVD patients | SNPs strongly associated with AA ($P < 1.0 \times 10^{-4}$) in both serum and erythrocytes. Significant associations were also observed for LA and αLNA ($P < 0.05$) |
| Tanaka et al. (2009) | 1210 + 1076 (replication) | 12–102 years | Both | GWAS | FA in plasma and blood lipids | Associations between rs174537 and AA ($P = 5.95 \times 10^{-46}$), TC ($P = 2.7 \times 10^{-2}$), LDL-C ($P = 1.1 \times 10^{-2}$) |
| Ameur et al. (2012) | 5652 genome-wide, 960 targeted resequencing | All ages | Both | GWAS | FA in blood phospholipid | FADS haplotype associated with lower LA ($P = 0.052$) and αLNA ($P = 0.024$) and higher EPA ($P = 1.1 \times 10^{-12}$), GLA ($P = 1.3 \times 10^{-18}$), DHA ($P = 8.3 \times 10^{-5}$), and AA ($P = 5.2 \times 10^{-18}$) |

*(Continued)*

**TABLE 12.1** Candidate Gene Studies and Genome-Wide Association Studies (Gwas): Associations Between *Fads* Snps, Fatty Acid Status, and Select Health Outcomes (*Cont.*)

| Study | Subjects | Age (Mean ± SD or range) | Sex | SNPs | Outcomes | Results (impact on fatty acid status and other significant findings) |
|---|---|---|---|---|---|---|
| Li et al. (2013) | 505 CAD, 510 controls | 33–85 years | Both | rs174460, rs174537, rs174550, rs174611, and rs174616 | Plasma FA | D6D activity (AA/LA) was higher in CAD patients ($P < 0.001$). rs174537 minor allele associated with lower risk of CAD [OR 0.743, 95% CI (0.624, 0.884), $P = 0.001$]. Carriers of the rs174460 minor allele were associated with a higher risk of CAD [OR 1.357, 95% CI (1.106, 1.665), $P = 0.003$] |
| Roke et al. (2013) | 878 | 20–29 years | Both | 19 SNPs in all subjects and 6 (rs174579, rs174593, rs174626, rs526126, rs968567, and rs17831757) were further analyzed | Plasma FA and hs-CRP | All 6 SNPs that were further analyzed significantly associated with AA levels and desaturase indices. Inverse association between *FADS1* desaturase index and hs-CRP ($P = 4.41 \times 10^{-6}$) |

| | | | | | | |
|---|---|---|---|---|---|---|
| Guan et al. (2014) | 8631 | GWAS | 60.3 years | Both | Plasma n-6 PUFA composition | *FADS* cluster associated with LA, AA, GLA, DGLA, and adrenic acid |
| Mozaffarian et al. (2015) | 8013 | GWAS | 45.8 ± 3.4–75.0 ± 5.1 years | Both | Phospholipid trans FA | 31 *FADS* SNPs associated with cis/trans-LA. No significant association with other TFAs |
| Li et al. (2016) | 872 | rs174450, rs174460, rs174537, and rs174616 | 59.3 ± 10.8 years | Both | Plasma FA and lipid composition T2D, CAD, both T2D and CAD, compared to healthy controls | T2D patients with rs174537 major allele were at risk of T2D and CAD (OR 1.763; 95% CI 1.143–2.718; $P = 0.010$), with elevated plasma LDL-C, AA, and desaturase activity |

*αLNA*, Alpha-linolenic acid; *AA*, arachidonic acid; *CAD*, coronary artery disease; *CVD*, cardiovascular disease; *D5D*, delta-5-desaturase; *D6D*, delta-6-desaturase; *DGLA*, dihomo-gamma-linolenic acid; *DHA*, docosahexaenoic acid; *EDA*, eicosadienoic acid; *EPA*, eicosapentaenoic acid; *FA*, fatty acids; *FADS*, fatty acid desaturase; *GLA*, gamma-linolenic acid; *GWAS*, genome wide association study; *hs-CRP*, high sensitivity C-reactive protein; *LA*, linoleic acid; *LD*, linkage disequilibrium; *LDL-C*, low density lipoprotein cholesterol; *MI*, myocardial infarction; *PC*, phosphatidylcholine; *PUFAs*, polyunsaturated fatty acids; *SD*, standard deviation; *T2D*, type 2 diabetes; *TC*, total cholesterol; *TFA*, trans fatty acid.

(2015) examined *FADS* SNPs, the lipidomic profile, and *FADS1–3* expression in liver samples ($n = 154$), and reported an associated with *FADS1* (but not *FADS2* and *3*) gene expression and protein levels, suggesting that the causal variant(s) may be located at *FADS1*. In addition, many of the highly linked SNPs were located in the transcription factor binding sites of the locus, which is consistent with the findings of Ameur et al. (2012) cited earlier and suggestive that *FADS* genotype mediates its effect by influencing transcription factor binding and activation of *FADS1*.

## IMPACT OF *FADS* GENOTYPE ON CARDIOVASCULAR AND OTHER HEALTH OUTCOMES

Variation across the *FADS* gene region and resulting LC PUFA status modulation has been shown to influence inflammation and associated chronic diseases such as CVD and dementia. Although carriers of the major alleles benefit from increased EPA and DHA, the majority of studies suggest that *FADS* minor alleles (associated with decreased desaturase activity) are associated with reduced inflammation, total cholesterol (TC), LDL-cholesterol (LDL-C), and coronary artery disease (CAD) risk (Aulchenko et al., 2009; Kwak et al., 2011; Martinelli et al., 2008). High levels of LA (relative to $\alpha$LNA), which are typical in Western diets, in conjunction with the presence of major alleles leads to increased conversion of LA to AA, which is a direct precursor of proinflammatory eicosanoids, such as prostaglandins and leukotrienes (Hester et al., 2014) (Fig. 12.1). Table 12.1 lists a number of studies that have examined both PUFA status and cardiovascular outcomes; however, other studies have also examined the impact of the FADS genotype on cardiovascular outcomes alone as previously reviewed by O'Neill and Minihane (2016). In the Verona Heart Study (2008), a CAD incidence of 84% versus 66% was evident in individuals with 6–7 versus 2–3 risk alleles, and a higher AA/LA ratio was an independent risk factor for CAD (Martinelli et al., 2008), which may be attributable to the greater production of proinflammatory eicosanoids. In addition to associations between *FADS* rs174537 and desaturase activity, Hester et al. (2014) also reported links between this SNP and the proinflammatory eicosanoids leukotriene B4 and 5-HETE (Fig. 12.1) in stimulated whole blood in adult females (Hester et al., 2014). In a recent finding, Vaittinen et al. (2016) reported significant associations between *FADS* SNPs (rs174547, rs174616) and adipose tissue inflammation following surgery-induced weight loss. However, a few studies have reported contradictory results (Lu et al., 2012; Qin et al., 2011; Song et al., 2013), which could be due to the ethnicity of the participants or differences in the n-6 to n-3 PUFA content of the habitual diet. For example, two studies in Chinese-Han populations reported the frequency of the rs174556 minor allele to be significantly higher in cases of both CAD and acute coronary syndrome compared to control groups (Qin et al., 2011; Song et al., 2013).

Less is known about the complex relationship between the *FADS* gene cluster, PUFA metabolism and cognitive function. *FADS* SNPs have been shown to modulate the effect of habitual fatty acid intake on cognitive outcomes in children (Lattka et al., 2012). As altered brain fatty acid status has been implicated in various cognitive disorders, Freemantle et al. investigated associations between *FADS* SNPs (rs174546, rs174548, rs174549, and rs174555) and fatty acid composition in cortical brain tissue in 61 adult males (Freemantle et al., 2012). A significant association between the minor haplotype with estimated fatty acid desaturase activity was evident with no impact of genotype on the expression of *FADS1* and *FADS2*. There was also a significant interaction between haplotype and age on LA and AA. Overall, the authors suggest that genetic variability in the *FADS* the gene cluster may affect FA composition in brain tissue but that this is unlikely due to an effect on local synthesis (Freemantle et al., 2012). Schuchardt et al. (2016) recently reported associations between 12 *FADS* SNPs and erythrocyte membrane LC-PUFAs in 111 patients with mild cognitive impairment. They found that minor allele carriers of several SNPs had higher LA and αLNA, and lower AA levels in erythrocyte membranes compared to the major allele carriers. However, this study lacked a healthy control group and was therefore unable to investigate the influence of *FADS* genotypes or LC-PUFA status on cognitive performance (Schuchardt et al., 2016). There do not appear to be any GWAS reporting an association between *FADS* genotype and cognitive function. It is clear that further research is warranted in this area.

## IMPACT OF OTHER GENOTYPES ON PUFA STATUS

In addition to the desaturases, elongase 2 and 5 (*ELOVL2* and 5) are involved in multiple steps in the biosynthesis of AA, EPA, and DHA (see Chapter 2). Less is known about the impact of variants in these loci (chromosome 6) on PUFA status. In the Cohorts for Heart and Aging Research in Genomic Epidemiology consortium, GWAS variant alleles of *ELOVL2* were associated with higher EPA and DPA and lower DHA. This suggests decreased elongase 2 activity, which is critical in the elongation of DPA to DHA. In addition to observational analysis, the impact of *ELOVL* variants on response to EPA and DHA supplementation was examined in the MARINA trial. After the 1.8 g/d dose, minor allele carriers had approximately 30% and 9% higher proportions of EPA and DHA than noncarriers (Alsaleh et al., 2014).

Apolipoprotein E (apoE) is involved in numerous stages of lipoprotein metabolism including dietary lipid absorption and chylomicron metabolism, the production of hepatic derived lipoproteins and the receptor mediated removal of lipoprotein remnants by the liver and other tissues (Minihane, 2016). Furthermore, it is the almost exclusive lipid transporter in the CNS. Two missense (resulting in changes in the amino acid in the protein) SNPs in the *APOE* gene on chromosome 19 (rs429358 and rs7412) result in three apoE protein isoforms, apoE2, apoE3, and apoE4. *APOE4* carriers have been inconsistently shown to be at higher risk of CVD and Alzheimer's disease, and *APOE* genotype has emerged as the strongest identified common genetic predictor of longevity (Minihane, 2016). In the Genetics of Healthy Ageing Study, the prevalence of the *APOE4* allele was 6.8% in nonagenarians (90–99 years old), compared to 12.7% in matched control (55–75 years old), with *APOE4* carriers having a 50% lower chance of reaching age 90 compared to noncarriers (Beekman et al., 2013).

Numerous mechanisms have been proposed to explain this association with health outcomes, including an impact of *APOE* genotype on LC n-3 PUFA status. Brain tissue is highly enriched in DHA, indicating its essentiality to neuronal function. A number of human studies have reported that the cognitive benefits associated with DHA/fish intake were absent or lower in *APOE4* carriers (Minihane, 2016). In the SATGENE intervention, in which adult males were supplemented with DHA (3.5 g/d) for 8 weeks, a 21% lower plasma phospholipid DHA enrichment was observed in overweight *APOE3/E4* males relative to those with the common *APOE3/E3* genotype (Chouinard-Watkins et al., 2015). Higher oxidation of DHA has been reported in *APOE4* carriers (Chouinard-Watkins et al., 2013), which may help explain the lower status. Therefore, although confirmation is needed and mechanisms are as yet poorly understood, available literature indicates defective LC n-3 PUFA metabolism in *APOE4* carriers (25% population) relative to noncarrier groups.

## GENOTYPE (COX, LOX, CYP450, AND SO ON) AND THE PRODUCTION OF BIOACTIVE LIPIDS

AA, EPA, and DHA are converted into a wide array of bioactive lipids including thromboxanes ($TXA_2$), prostaglandins, leukotrienes, lipoxins, resolvins, protectins, and other CYP450 derived oxylipins (Fig. 12.1) (Buczynski et al., 2009; Funk, 2001; Serhan et al., 2008; Stables and Gilroy, 2011), which modulate inflammation and vascular function. Given that these compounds are several-fold more potent than their parent PUFAs, with AA metabolites tending to be more inflammatory than their EPA and DHA derived equivalents, in addition to PUFA status, the rate of biosynthesis of these bioactive lipids is likely to be an important determinant of PUFA-health outcome associations. As indicated in Fig. 12.1, cyclooxygenases (COX), lipoxygenases (LOX), cytochrome P450s (CYP450), and soluble epoxide hydrolase (sEH) are the main enzyme groups involved in their synthesis. Although as yet relatively unknown, variants in these genes, which impact on enzyme concentration, structure, and function, are likely to determine tissue concentration of this compound group.

Variation in the *PTGS2* gene, which encodes for COX-2, has been associated with cardiovascular health and cancer risk (Ross et al., 2014; Shahedi et al., 2006), likely in part due to differences in the production of PUFA derived inflammatory mediators. In a prospective analysis using data from six patient groups ($n = 49{,}232$), the minor allele of the rs20417 SNP was associated with a decreased risk of CVD outcome (Ross et al., 2014). In 117 healthy controls, carriers had significantly lower urinary levels of select $TXA_2$ and prostacyclins ($PGF_{1\alpha}$) compared with noncarriers (Ross et al., 2014). Although evidence is currently limited, variation in 5-LOX and 12-LOX has been associated with myocardial infarction (Gammelmark et al., 2016) and essential hypertension (Quintana et al., 2006) and incidence, although no associations between 1348 LOX5 SNPs and carotid intima-media thickness as well as incident coronary heart disease was observed in the MESA study (Tsai et al., 2016). Quintana et al. (2006) reported that the 12-LOX R261Q polymorphism (which results in an amino acid arginine to glutamine change in the protein) was associated with urinary 12-(S)hydroxyeicosatetraenoic acid (12(S)-HETE).

CYP450 variants have been linked with health end-points (Gervasini et al., 2016; Yan et al., 2013; Zordoky and El-Kadi, 2010) and the bioactive lipid profile (Schwarz et al., 2005). A genetic variant in the CYP4F2 gene, the main gene involved in 20-HETE synthesis, was associated with the risk for posttransplant diabetes mellitus (PTDM) in kidney recipients (Gervasini et al., 2016). In in vitro catalytic assays the CYP1A1.2 I462V variant affected the production of a number of EPA derived epoxide metabolites (Schwarz et al., 2005).

## SUMMARY AND CONCLUSION

Animal products and oily fish are the richest dietary sources of AA and EPA/DHA respectively, and, in omnivores, intakes of these products are the major variable determining tissue status. Common gene variants are thought to account for at least half the variability in the concentration of these LC-PUFA, and are likely to make an even greater contribution in vegetarians and vegans where endogenous biosynthesis is the almost exclusive source. Although variation in the *FADS* locus have been confirmed as an important modulator of LC-PUFA status, the functional variants have not been identified and the impact of variables such as sex, ethnicity, and habitual diet on its penetrance is relatively unquantified. Although the size effect has not been fully established, and is likely to be variable among population subgroups, the *FADS* genotype is likely to have health implications. For example, for EPA + DHA, differences in concentrations of up to 30% have been reported between *FADS* genotype/haplotype groups (Ameur et al., 2012; Li et al., 2016) (Table 12.1), which influence the risk of CVDs (Albert et al., 2002).

Modest dietary intakes of EPA and DHA could overcome this genotype effect; supplementation of 300 mg EPA and DHA or 90 g of salmon per week has been shown to increase combined plasma EPA plus DHA by about 30% (Flock et al., 2013; Raatz et al., 2013).

In addition to determining PUFA status, common genotypes are likely to modulate the synthesis of a whole host of bioactive lipids, which mediate the cell and tissue actions of the LC PUFAs. However, research to identify functional variants in rate limiting COX, LOX, CYP450, and sEH enzymes, which modulate PUFA metabolism, is in its relative infancy.

In an era of precision, preventive and therapeutic medicine, *FADS* and other genotypes could contribute to the future stratification and targeting of dietary PUFA recommendations, and in particular in subgroups with a genetic predisposition towards low LC PUFA status.

## References

Albert, C.M., Ma, J., Rifai, N., Stampfer, M.J., Ridker, P.M., 2002. Prospective study of C-reactive protein, homocysteine, and plasma lipid levels as predictors of sudden cardiac death. Circulation 105 (22), 2595–2599.

Alsaleh, A., Maniou, Z., Lewis, F.J., Hall, W.L., Sanders, T.A., O'Dell, S.D., 2014. ELOVL2 gene polymorphisms are associated with increases in plasma eicosapentaenoic and docosahexaenoic acid proportions after fish oil supplement. Genes Nutr. 9 (1), 362.

Ameur, A., Enroth, S., Johansson, A., Zaboli, G., Igl, W., Johansson, A.C., Rivas, M.A., Daly, M.J., Schmitz, G., Hicks, A.A., Meitinger, T., Feuk, L., van Duijn, C., Oostra, B., Pramstaller, P.P., Rudan, I., Wright, A.F., Wilson, J.F., Campbell, H., Gyllensten, U., 2012. Genetic adaptation of fatty-acid metabolism: a human-specific haplotype increasing the biosynthesis of long-chain omega-3 and omega-6 fatty acids. Am. J. Hum. Genet. 90 (5), 809–820.

Aulchenko, Y.S., Ripatti, S., Lindqvist, I., Boomsma, D., Heid, I.M., Pramstaller, P.P., Penninx, B.W.J.H., Janssens, A.C.J.W., Wilson, J.F., Spector, T., Martin, N.G., Pedersen, N.L., Kyvik, K.O., Kaprio, J., Hofman, A., Freimer, N.B., Jarvelin, M.-R., Gyllensten, U., Campbell, H., Rudan, I., Johansson, A., Marroni, F., Hayward, C., Vitart, V., Jonasson, I., Pattaro, C., Wright, A., Hastie, N., Pichler, I., Hicks, A.A., Falchi, M., Willemsen, G., Hottenga, J.-J., de Geus, E.J.C., Montgomery, G.W., Whitfield, J., Magnusson, P., Saharinen, J., Perola, M., Silander, K., Isaacs, A., Sijbrands, E.J.G., Uiterlinden, A.G., Witteman, J.C.M., Oostra, B.A., Elliott, P., Ruokonen, A., Sabatti, C., Gieger, C., Meitinger, T., Kronenberg, F., Döring, A., Wichmann, H.E., Smit, J.H., McCarthy, M.I., van Duijn, C.M., Peltonen, L., Consortium, E., 2009. Loci influencing lipid levels and coronary heart disease risk in 16 European population cohorts. Nat. Genet. 41 (1), 47–55.

Auton, A., Brooks, L.D., Durbin, R.M., Garrison, E.P., Kang, H.M., Korbel, J.O., Marchini, J.L., McCarthy, S., McVean, G.A., Abecasis, G.R., 2015. A global reference for human genetic variation. Nature 526 (7571), 68–74.

Baylin, A., Ruiz-Narvaez, E., Kraft, P., Campos, H., 2007. α-Linolenic acid, Δ6-desaturase gene polymorphism, and the risk of nonfatal myocardial infarction. Am. J. Clin. Nutr. 85 (2), 554–560.

Beekman, M., Blanche, H., Perola, M., Hervonen, A., Bezrukov, V., Sikora, E., Flachsbart, F., Christiansen, L., De Craen, A.J., Kirkwood, T.B., Rea, I.M., Poulain, M., Robine, J.M., Valensin, S., Stazi, M.A., Passarino, G., Deiana, L., Gonos, E.S., Paternoster, L., Sorensen, T.I., Tan, Q., Helmer, Q., van den Akker, E.B., Deelen, J., Martella, F., Cordell, H.J., Ayers, K.L., Vaupel, J.W., Tornwall, O., Johnson, T.E., Schreiber, S., Lathrop, M., Skytthe, A., Westendorp, R.G., Christensen, K., Gampe, J., Nebel, A., Houwing-Duistermaat, J.J., Slagboom, P.E., Franceschi, C., 2013. Genome-wide linkage analysis for human longevity: genetics of healthy aging study. Aging Cell 12 (2), 184–193.

Buczynski, M.W., Dumlao, D.S., Dennis, E.A., 2009. Thematic review series: proteomics. An integrated omics analysis of eicosanoid biology. J. Lipid Res. 50 (6), 1015–1038.

Burdge, G.C., 2006. Metabolism of $\alpha$-linolenic acid in humans. Prostaglandins Leukot. Essent. Fatty Acids 75 (3), 161–168.

Cho, H.P., Nakamura, M., Clarke, S.D., 1999a. Cloning, expression, and fatty acid regulation of the human delta-5 desaturase. J. Biol. Chem. 274 (52), 37335–37339.

Cho, H.P., Nakamura, M.T., Clarke, S.D., 1999b. Cloning, expression, and nutritional regulation of the mammalian delta-6 desaturase. J. Biol. Chem. 274 (1), 471–477.

Chouinard-Watkins, R., Rioux-Perreault, C., Fortier, M., Tremblay-Mercier, J., Zhang, Y., Lawrence, P., Vohl, M.C., Perron, P., Lorrain, D., Brenna, J.T., Cunnane, S.C., Plourde, M., 2013. Disturbance in uniformly 13C-labelled DHA metabolism in elderly human subjects carrying the apoE ε4 allele. Br. J. Nutr. 110 (10), 1751–1759.

Chouinard-Watkins, R., Conway, V., Minihane, A.M., Jackson, K.G., Lovegrove, J.A., Plourde, M., 2015. Interaction between BMI and APOE genotype is associated with changes in the plasma long-chain-PUFA response to a fish-oil supplement in healthy participants. Am. J. Clin. Nutr. 102 (2), 505–513.

Flock, M.R., Skulas-Ray, A.C., Harris, W.S., Etherton, T.D., Fleming, J.A., Kris-Etherton, P.M., 2013. Determinants of erythrocyte omega-3 fatty acid content in response to fish oil supplementation: a dose–response randomized controlled trial. J. Am. Heart Assoc. 2 (6).

Freemantle, E., Lalovic, A., Mechawar, N., Turecki, G., 2012. Age and haplotype variations within FADS1 interact and associate with alterations in fatty acid composition in human male cortical brain tissue. PLoS One 7 (8), e42696.

Funk, C.D., 2001. Prostaglandins and leukotrienes: advances in eicosanoid biology. Science 294 (5548), 1871–1875.

Gammelmark, A., Nielsen, M.S., Lundbye-Christensen, S., Tjonneland, A., Schmidt, E.B., Overvad, K., 2016. Common polymorphisms in the 5-lipoxygenase pathway and risk of incident myocardial infarction: a Danish case-cohort study. PLoS One 11 (11), e0167217.

Gervasini, G., Luna, E., Garcia-Cerrada, M., Garcia-Pino, G., Cubero, J.J., 2016. Risk factors for post-transplant diabetes mellitus in renal transplant: role of genetic variability in the CYP450-mediated arachidonic acid metabolism. Mol. Cell. Endocrinol. 419, 158–164.

Gieger, C., Geistlinger, L., Altmaier, E., Hrabé de Angelis, M., Kronenberg, F., Meitinger, T., Mewes, H.-W., Wichmann, H.E., Weinberger, K.M., Adamski, J., Illig, T., Suhre, K., 2008. Genetics meets metabolomics: a genome-wide association study of metabolite profiles in human serum. PLoS Genet. 4 (11), e1000282.

Guan, W., Steffen, B.T., Lemaitre, R.N., Wu, J.H., Tanaka, T., Manichaikul, A., Foy, M., Rich, S.S., Wang, L., Nettleton, J.A., Tang, W., Gu, X., Bandinelli, S., King, I.B., McKnight, B., Psaty, B.M., Siscovick, D., Djousse, L., Ida Chen, Y.D., Ferrucci, L., Fornage, M., Mozaffarian, D., Tsai, M.Y., Steffen, L.M., 2014. Genome-wide association study of plasma N6 polyunsaturated fatty acids within the cohorts for heart and aging research in genomic epidemiology consortium. Circ. Cardiovasc. Genet. 7 (3), 321–331.

Hester, A.G., Murphy, R.C., Uhlson, C.J., Ivester, P., Lee, T.C., Sergeant, S., Miller, L.R., Howard, T.D., Mathias, R.A., Chilton, F.H., 2014. Relationship between a common variant in the fatty acid desaturase (FADS) cluster and eicosanoid generation in humans. J. Biol. Chem. 289 (32), 22482–22489.

Janssen, C.I., Kiliaan, A.J., 2014. Long-chain polyunsaturated fatty acids (LCPUFA) from genesis to senescence: the influence of LCPUFA on neural development, aging, and neurodegeneration. Prog. Lipid. Res. 53, 1–17.

Kwak, J.H., Paik, J.K., Kim, O.Y., Jang, Y., Lee, S.H., Ordovas, J.M., Lee, J.H., 2011. FADS gene polymorphisms in Koreans: association with omega6 polyunsaturated fatty acids in serum phospholipids, lipid peroxides, and coronary artery disease. Atherosclerosis 214 (1), 94–100.

Lander, E.S., 2011. Initial impact of the sequencing of the human genome. Nature 470 (7333), 187–197.

Lattka, E., Klopp, N., Demmelmair, H., Klingler, M., Heinrich, J., Koletzko, B., 2012. Genetic variations in polyunsaturated fatty acid metabolism—implications for child health? Ann. Nutr. Metab. 60 (Suppl. 3), 8–17.

Lemaitre, R.N., Siscovick, D.S., Berry, E.M., Kark, J.D., Friedlander, Y., 2008. Familial aggregation of red blood cell membrane fatty acid composition: the Kibbutzim Family Study. Metabolism 57 (5), 662–668.

Li, S.-W., Lin, K., Ma, P., Zhang, Z.-L., Zhou, Y.-D., Lu, S.-Y., Zhou, X., Liu, S.-M., 2013. FADS gene polymorphisms confer the risk of coronary artery disease in a Chinese Han population through the altered desaturase activities: based on high-resolution melting analysis. PLoS One 8 (1), e55869.

Li, S.-W., Wang, J., Yang, Y., Liu, Z.-J., Cheng, L., Liu, H.-Y., Ma, P., Luo, W., Liu, S.-M., 2016. Polymorphisms in FADS1 and FADS2 alter plasma fatty acids and desaturase levels in type 2 diabetic patients with coronary artery disease. J. Transl. Med. 14 (1), 1–9.

Lu, Y., Vaarhorst, A., Merry, A.H., Dolle, M.E., Hovenier, R., Imholz, S., Schouten, L.J., Heijmans, B.T., Muller, M., Slagboom, P.E., van den Brandt, P.A., Gorgels, A.P., Boer, J.M., Feskens, E.J., 2012. Markers of endogenous desaturase activity and risk of coronary heart disease in the CAREMA cohort study. PLoS One 7 (7), e41681.

Malerba, G., Schaeffer, L., Xumerle, L., Klopp, N., Trabetti, E., Biscuola, M., Cavallari, U., Galavotti, R., Martinelli, N., Guarini, P., Girelli, D., Olivieri, O., Corrocher, R., Heinrich, J., Pignatti, P.F., Illig, T., 2008. SNPs of the FADS gene cluster are associated with polyunsaturated fatty acids in a cohort of patients with cardiovascular disease. Lipids 43 (4), 289–299.

Marquardt, A., Stohr, H., White, K., Weber, B.H., 2000. cDNA cloning, genomic structure, and chromosomal localization of three members of the human fatty acid desaturase family. Genomics 66 (2), 175–183.

Martinelli, N., Girelli, D., Malerba, G., Guarini, P., Illig, T., Trabetti, E., Sandri, M., Friso, S., Pizzolo, F., Schaeffer, L., Heinrich, J., Pignatti, P.F., Corrocher, R., Olivieri, O., 2008. FADS genotypes and desaturase activity estimated by the ratio of arachidonic acid to linoleic acid are associated with inflammation and coronary artery disease. Eur. J. Clin. Nutr. 88 (4), 941–949.

Merino, D.M., Johnston, H., Clarke, S., Roke, K., Nielsen, D., Badawi, A., El-Sohemy, A., Ma, D.W., Mutch, D.M., 2011. Polymorphisms in FADS1 and FADS2 alter desaturase activity in young Caucasian and Asian adults. Mol. Genet. Metab. 103 (2), 171–178.

Minihane, A.M., 2016. Impact of genotype on EPA and DHA status and responsiveness to increased intakes. Nutrients 8 (3), 123.

Mozaffarian, D., Wu, J.H., 2011. Omega-3 fatty acids and cardiovascular disease: effects on risk factors, molecular pathways, and clinical events. J. Am. Coll. Cardiol. 58 (20), 2047–2067.

Mozaffarian, D., Kabagambe, E.K., Johnson, C.O., Lemaitre, R.N., Manichaikul, A., Sun, Q., Foy, M., Wang, L., Wiener, H., Irvin, M.R., Rich, S.S., Wu, H., Jensen, M.K., Chasman, D.I., Chu, A.Y., Fornage, M., Steffen, L., King, I.B., McKnight, B., Psaty, B.M., Djousse, L., Chen, I.Y., Wu, J.H., Siscovick, D.S., Ridker, P.M., Tsai, M.Y., Rimm, E.B., Hu, F.B., Arnett, D.K., 2015. Genetic loci associated with circulating phospholipid trans fatty acids: a meta-analysis of genome-wide association studies from the CHARGE Consortium. Am. J. Clin. Nutr. 101 (2), 398–406.

O'Neill, C.M., Minihane, A.M., 2016. The impact of fatty acid desaturase genotype on fatty acid status and cardiovascular health in adults. Proc. Nutr. Soc., 1–12.

Qin, L., Sun, L., Ye, L., Shi, J., Zhou, L., Yang, J., Du, B., Song, Z., Yu, Y., Xie, L., 2011. A case-control study between the gene polymorphisms of polyunsaturated fatty acids metabolic rate-limiting enzymes and coronary artery disease in a Chinese Han population. Prostaglandins Leukot. Essent. Fatty Acids 85 (6), 329–333.

Quintana, L.F., Guzman, B., Collado, S., Claria, J., Poch, E., 2006. A coding polymorphism in the 12-lipoxygenase gene is associated to essential hypertension and urinary 12(S)-HETE. Kidney Int. 69 (3), 526–530.

Raatz, S.K., Rosenberger, T.A., Johnson, L.K., Wolters, W.W., Burr, G.S., Picklo, M.J., 2013. Dose-dependent consumption of farmed Atlantic salmon (*Salmo salar*) increases plasma phospholipid n-3 fatty acids differentially. J. Acad. Nutr. Diet. 113 (2), 282–287.

Riserus, U., Willett, W.C., Hu, F.B., 2009. Dietary fats and prevention of type 2 diabetes. Prog. Lipid Res. 48 (1), 44–51.

Roke, K., Ralston, J.C., Abdelmagid, S., Nielsen, D.E., Badawi, A., El-Sohemy, A., Ma, D.W., Mutch, D.M., 2013. Variation in the FADS1/2 gene cluster alters plasma n-6 PUFA and is weakly associated with hsCRP levels in healthy young adults. Prostaglandins Leukot. Essent. Fatty Acids 89 (4), 257–263.

Ross, S., Eikelboom, J., Anand, S.S., Eriksson, N., Gerstein, H.C., Mehta, S., Connolly, S.J., Rose, L., Ridker, P.M., Wallentin, L., Chasman, D.I., Yusuf, S., Pare, G., 2014. Association of cyclo-oxygenase-2 genetic variant with cardiovascular disease. Eur. Heart J. 35 (33), 2242–2248a.

Schaeffer, L., Gohlke, H., Muller, M., Heid, I.M., Palmer, L.J., Kompauer, I., Demmelmair, H., Illig, T., Koletzko, B., Heinrich, J., 2006. Common genetic variants of the FADS1 FADS2 gene cluster and their reconstructed haplotypes are associated with the fatty acid composition in phospholipids. Hum. Mol. Genet. 15 (11), 1745–1756.

Schuchardt, J.P., Köbe, T., Witte, V., Willers, J., Gingrich, A., Tesky, V., Pantel, J., Rujescu, D., Illig, T., Flöel, A., Hahn, A., 2016. Genetic variants of the FADS gene cluster are associated with erythrocyte membrane LC PUFA levels in patients with mild cognitive impairment. J. Nutr. Health Aging, 1–10.

Schwarz, D., Kisselev, P., Chernogolov, A., Schunck, W.H., Roots, I., 2005. Human CYP1A1 variants lead to differential eicosapentaenoic acid metabolite patterns. Biochem. Biophys. Res. Commun. 336 (3), 779–783.

Serhan, C.N., Yacoubian, S., Yang, R., 2008. Anti-inflammatory and proresolving lipid mediators. Annu. Rev. Pathol. 3, 279–312.

Shahedi, K., Lindstrom, S., Zheng, S.L., Wiklund, F., Adolfsson, J., Sun, J., Augustsson-Balter, K., Chang, B.L., Adami, H.O., Liu, W., Gronberg, H., Xu, J., 2006. Genetic variation in the COX-2 gene and the association with prostate cancer risk. Int. J. Cancer 119 (3), 668–672.

Song, Z., Cao, H., Qin, L., Jiang, Y., 2013. A case-control study between gene polymorphisms of polyunsaturated fatty acid metabolic rate-limiting enzymes and acute coronary syndrome in Chinese Han population. BioMed Res. Int. 2013, 928178.

Stables, M.J., Gilroy, D.W., 2011. Old and new generation lipid mediators in acute inflammation and resolution. Prog. Lipid Res. 50 (1), 35–51.

Tanaka, T., Shen, J., Abecasis, G.R., Kisialiou, A., Ordovas, J.M., Guralnik, J.M., Singleton, A., Bandinelli, S., Cherubini, A., Arnett, D., Tsai, M.Y., Ferrucci, L., 2009. Genome-wide association study of plasma polyunsaturated fatty acids in the InCHIANTI Study. PLoS Genet. 5 (1), e1000338.

Tsai, M.Y., Cao, J., Steffen, B.T., Weir, N.L., Rich, S.S., Liang, S., Guan, W., 2016. 5-Lipoxygenase gene variants are not associated with atherosclerosis or incident coronary heart disease in the Multi-Ethnic Study of Atherosclerosis Cohort. J. Am. Heart Assoc. 5 (3), e002814.

Vaittinen, M., Walle, P., Kuosmanen, E., Mannisto, V., Kakela, P., Agren, J., Schwab, U., Pihlajamaki, J., 2016. FADS2 genotype regulates delta-6 desaturase activity and inflammation in human adipose tissue. J. Lipid Res. 57 (1), 56–65.

Wang, L., Athinarayanan, S., Jiang, G., Chalasani, N., Zhang, M., Liu, W., 2015. Fatty acid desaturase 1 gene polymorphisms control human hepatic lipid composition. Hepatology 61 (1), 119–128.

Yan, H.C., Liu, J.H., Li, J., He, B.X., Yang, L., Qiu, J., Li, L., Ding, D.P., Shi, L., Zhao, S.J., 2013. Association between the CYP4A11 T8590C variant and essential hypertension: new data from Han Chinese and a meta-analysis. PLoS One 8 (11), e80072.

Zordoky, B.N., El-Kadi, A.O., 2010. Effect of cytochrome P450 polymorphism on arachidonic acid metabolism and their impact on cardiovascular diseases. Pharmacol. Ther. 125 (3), 446–463.

# Interactions Between Polyunsaturated Fatty Acids and the Epigenome

*Karen A. Lillycrop\*, Graham C. Burdge\*\**

\*Centre for Biological Sciences, University of Southampton, Southampton, United Kingdom; \*\*University of Southampton, Southampton, United Kingdom

## INTRODUCTION

Polyunsaturated fatty acids (PUFA) can influence cell function through several well-established mechanisms including membrane fluidity, by acting as substrates for the production of a large number of second messengers and by the activation of transcription factors, primarily peroxisomal proliferator-activated receptors (PPARs). In addition, PUFA biosynthesis can be modified by hormones, dietary fat intake and by polymorphisms

Polyunsaturated Fatty Acid Metabolism. http://dx.doi.org/10.1016/B978-0-12-811230-4.00013-2

in genes that encode the major enzymes that catalyse this pathway. There is emerging evidence that PUFA can alter cell function by modifying the epigenetic regulation of genes and, in addition, that PUFA biosynthesis is regulated by epigenetic processes. This chapter discusses the evidence for interactions between PUFA and the epigenome, in terms of the role of epigenetics in regulating PUFA biosynthesis and the effects of PUFA on epigenetic processes.

## An Overview of Epigenetics

Epigenetics refers to a group of closely interrelated processes that regulate transcription; methylation at the 5' position of cytosine bases in CpG dinucleotide pairs, covalent modifications of histones (including acetylation and methylation of lysine residues) and the activities of noncoding RNA species (Bird, 2002). Such processes, in particular DNA methylation, are central to the induction during embryogenesis and maintenance of cellular phenotype and for mediating, via imprinting, parental influence on growth and development (Bird, 2002). DNA methylation marks are induced by the activities of DNA methyltransferases (Dnmts) 3a and 3b, and maintained by Dnmt 1. Both imprinted and cell-type specific epigenetic marks are retained throughout the life course. However, there is evidence that DNA methylation at some loci can retain plasticity and respond to environmental inputs (Szyf, 2007). Such plasticity has been associated with disease traits such as cancer (Lillycrop and Burdge, 2014) and with cellular response to environmental signals, for example T cell maturation in the Th1 or Th2 phenotypes (Leoni et al., 2015). One implication of epigenetic plasticity is that although variations in the epigenetic control of genes involved in nutrient metabolism may contribute to individual differences in nutrient requirements, dietary choices may modify the epigenome and hence influence the levels of nutrient intakes that are needed for health (Burdge et al., 2012). Methylation of CpG loci is often associated with repression of transcription, whereas unmethylated CpG dinucleotides facilitate transcriptional activation, although there are exceptions (Hu et al., 2013). Covalent modifications of histones may vary more dynamically than DNA methylation marks. Changes in the acetyl or methyl status of individual lysine residences can alter the conformation of the associated chromatin. Acetylated lysines facilitate a potentially transcriptionally active chromatin structure, whereas deactylation can induce a condensed transcriptionally inactive structure. Methylation of lysine residues may facilitate or repress transcription in a manner dependent on the residue histone and a number of methyl groups (mono-, di-, or trimethyl). Noncoding RNAs may modify gene activity via RNA interference leading to degradation of specific RNA transcripts, modifying DNA methylation, repression of transposons, altering transcriptional activation and by regulating RNA polymerase II activity.

# The Effect of PUFA on Epigenetic Marks in Cultured Cells

Perhaps surprisingly, there have been relatively few studies of the effects of PUFA, or other fatty acid classes, on epigenetic processes in cultured cells. Treatment of M17 neuroblastoma cells with docosahexaenoic acid (DHA) (10 μmol/L) for 48 h induced increased global H3K9 acetylation and decreased global dimethyl H3K4, dimethyl H3K9, dimethyl H3K27, dimethyl H3K36, and dimethyl H3K79 levels, which suggests a genome-wide increase in potentially transcriptionally active genes (Sadli et al., 2012). This was accompanied by decreased protein expression of histone deacetylases (HDACs) 1, 2, and 3, although the effect of DHA on HDAC activity was not tested. Unfortunately, this study did not test whether such effects were specific to DHA or whether they were dependent on DHA concentration. Treatment of U937 leukemic cells with eicosapentaenoic acid (EPA) (100 μmol/L) increased mRNA expression of the CCAAT/enhancer-binding proteins C/EBP-$\beta$ and C/EBP-$\delta$, PU.1 and c-Jun (Ceccarelli et al., 2011). This was accompanied by demethylation of a single CpG locus the C/EBP-$\delta$ promoter in all clones compared to untreated cells. In contrast, this locus was unmethylated in 2/7 clones showed cells treated with oleic acid. Furthermore, the same groups showed that treatment of U937 cells with EPA also induced almost complete demethylation of the H-Ras intron 1 CpG island, which was accompanied by increased H-Ras protein expression compared to untreated cells or cells treated with oleic acid (Ceccarelli et al., 2014). To date there have not been any studies on the effects of n-6 PUFA on epigenetic processes in vitro; however, one study has shown that treatment with arachidonoylethanolamine reduced the expression of differentiation-associated markers in HaCaT keratinocytes (Paradisi et al., 2008) induced increased methylation of the keratin 10 genes via a putative mechanism involving type-1 cannabinoid receptor-mediated induction of DNA methyltransferase activity.

Together these findings show that treatment of cultured cells with PUFA can induce changes in the epigenetic processes in a manner and which exhibits specificity in terms of target epigenetic mark and the genes that are affected. Furthermore, epigenetic changes can also be induced by metabolites of PUFA.

# The Effect of Dietary PUFA on Epigenetic Processes in Pregnant Animal Models and Their Offspring

The majority requirements of developing animals are met primarily by the maternal diet and by adaptions to maternal metabolism (Burdge et al., 1994; Postle et al., 1995). Animals that are born at a relatively early stage of development, such as rodents, tend to accumulate fatty acids in adipose tissue and in specific organs such as the brain after birth (Sinclair and Crawford, 1972). In these species, lactation is the primary route of fatty acids supply. Other species, including humans and guinea pigs, tend to develop substantial adipose tissue reserves (Tanner, 1989) and incorporate fatty acids into specific organs, for example the brain (Innis, 1991), before birth. In these animals, both placenta fatty acid transport and lactation play important roles in providing fatty acids during development. However, whether fatty acids are supplied primarily through lactation or by a combination of placental transfer and lactation, pregnant females undergo metabolic adaptations to facilitate supply of fatty acids either via the placenta or milk including hyperlipidaemia, insulin resistance in late gestation (Herrera et al., 2006) and changes in phospholipid and PUFA metabolism (Burdge et al., 1994; Postle et al., 1994). Thus the outcome of interventions with PUFA during pregnancy can be modified by maternal metabolism and the stage of development of the offspring and the species of animal, which in turn have implications for studies involving the effect of maternal PUFA on epigenetic processes in the offspring.

Feeding diets with different ratios of linoleic acid (LA) to $\alpha$-linolenic acid (ALNA) during pregnancy and lactation has been shown to alter DNA methylation in the liver of the offspring at weaning. In one study, female mice were fed diets containing either a LA:ALNA ratio of 8:1 or of 55:1 for 30 days before pregnancy and throughout pregnancy followed by either the same diet as during pregnancy or an ALNA-supplemented (LA:ALNA, 0.3:1) (Niculescu et al., 2013). There was no significant effect of the diets fed during pregnancy on the mean methylation level of the *Fads2* promoter and of intron 1 in the liver of the dams at the end of lactation (Niculescu et al., 2013). However, *Fads2* promoter, but not intron 1, average methylation was approximately 2% higher in the liver of dams fed the ALNA-supplemented diet during lactation irrespective of the diet fed during pregnancy, which explained 6% of the variation in the level of the *Fads2* transcript. Variation in the expression of Dnmts explained between 3% and 16% of the variation in *Fads2* methylation. The ALNA supplemented diet was also associated with higher DHA concentration in maternal liver and serum. Similarly, the average methylation of the *Fads2* promoter was higher in offspring of dams fed the ALNA-supplemented diet than those fed the control or deficient diets throughout pregnancy and lactation (Niculescu et al., 2013). There was no difference in DHA concentration in the brain of the offspring between the maternal dietary groups. These findings suggest that differences in capacity of the dams to synthesise DHA from ALNA reflected small variations in the epigenetic regulation of *Fads2*. However, although the magnitude of differences in DHA synthesis was associated with higher status, this did not appear to affect DHA accumulation in the offspring. However, ALA supplementation during lactation increased neurogenesis in the dentate gyrus of the offspring, which was prevented in offspring of dams fed the diet with LA:ALNA ratio of 55:1 diet during pregnancy (Niculescu et al., 2011).

In order to test whether the magnitude of induced changes in DNA methylation was related to the amount or type of dietary fat, female rats were fed diets containing either 3.5%, 7%, or 21% (w/w) saturated and monounsaturated fatty acids (butter) or in EPA + DHA (fish oil) (Hoile et al., 2013). The offspring were fed 4% (w/w) soybean oil. The proportions of arachidonic acid (AA) and DHA in liver and plasma phospholipids of the adult offspring were associated inversely with the amount of fat in the maternal diet irrespective of the type of fat consumed. *Fads*2 mRNA expression in the liver of the adult offspring was also associated negatively with the amount of fat in the maternal diet intake. The methylation status of some CpG loci in the *Fads*2 promoter was 20% higher in the offspring of dams fed 21% fat compared to those fed 3.5% fat. A second study by the same group tested the effects of feeding dams diets containing 7% and 21% (w/w) safflower oil, enriched in LA, or hydrogenated soybean oil, enriched in *trans* fatty acids (Kelsall et al., 2012). The proportion of AA, but not DHA, was lower in the aorta from the offspring of dams fed 21% fat compared to those of dams fed 7% fat irrespective of the fatty acid content of the diet. This was associated with a reciprocal change in the mRNA expression of *Fads*1 and *Fads*2. Maternal diet induced specific CpG locus-specific hypermethylation of the Fads2 promoter, whereas the level of DNA methylation of the *Fads*1 promoter was not altered. Mutation of one CpG locus at position $-394$ (relative to the transcription start site) that was hypermethylated in offspring aorta of dams fed 21% fat and which was located within an estrogen receptor response element reduced the transcription of the gene in a luciferase reporter vector. This suggests that, at least for this locus, altered methylation may be involved directly in changes in transcriptional activity.

Together these findings suggest that, at least in rodents, the amount of maternal dietary fat appears to exert a greater influence on the DNA methylome of the offspring than the fatty acid composition of the diet. Furthermore, maternal dietary fat induced changes in DNA methylation in the adult offspring. Unfortunately, the experimental designs did not allow investigation as to whether the loci that were altered in adults were also differentially methylated in neonatal or juvenile offspring. Answering this question has implications for understanding the long-term impact of maternal diet on epigenetic processes in the offspring.

## The Effect of Dietary PUFA on Epigenetic Processes in Adult Animal Models

Feeding adult, nonpregnant female rats a fish oil (FO)-enriched diet for 9 weeks induced reduced *Fads*2 mRNA expression, a lower proportion of AA in hepatic phospholipids and increased methylation of specific

CpG loci in the *Fads2* promoter compared to those fed soy bean oil (Hoile et al., 2013). These *Fads2* loci were the same as those that were hypermethylated in the liver of the adult offspring of dams fed a higher FO diet (Hoile et al., 2013). However, these changes in DNA methylation were lost when the diet was switched from FO to soybean oil for a further 4 weeks. These findings imply that although dietary PUFA can alter epigenetic processes, the epigenome may revert to the state before the dietary intervention when the dietary input is removed. One implication of these findings is that dietary inputs during development appear to be persistent, whereas those during adulthood may be transient. This apparent difference has implications for strategies to modify epigenetic marks in adulthood that are associated with disease traits.

Dietary supplementation with FO, together with aerobic exercise, has been shown to reduce adiposity and improve cardio-metabolic health (Hill et al., 2007). To investigate the underlying mechanism, Fan et al. (2011) investigated the effect of a mixed soybean oil and FO diet on the DNA methylation status CpG islands within the of the leptin, and pro-opiomelanocortin (POMC) genes in the adipose tissue or brain, respectively, form juvenile male and young pregnant female mice. As expected, the n-3 PUFA diet reversed the effects of the obesogenic diet on the expression of these genes. However, there was no significant effect of the obesogenic diet, FO or the FO/soybean oil blend on the methylation status of leptin or POMC. These findings suggest that neither of these diets alter appetite control through an epigenetic mechanism, although the DNA methylation status of POMC has been shown be modified by dietary energy (Plagemann et al., 2009). However, it is possible that the limited sequence covered by the analysis (leptin 295 bp; POMC 607 bp) may have missed altered CpG loci. In contrast, the same group showed more recently that feeding juvenile male mice an obesogenic diet induced hypermethylation of specific CpG loci in the leptin promoter in adipose tissue and that this was presented by addition of FO to the obesogenic diet (Shen et al., 2014). The reason for these conflicting results is not clear.

## THE EFFECT OF PUFA INTAKE DURING HUMAN PREGNANCY ON THE EPIGENOME OF THE OFFSPRING

The relatively few studies of the effect of dietary supplementation with EPA and DHA during pregnancy on the DNA methylome of the offspring have produced inconsistent findings. Lee et al. (2013) investigated the effect of supplementation of the diet of pregnant women (18–22 weeks gestation) with DHA (400 mg DHA/d) or olive oil on the DNA methylation status of genes involved in immune function in umbilical cord blood. There were no significant differences in the average methylation in any of the genes measured between groups, even when the groups were stratified for smoking and the results adjusted major confounders. However, average methylation of long-interspersed repetitive sequence (LINE)-1 was approximately 1% higher in cord blood from pregnancies in which women smoked in the DHA group compared to placebo group, although this effect disappeared when adjusted for confounders. However, some concerns have been raised about the study protocol (Burdge, 2013). The methylation status of CpG loci in the insulin-like growth factor receptor-2 and H19 genes was also analysed in samples from this study (Lee et al., 2014). One CpG locus in the IGF2 promoter-3 had 0.9% higher methylation in the DHA supplemented group compared to the olive oil supplemented group. There were no significant differences between groups in the methylation status of known differentially methylated regions in the IGF2 and H19 genes.

Amarasekera et al. (2014) investigated the effect of supplementing the diet of pregnant atopic women either 1 g EPA plus 2.1 g DHA or an undisclosed placebo from the second trimester until delivery on the DNA methylome of umbilical cord $CD4^+$ T cells using the Ilumina 450 k array. There were no significant differences between groups in any of the CpG loci or regions covered by the array and the effect sizes were consistently less than 5% difference. In contrast, Lind et al. (2015) have shown that supplementing the diet of 9 month old infants with FO or sunflower oil (both 3.8 g/d) altered the methylation status of 43 CpG loci in peripheral blood leukocytes by more than 10%.

Overall, there appears to be no evidence to support an effect of maternal supplementation with n-3 PUFA on the DNA methylation status of genes in umbilical blood cells. However, the possibility that the amount of PUFA supplement was too small to induce changes in the offspring remains after partitioning of EPA and DHA into various maternal, fetal, and placental lipid pools. To date there have not been any studies that have investigated whether there is a dose-dependent effect of the amount of supplement consumed by pregnant women and the status of the DNA methylome of their infants. The study by Lind et al. (2015) suggests that, at least in leukocytes, the DNA methylome of infants retains plasticity with respect to dietary PUFA intake and so may be amenable to modification by dietary intervention.

## THE EFFECT OF PUFA INTAKE ON THE EPIGENOME OF CHILDREN AND ADULT HUMANS

Recent studies have reported the effects of increasing dietary fat intake on epigenetic marks in adult humans, including feeding young men either a diet in which 60% of energy was derived from fat, one-third from each of the monounsatured fatty acids, PUFA (the fatty acid composition was not disclosed), and saturated fatty acids or a lower fat diet (35% of the energy from fat) for 5 days in a crossover study (Brons et al., 2009). 7909 CpG loci in 6508 genes were differentially methylated in skeletal muscle (Jacobsen et al., 2012). The median difference in methylation of hypermethylated CpG dinucleotides was 2%–3% (1979 loci) and was associated with pathways related to cancer, reproductive system disease, and gastrointestinal disease. Eight loci differed by more than 10%. CpG loci that were hypomethylated had a median difference of 4%–5% (214 loci) and were associated with inflammatory disease, inflammatory response, and ophthalmic disease. Switching from the high fat to the lower fat diet reversed the methylation changes in all but 5% of the altered genes in which the level of methylation was significantly altered in the opposite direction to that induced by the high fat diet. These findings are consistent with the observation in rats that the effect of dietary interventions on DNA methylation in adults may be transient and lost once the dietary input has been removed (Hoile et al., 2013).

Supplementation of the diet of patients with moderate renal disease with 4 g daily of EPA plus docosapentaenoic acid (DPA) plus DHA ethyl esters or olive oil for 8 weeks induced the level of methylation of individual CpG loci in the 5′ regulatory regions of genes involved in PUFA biosynthesis (Hoile et al., 2014). Olive oil and n-3 PUFA induced differential changes (≥10% difference in methylation) in specific CpG loci in *FADS2* and *ELOVL5* contingent on sex, but not *FADS1* or *ELOVL2*. The methylation status of the altered CpG loci was associated negatively with the level of their transcripts. These findings were replicated in healthy men and women. One implication of these findings is that even modest fatty acid supplementation can modify the DNA methylation of specific CpG loci in adult humans. Whether these effects persisted beyond the period of supplementation was not tested.

Lupu et al. (2015) investigated the relationship between maternal FADS2 polymorphism rs174575 and DNA methylation status and PUFA concentrations and leukocyte DNA methylation levels in their 16-month-old children. Maternal FADS2 DNA methylation accounted for 20% of the variation in ALNA concentration in the children irrespective of sex, and also accounted for 26% of the variation in AA concentration in boys, but was not associated with AA in girls. One CpG locus in mothers accounted for 5% of the variation in methylation at that CpG dinucleotide in their children, while stratification for sex showed that maternal DNA methylation accounted for 11% of the variation in girls, but was not related DNA methylation at that locus in boys. There were no sex differences in DNA methylation in the children.

## The Effect of Variations in DNA Methylation of Genes Involved in PUFA Synthesis on Health-related Outcomes

Studies in rodent models and in human subjects have shown that the DNA methylation status of their 5′-regulatory regions can modify the activity of the genes in the PUFA synthesis pathway a manner associated with variations in PUFA status (Hoile et al., 2013; Kelsall et al., 2012). Two studies have suggested that such epigenetic variation has health implications. Cui et al. (2016) have shown that the level of methylation of cg27386326 in the FADS gene cluster was associated negatively with the concentration of AA in prostate cancer tumour tissue. Since eicosanoids derived from AA have been implicated in prostate cancer (Locke et al., 2010; Tawadros et al., 2012), DNA methylation at the cg27386326 locus may have implications for risk or progression of the disease.

Haghighi et al. (2015) investigated the relationship between the methylation status of *FADS*1 and 2, and *ELOVL5* in leukocytes and risk of depression or suicide. The findings showed individuals diagnosed with major depressive disorder (MDD) had lower methylation of a region upstream of the transcription start site of *FADS2* and higher methylation of an upstream region of *ELOVL5* compared to healthy controls. The precise regions that were analysed were not disclosed. Patients with MDD who had attempted suicide had lower levels of methylation in a downstream region of ELOVL5, but the higher methylation of an upstream region compared to those who has not attempted suicide. Low long chain n-3 PUFA status has been associated with increased risk of MDD and suicide (Hibbeln, 2009). However, the mechanism underlying the association between DNA methylation in leukocytes and depression is not known, although it is possible that it may represent a proxy measure of capacity for long chain n-3 PUFA synthesis in the liver or brain.

## POSSIBLE MECHANISMS

The mechanism by which PUFA modify the epigenome is currently not known. There is substantial evidence that short chain fatty acids can inhibit HDAC activity (Davie, 2003). Variation in energy intake leading to changes in cellular $NAD^+/NADH$ may alter histone acetylation by modulating the activities of the HDAC sirtuin (Imai et al., 2000). Such effects may be involved in epigenetic changes induced by high fat diets, but they are less likely to explain the impact of modest changes in fatty acid intake or the differential effects of different dietary lipids on specific epigenetic marks.

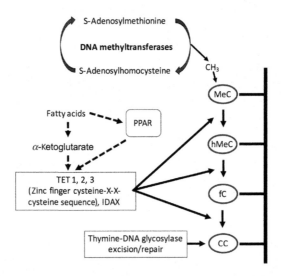

FIGURE 13.1   **A model of a putative mechanism for induction of altered DNA methylation by PUFA.** Methylation of CpG dinucleotides is mediated by DNA methyltransferases, which transfer a methyl group to the 5-posiiton of cytosine. Demethylation has been suggested to involve sequential oxidation of methylcytosine (MeC) to hydroxymethylcytosine (hMeC), formylcytosine (fC), and crboxycytosine (CC) by the activities of ten-eleven translocase (TET) proteins. TET activity may be up-regulated by metabolism of PUFA to α-ketoglutarate and it has been suggested that sequence specificity could be mediated by the TET chromatin binding domain and/or via a PPAR. Removal of CC is via thymine glycosylase excision/repair.

Feeding pregnant and lactating mice a high-fat diet induced lower expression of a number of noncoding RNAs in the adult offspring (Zhang et al., 2009). Noncoding RNAs have been shown to modify DNA methylation by changing the activity of Dnmts 3a and 3b (Fabbri et al., 2007), and by altering chromatin structure by modulating methyl CpG binding protein-2 activity (Klein et al., 2007) and by altering the expression of the histone methyltransferase enhancer of Zeste homolog-2 (Varambally et al., 2008). Thus specific noncoding RNA species represent a putative mechanism by which PUFA could modify specific epigenetic regulatory processes. However, such mechanisms await empirical evidence.

It has been suggested that the level of DNA methylation is the product of the equilibrium between methylation and demethylation reactions (Szyf, 2007) (Fig. 13.1). Methylation of CpG dinucleotides by Dnmts is well described. However, less is known about DNA demethylation, although several candidate mechanisms have been proposed (Wu and Zhang, 2014). There is growing evidence for the involvement of ten-eleven translocation proteins (TET) in DNA demethylation, TET

1, 2, and 3 demethylate 5-methylcytosine by sequential oxidation, via 5-hydroymethylcytosine and 5-formylcytosine, to form 5-carboxylcytosine, followed by base excision of formylcytosine or carboxylcytosine by thymine-DNA glycosylase (Wu and Zhang, 2014) (Fig. 13.1). TET activities are positively regulated by $\alpha$-ketoglutaric acid (Kaelin and McKnight, 2013). This suggests that increased dietary fat may lead to up-regulation of DNA demethylation by increasing flux of acetyl-CoA derived from fatty acid $\beta$-oxidation to Kreb's cycle leading to a rise in $\alpha$-ketoglutaric acid concentration. Sequence specificity may be conferred by specific TET DNA binding domains. TET1 and TET3, but not TET2, encode an N-terminal domain zinc finger cysteine-X-X-cysteine sequence, which interacts with chromatin (Long et al., 2013). The interaction of TET2 with DNA occurs via binding to the Inhibition of the Dvl and Axin complex (IDAX), which contains a zinc finger cysteine-X-X-cysteine sequence (Ko et al., 2013). However, unlike other chromatin binding proteins, TET 1 and 3, and IDAX recognize both methylated and unmethylated CpG loci (Williams et al., 2011; Wu et al., 2011; Wu and Zhang, 2014; Xu et al., 2011, 2012; Zhang et al., 2010). Thus, one possible mechanism to explain how PUFA could induce sequence-specific changes in DNA methylation is by altering the relative activities of Dnmts and TET proteins. This mechanism could explain how fatty acids can induce directionally opposite changes in DNA methylation as well as targeting of individual loci. Sequence specificity may also be conferred by poly(ADP-ribosyl)ation (PARylation) of transcription factors. PARylation of PPAR® has been shown to recruit TET activity and induce demethylation of CpG loci that are proximal to the PPAR® response element, although this has yet to be tested using fatty acids (Fujiki et al., 2013).

## CONCLUSIONS

There is increasing evidence that PUFA can modify DNA methylation, histone modifications, and noncoding RNAs. However, there are several important limitations to progress in this area. A number of studies have shown relatively small changes, in particular in DNA methylation, and it remains to be determined whether such effects are sufficient to alter transcription or the activity of metabolic pathways. In addition, it is not known whether epigenetic changes induced by PUFA in leukocytes reflect the effects of these fatty acids on less accessible tissues such as the liver or brain. Nevertheless, this field of research has potential to provide novel insights into the mechanisms by which PUFA influence development, metabolism, and risk of disease.

# References

Amarasekera, M., Noakes, P., Strickland, D., Saffery, R., Martino, D.J., Prescott, S.L., 2014. Epigenome-wide analysis of neonatal CD4$^{(+)}$ T-cell DNA methylation sites potentially affected by maternal fish oil supplementation. Epigenetics 9, 1570–1576.

Bird, A., 2002. DNA methylation patterns and epigenetic memory. Genes Dev. 16, 6–21.

Brons, C., Jensen, C.B., Storgaard, H., Hiscock, N.J., White, A., Appel, J.S., Jacobsen, S., Nilsson, E., Larsen, C.M., Astrup, A., Quistorff, B., Vaag, A., 2009. Impact of short-term high-fat feeding on glucose and insulin metabolism in young healthy men. J. Physiol. 587, 2387–2397.

Burdge, G.C., Calder, P.C., 2013. Does early n-3 fatty acid exposure alter DNA methylation in the developing human immune system. Clin. Lipidol. 8, 505–508.

Burdge, G.C., Hoile, S.P., Lillycrop, K.A., 2012. Epigenetics: are there implications for personalised nutrition? Curr. Opin. Clin. Nutr. Metab. Care 15, 442–447.

Burdge, G.C., Hunt, A.N., Postle, A.D., 1994. Mechanisms of hepatic phosphatidylcholine synthesis in adult rat: effects of pregnancy. Biochem. J. 303 (Pt 3), 941–947.

Ceccarelli, V., Nocentini, G., Billi, M., Racanicchi, S., Riccardi, C., Roberti, R., Grignani, F., Binaglia, L., Vecchini, A., 2014. Eicosapentaenoic acid activates RAS/ERK/C/EBPbeta pathway through H-Ras intron 1 CpG island demethylation in U937 leukemia cells. PLoS One 9, e85025.

Ceccarelli, V., Racanicchi, S., Martelli, M.P., Nocentini, G., Fettucciari, K., Riccardi, C., Marconi, P., Di, N.P., Grignani, F., Binaglia, L., Vecchini, A., 2011. Eicosapentaenoic acid demethylates a single CpG that mediates expression of tumor suppressor CCAAT/enhancer-binding protein delta in U937 leukemia cells. J. Biol. Chem. 286, 27092–27102.

Cui, T., Hester, A.G., Seeds, M.C., Rahbar, E., Howard, T.D., Sergeant, S., Chilton, F.H., 2016. Impact of genetic and epigenetic variations within the FADS cluster on the composition and metabolism of polyunsaturated fatty acids in prostate cancer. Prostate 76, 1182–1191.

Davie, J.R., 2003. Inhibition of histone deacetylase activity by butyrate. J. Nutr. 133, 2485S–2493S.

Fabbri, M., Garzon, R., Cimmino, A., Liu, Z., Zanesi, N., Callegari, E., Liu, S., Alder, H., Costinean, S., Fernandez-Cymering, C., Volinia, S., Guler, G., Morrison, C.D., Chan, K.K., Marcucci, G., Calin, G.A., Huebner, K., Croce, C.M., 2007. MicroRNA-29 family reverts aberrant methylation in lung cancer by targeting DNA methyltransferases 3A and 3B. Proc. Natl. Acad. Sci. USA 104, 15805–15810.

Fan, C., Liu, X., Shen, W., Deckelbaum, R.J., Qi, K., 2011. The regulation of leptin, leptin receptor and pro-opiomelanocortin expression by N-3 PUFAs in diet-induced obese mice is not related to the methylation of their promoters. Nutr. Metab. (Lond) 8, 31.

Fujiki, K., Shinoda, A., Kano, F., Sato, R., Shirahige, K., Murata, M., 2013. PPARgamma-induced PARylation promotes local DNA demethylation by production of 5-hydroxymethylcytosine. Nat. Commun. 4, 2262.

Haghighi, F., Galfalvy, H., Chen, S., Huang, Y.Y., Cooper, T.B., Burke, A.K., Oquendo, M.A., Mann, J.J., Sublette, M.E., 2015. DNA methylation perturbations in genes involved in polyunsaturated fatty acid biosynthesis associated with depression and suicide risk. Front. Neurol. 6, 92.

Herrera, E., Amusquivar, E., Lopez-Soldado, I., Ortega, H., 2006. Maternal lipid metabolism and placental lipid transfer. Horm. Res. 65 (Suppl. 3), 59–64.

Hibbeln, J.R., 2009. Depression, suicide and deficiencies of omega-3 essential fatty acids in modern diets. World Rev. Nutr. Diet. 99, 17–30.

Hill, A.M., Buckley, J.D., Murphy, K.J., Howe, P.R., 2007. Combining fish-oil supplements with regular aerobic exercise improves body composition and cardiovascular disease risk factors. Am. J. Clin. Nutr. 85, 1267–1274.

Hoile, S.P., Clarke-Harris, R., Huang, R.C., Calder, P.C., Mori, T.A., Beilin, L.J., Lillycrop, K.A., Burdge, G.C., 2014. Supplementation with N-3 long-chain polyunsaturated fatty acids or olive oil in men and women with renal disease induces differential changes in the DNA methylation of FADS2 and ELOVL5 in peripheral blood mononuclear cells. PLoS One 9, e109896.

Hoile, S.P., Irvine, N.A., Kelsall, C.J., Sibbons, C., Feunteun, A., Collister, A., Torrens, C., Calder, P.C., Hanson, M.A., Lillycrop, K.A., Burdge, G.C., 2013. Maternal fat intake in rats alters 20:4n-6 and 22:6n-3 status and the epigenetic regulation of Fads2 in offspring liver. J. Nutr. Biochem. 24, 1213–1220.

Hu, S., Wan, J., Su, Y., Song, Q., Zeng, Y., Nguyen, H.N., Shin, J., Cox, E., Rho, H.S., Woodard, C., Xia, S., Liu, S., Lyu, H., Ming, G.L., Wade, H., Song, H., Qian, J., Zhu, H., 2013. DNA methylation presents distinct binding sites for human transcription factors. eLife 2, e00726.

Imai, S., Armstrong, C.M., Kaeberlein, M., Guarente, L., 2000. Transcriptional silencing and longevity protein Sir2 is an NAD-dependent histone deacetylase. Nature 403, 795–800.

Innis, S.M., 1991. Essential fatty acids in growth and development. Prog. Lipid Res. 30, 39–103.

Jacobsen, S.C., Brons, C., Bork-Jensen, J., Ribel-Madsen, R., Yang, B., Lara, E., Hall, E., Calvanese, V., Nilsson, E., Jorgensen, S.W., Mandrup, S., Ling, C., Fernandez, A.F., Fraga, M.F., Poulsen, P., Vaag, A., 2012. Effects of short-term high-fat overfeeding on genome-wide DNA methylation in the skeletal muscle of healthy young men. Diabetologia 55, 3341–3349.

Kaelin, Jr., W.G., McKnight, S.L., 2013. Influence of metabolism on epigenetics and disease. Cell 153, 56–69.

Kelsall, C.J., Hoile, S.P., Irvine, N.A., Masoodi, M., Torrens, C., Lillycrop, K.A., Calder, P.C., Clough, G.F., Hanson, M.A., Burdge, G.C., 2012. Vascular dysfunction induced in offspring by maternal dietary fat involves altered arterial polyunsaturated fatty acid biosynthesis. PLoS One 7, e34492.

Klein, M.E., Lioy, D.T., Ma, L., Impey, S., Mandel, G., Goodman, R.H., 2007. Homeostatic regulation of MeCP2 expression by a CREB-induced microRNA. Nat. Neurosci. 10, 1513–1514.

Ko, M., An, J., Bandukwala, H.S., Chavez, L., Aijo, T., Pastor, W.A., Segal, M.F., Li, H., Koh, K.P., Lahdesmaki, H., Hogan, P.G., Aravind, L., Rao, A., 2013. Modulation of TET2 expression and 5-methylcytosine oxidation by the CXXC domain protein IDAX. Nature 497, 122–126.

Lee, H.S., Barraza-Villarreal, A., Biessy, C., Duarte-Salles, T., Sly, P.D., Ramakrishnan, U., Rivera, J., Herceg, Z., Romieu, I., 2014. Dietary supplementation with polyunsaturated fatty acid during pregnancy modulates DNA methylation at IGF2/H19 imprinted genes and growth of infants. Physiol. Genomics 46, 851–857.

Lee, H.S., Barraza-Villarreal, A., Hernandez-Vargas, H., Sly, P.D., Biessy, C., Ramakrishnan, U., Romieu, I., Herceg, Z., 2013. Modulation of DNA methylation states and infant immune system by dietary supplementation with omega-3 PUFA during pregnancy in an intervention study. Am. J. Clin. Nutr. 98 (2), 480–487.

Leoni, C., Vincenzetti, L., Emming, S., Monticelli, S., 2015. Epigenetics of T lymphocytes in health and disease. Swiss Med. Wkly. 145, w14191.

Lillycrop, K.A., Burdge, G.C., 2014. Breast cancer and the importance of early life nutrition. Cancer Treat. Res. 159, 269–285.

Lind, M.V., Martino, D., Harslof, L.B., Kyjovska, Z.O., Kristensen, M., Lauritzen, L., 2015. Genome-wide identification of mononuclear cell DNA methylation sites potentially affected by fish oil supplementation in young infants: a pilot study. Prostaglandins Leukot. Essent. Fatty Acids 101, 1–7.

Locke, J.A., Guns, E.S., Lehman, M.L., Ettinger, S., Zoubeidi, A., Lubik, A., Margiotti, K., Fazli, L., Adomat, H., Wasan, K.M., Gleave, M.E., Nelson, C.C., 2010. Arachidonic acid activation of intratumoral steroid synthesis during prostate cancer progression to castration resistance. Prostate 70, 239–251.

Long, H.K., Blackledge, N.P., Klose, R.J., 2013. ZF-CxxC domain-containing proteins, CpG islands and the chromatin connection. Biochem. Soc. Trans. 41, 727–740.

Lupu, D.S., Cheatham, C.L., Corbin, K.D., Niculescu, M.D., 2015. Genetic and epigenetic transgenerational implications related to omega-3 fatty acids. Part I: maternal FADS2 genotype and DNA methylation correlate with polyunsaturated fatty acid status in toddlers: an exploratory analysis. Nutr. Res. 35, 939–947.

Niculescu, M.D., Lupu, D.S., Craciunescu, C.N., 2011. Maternal alpha-linolenic acid availability during gestation and lactation alters the postnatal hippocampal development in the mouse offspring. Int. J. Dev. Neurosci. 29, 795–802.

Niculescu, M.D., Lupu, D.S., Craciunescu, C.N., 2013. Perinatal manipulation of alpha-linolenic acid intake induces epigenetic changes in maternal and offspring livers. FASEB J. 27, 350–358.

Paradisi, A., Pasquariello, N., Barcaroli, D., Maccarrone, M., 2008. Anandamide regulates keratinocyte differentiation by inducing DNA methylation in a CB1 receptor-dependent manner. J. Biol. Chem. 283, 6005–6012.

Plagemann, A., Harder, T., Brunn, M., Harder, A., Roepke, K., Wittrock-Staar, M., Ziska, T., Schellong, K., Rodekamp, E., Melchior, K., Dudenhausen, J.W., 2009. Hypothalamic proopiomelanocortin promoter methylation becomes altered by early overfeeding: an epigenetic model of obesity and the metabolic syndrome. J. Physiol. 587, 4963–4976.

Postle, A.D., Al, M.D., Burdge, G.C., Hornstra, G., 1995. The composition of individual molecular species of plasma phosphatidylcholine in human pregnancy. Early Hum. Dev. 43, 47–58.

Postle, A.D., Burdge, G.C., Al, M.D., 1994. Molecular species composition of plasma phosphatidylcholine in human pregnancy. World Rev. Nutr. Diet. 75, 109–111.

Sadli, N., Ackland, M.L., De Mel, D., Sinclair, A.J., Suphioglu, C., 2012. Effects of zinc and DHA on the epigenetic regulation of human neuronal cells. Cell. Physiol. Biochem. 29, 87–98.

Shen, W., Wang, C., Xia, L., Fan, C., Dong, H., Deckelbaum, R.J., Qi, K., 2014. Epigenetic modification of the leptin promoter in diet-induced obese mice and the effects of N-3 polyunsaturated fatty acids. Sci. Rep. 4, 5282.

Sinclair, A.J., Crawford, M.A., 1972. The accumulation of arachidonate and docosahexaenoate in the developing rat brain. J. Neurochem. 19, 1753–1758.

Szyf, M., 2007. The dynamic epigenome and its implications in toxicology. Toxicol. Sci. 100, 7–23.

Tanner, E.M., 1989. Foetus Into Man. Castlemead Publications, Ware, United Kingdom.

Tawadros, T., Brown, M.D., Hart, C.A., Clarke, N.W., 2012. Ligand-independent activation of EphA2 by arachidonic acid induces metastasis-like behaviour in prostate cancer cells. Br. J. Cancer 107, 1737–1744.

Varambally, S., Cao, Q., Mani, R.S., Shankar, S., Wang, X., Ateeq, B., Laxman, B., Cao, X., Jing, X., Ramnarayanan, K., Brenner, J.C., Yu, J., Kim, J.H., Han, B., Tan, P., Kumar-Sinha, C., Lonigro, R.J., Palanisamy, N., Maher, C.A., Chinnaiyan, A.M., 2008. Genomic loss of microRNA-101 leads to overexpression of histone methyltransferase EZH2 in cancer. Science 322, 1695–1699.

Williams, K., Christensen, J., Pedersen, M.T., Johansen, J.V., Cloos, P.A., Rappsilber, J., Helin, K., 2011. TET1 and hydroxymethylcytosine in transcription and DNA methylation fidelity. Nature 473, 343–348.

Wu, H., D'Alessio, A.C., Ito, S., Xia, K., Wang, Z., Cui, K., Zhao, K., Sun, Y.E., Zhang, Y., 2011. Dual functions of Tet1 in transcriptional regulation in mouse embryonic stem cells. Nature 473, 389–393.

Wu, H., Zhang, Y., 2014. Reversing DNA methylation: mechanisms, genomics, and biological functions. Cell 156, 45–68.

Xu, Y., Wu, F., Tan, L., Kong, L., Xiong, L., Deng, J., Barbera, A.J., Zheng, L., Zhang, H., Huang, S., Min, J., Nicholson, T., Chen, T., Xu, G., Shi, Y., Zhang, K., Shi, Y.G., 2011. Genome-wide regulation of 5hmC, 5mC, and gene expression by Tet1 hydroxylase in mouse embryonic stem cells. Mol. Cell 42, 451–464.

Xu, Y., Xu, C., Kato, A., Tempel, W., Abreu, J.G., Bian, C., Hu, Y., Hu, D., Zhao, B., Cerovina, T., Diao, J., Wu, F., He, H.H., Cui, Q., Clark, E., Ma, C., Barbara, A., Veenstra, G.J., Xu, G., Kaiser, U.B., Liu, X.S., Sugrue, S.P., He, X., Min, J., Kato, Y., Shi, Y.G., 2012. Tet3 CXXC domain and dioxygenase activity cooperatively regulate key genes for Xenopus eye and neural development. Cell 151, 1200–1213.

Zhang, H., Zhang, X., Clark, E., Mulcahey, M., Huang, S., Shi, Y.G., 2010. TET1 is a DNA-binding protein that modulates DNA methylation and gene transcription via hydroxylation of 5-methylcytosine. Cell Res. 20, 1390–1393.

Zhang, J., Zhang, F., Didelot, X., Bruce, K.D., Cagampang, F.R., Vatish, M., Hanson, M., Lehnert, H., Ceriello, A., Byrne, C.D., 2009. Maternal high fat diet during pregnancy and lactation alters hepatic expression of insulin like growth factor-2 and key microRNAs in the adult offspring. BMC Genomics 10, 478.

# Index

Printed in the United States
By Bookmasters